Contextualizing Systems Biology

Martin Döring • Imme Petersen
Anne Brüninghaus • Regine Kollek

Contextualizing Systems Biology

Presuppositions and Implications of a New Approach in Biology

Springer

Martin Döring
University of Hamburg
Hamburg, Germany

Anne Brüninghaus
University of Hamburg
Hamburg, Germany

Imme Petersen
University of Hamburg
Hamburg, Germany

Regine Kollek
University of Hamburg
Hamburg, Germany

ISBN 978-3-319-17105-0 ISBN 978-3-319-17106-7 (eBook)
DOI 10.1007/978-3-319-17106-7

Library of Congress Control Number: 2015944085

Springer Cham Heidelberg New York Dordrecht London

Printed on acid-free paper

Springer International Publishing AG Switzerland is part of Springer Science+Business Media (www.springer.com)

Preface

This book explores the development of systems biology as a new approach to life. It is based on an empirical social study of science and analyzes the epistemic preconditions, infrastructural requirements, innovative potentials, and policy implications of emerging and expanding concepts and practices of systems biology. In conducting the research that provided the basis for this publication we were interested not only in systems biology's capacity to give rise to a better understanding of complex biological entities such as cells or organisms but also in its cognitive, social, and policy framings and contexts. The results of the study show that systems biology is as complex as its objects of research, and that it also is an interdisciplinary enterprise which will most likely have a profound impact on our perception of life as well as on science itself.

The overall aim of this book is to contribute to a better understanding of the implications nestling in the current shift in molecular biology towards a systems-oriented perspective for science and society. It was written for specialists of different academic disciplines as well as for experts coming from nonacademic fields. Talking about experts from academia, we do not only think about those from biology, informatics, physics or other natural sciences, or medicine but also think about scholars from the social and cultural studies of science, from history and philosophy of science, or from linguistics. And when we talk about experts from nonacademic fields, we mean anyone interested in scientific developments such as systems biology coming from science policy, science administration, or the media reporting about science. It is our mission to make science, its presuppositions and preconditions, as well as its implications, as transparent and accountable as possible. Therefore we tried—and we hope succeeded—to use a language that makes a complex, but nevertheless highly topical and important subject accessible to all of those who are interested in a more than superficial understanding of science and of how it shapes and is shaped by us, by society, and by culture.

Hamburg, Germany

Martin Döring
Imme Petersen
Anne Brüninghaus
Regine Kollek

Acknowledgments

The study presented here and the research conducted would not have been possible without the participation of the many individuals who encouraged and supported our work. We are extremely grateful to all of them. First of all, we would like to thank scientists active in systems biology as well as experts from politics, administration, and the media, who were available for interviews and patiently answered our questions. Furthermore, we also thank the participants of the feedback-workshops at Charité Universitätsmedizin in Berlin and Technische Universität Braunschweig in Spring 2014 for open and inspiring discussions.

The work presented here was conducted in the context of the research project *Towards a holistic conception of life? Epistemic presumptions and socio-cultural implications of systems biology* (THCL; www.thcl.de), which was funded by the German Federal Ministry of Education and Research (BMBF) (May 2010 to July 2014). THCL was part in the funding line Ethical, Legal, and Social Aspects of Genome Research (ELSA-GEN). We would like to thank Marina Schindel, Detlef Böcking, and Anja Heintze at Deutsches Zentrum für Luft- und Raumfahrt (DLR) for taking care of all administrative and financial issues and providing help whenever necessary.

THCL was carried out by the Research Centre for Biotechnology, Society and the Environment (FSP BIOGUM)—Research Group Medicine and Neurosciences at Universität Hamburg. The principle investigator of the project was Regine Kollek. Martin Döring (May 2010 to September 2012) and Imme Petersen (October 2012 to July 2014) were responsible for managing the project. Anne Brüninghaus, Martin Döring, and Imme Petersen conducted the empirical research presented in this book. Most of it was done in Germany, but partially also in Austria. This resulted inter alia in a comparative analysis of media representations of systems biology in Austria and Germany, which is included in this book. Part of the project was carried out in cooperation with Karen Kastenhofer and Helge Torgersen at the Institute for Technology Assessment (ITA) at the Österreichische Akademie der Wissenschaften in Vienna. However, this book presents the results of the research group at BIOGUM whereas the results of our cooperating partners at ITA have been published elsewhere. The close and continuous exchange with Karen Kastenhofer and Helge Torgersen and the joint meetings and discussions in Vienna and Hamburg were extremely inspiring and fruitful; we thank both of them for this wonderful cooperation.

We also would like to express our gratitude to those of our former colleagues who cooperated with us in the course of the EU project *ACGT—Advancing Clinico-Genomic Clinical Trials on Cancer* in the 6th Framework Program and who agreed to participate in the interviews for the case study reported in this book. In particular, we would like to thank Georgios Stamatakos (Institute of Communication and Computer Systems, National Technical University of Athens) for advice and assistance in understanding the function of the oncosimulator and for proofreading Sect. 4.2.

The project was expertly advised by a scientific board whose members were Paul Martin (Department of Sociological Studies, University of Sheffield), Roger Strand (Centre for the Study of the Sciences and the Humanities, University of Bergen), Hans Westerhoff (Centre for Integrative Systems Biology, University of Manchester), and Ana Cãno-Delgado (Centre for Research in Agricultural Genomics, Barcelona). We would like to thank all of them for advice and support and for participating in the inspiring project workshop on *Different Forms of Life? Comparative Perspectives on Systems and Synthetic Biology* on January 19–20, 2012 in Hamburg.

The interviews were transcribed by Rita Stark (Köln) and Angela Wiedl (East Aurora), and the translation of German interview citations was done by Sandra H. Lustig (Hamburg). Many thanks to all of them for their work! Last, but not least, we want to give thanks to our colleagues at the FSP BIOGUM, especially to Birgit Sonntag for giving excellent organizational and administrative support as well as for taking care of a constant supply of coffee, tea, and biscuits during long meetings and workshops and Moritz Hettrich for arranging the literature, formatting, and proofreading the manuscript.

Contents

Contributors

Anne Brüninghaus Research Centre for Biotechnology, Society and Environment (FSP BIOGUM), University of Hamburg, Hamburg, Germany

Martin Döring Research Centre for Biotechnology, Society and Environment (FSP BIOGUM), University of Hamburg, Hamburg, Germany

Regine Kollek Research Centre for Biotechnology, Society and Environment (FSP BIOGUM), University of Hamburg, Hamburg, Germany

Imme Petersen Research Centre for Biotechnology, Society and Environment (FSP BIOGUM), University of Hamburg, Hamburg, Germany

Chapter 1
Understanding Systems Biology: A Place for Social Science Analysis

Regine Kollek, Imme Petersen, Anne Brüninghaus, and Martin Döring

Abstract Systems biology is a new approach in the life sciences aiming at a more holistic understanding of biological entities. Developing out of the shortcomings of molecular biology in explaining complex and dynamic features of living systems, it can have far-reaching implications for science and society, as well as for our understanding of life. In order to understand the potentials and impacts of systems biology, a broad analysis of this new approach was undertaken from the social and cultural studies of science perspective, the results of which are laid down in this book. This chapter introduces the subject and our research concept and outlines the scope and the aims of the book. The first section describes the rise of systems biology, its definitions and main aims. In a second step the conceptual approach of our analysis is laid out. We introduce the concept of context and context analysis and outline the cultural, practice-related, and societal environments, which were considered in our exploration of systems biology. The second section first describes the methodical approaches applied in our study before it depicts the goals and hypotheses of this book and provides a short synopsis of the following chapters.

Keywords Systems biology • Context • Social and cultural analysis of science • Philosophy of science • Holism • Implications for science and society

Systems biology is a new approach in the life sciences aiming at a more holistic understanding of life. During the second part of the last century, activities in the life sciences focused primarily on the molecular constituents of cells, their structure and immediate biochemical functions. After successfully sequencing the human genome and that of many other organisms, scientific attention shifted towards complexity and dynamics of biological processes and entities. This shift brought a systems approach to the fore that finally led to the advent of systems biology by the turn of the last century. Still in its infancy today, systems biology wants to explain

---

R. Kollek (✉) • I. Petersen • A. Brüninghaus • M. Döring
Research Centre for Biotechnology, Society and Environment (FSP BIOGUM), University of Hamburg, Hamburg, Germany
e-mail: kollek@uni-hamburg.de

M. Döring et al., *Contextualizing Systems Biology*,
DOI 10.1007/978-3-319-17106-7_1

biological entities such as cells, organs, or organisms at the systems level. It wants to understand how emergent properties of such entities arise from the interactions among their components and how such properties are influenced by the requirements of higher organizational levels. The turn towards a systems approach in biology, however, demands not only a much stronger formalization of biological processes compared to earlier developments in (molecular) biology, but also an extended interdisciplinary cooperation and philosophical reflection of old and new concepts and presumptions. Such changes in concepts, methods, practices, and disciplinary structures bear the potential to fundamentally transform biology. Such a transformation may not only deeply affect our understanding of life, but also the way research is undertaken and organized. Together these developments may have far-reaching consequences for biology in particular, but also for the life sciences in general. Moreover, concepts and insights from systems biology will also trigger translational research, such as modeling and prediction of disease processes, the production of high-value biotechnological substances, or the development of new targeted therapies. This translation of systems approaches towards applications in different fields (e.g., bioengineering, systems medicine) may yield its own ethical, legal, and social consequences that go beyond those of earlier concepts and practices in molecular biology and genetics. These developments and their possible implications for science and society need to be analyzed more closely. This is especially true if we want to gain a detailed and sophisticated understanding of systems biology and its potential for innovation in current governmental and public contexts.

The first chapter introduces the subject of systems biology, outlines our research concept and provides an overview of the methods applied and the content of the book. It consists of two main sections. The first is dedicated to systems biology as a new approach in the life sciences. Here we first provide a detailed overview of the development of systems biology and explore the different understandings and definitions that currently exist. In a second step the conceptual approach of our analysis is laid out. We introduce the concept of context and context analysis and outline the different cultural, practice-related, and societal environments that we included in our analysis. The second section first describes the methodical approaches applied in our study before it depicts the goals and hypotheses of this book and provides a short synopsis of the following chapters.

1.1 Systems Biology: A New Approach in the Life Sciences

1.1.1 From Parts to Wholes: The Rise of Systems Biology

Living organisms are complex and dynamic entities: they exist in space and time. Philosophers and naturalists have been curious about their nature, functions, and behavior for as long as they have started to reflect on it. The modern sciences, especially molecular biology, chemistry, and physics provided researchers with systematic approaches to analyze organisms (including humans) in detail. By a multiplicity of different strategies, procedures, and methods, organisms and cells could be dissected into millions of components, structures, and processes. The sequence of the human genome, the draft of which was published in February 2001, has possibly been one of the most publicly visible, but also only one of the last in a long row of important achievements in the history of the life sciences.

The more we know about the molecular elements and processes, however, the more difficult it seems to understand how they interact not only on the molecular, but also on the cellular level and beyond in order to create specific biological functions, structures, and other high-level phenomena of living entities. Molecular biology, the leading discipline during the second half of the twentieth century, was enormously successful in analyzing the molecular parts of the cell. However, it did not provide convincing theories or concepts necessary for making sense of the ever-growing amount of molecular data (O'Malley and Dupré 2005). Even Omics approaches which aim at large-scale analysis and measurements of genes (genomics), transcribed sequences (transcriptomics), or proteins (proteomics) did by themselves not lead to a comprehensive understanding of the interaction and integrated functioning of identified parts (O'Malley et al. 2014). As first and foremost reductionist enterprises, they aim at the analytical dissection of cellular components and result primarily in ever-growing lists of parts. Such large-scale measurements have driven the resolution of cells and their constituents to an unprecedented high definition. Nevertheless, this did not provide us with an understanding of biological systems as systems (Kitano 2000, 200). Hence, what was needed around the turn of the last century was an approach able to integrate the multitude of data about isolated parts and processes into an overall perspective. Such integration is thought to be the key to decoding what forms the basis of the functions of living systems (Pesce et al. 2013; Green and Wolkenhauer 2012). Furthermore, it most likely would considerably speed up hypothesis generation in biology and finally yield deep insights into regular and pathological processes of the cell and of organisms (Gomez-Cabrero et al. 2014).

At this point, modern systems biology enters the stage. Systems biology is a new field in biology that aims at understanding biological systems such as cells and organisms from a system's perspective (Kitano 2000, 2001). It emerged more or less at the time when the working draft of the human genome became available. Although not a completely new perspective in biology, it seems to have hit the nerve of the life sciences. After decades of dissecting biological phenomena down to the molecular level and extensive analysis of DNA-sequences and expression patterns in the context of Omics research, the time had come for stepping back and reflecting on the meaning of accumulated data and information and searching for an integrative perspective. The new systems biology promises to provide such perspective. It wants to understand intra- and intercellular processes in order to gain a comprehensive idea of all interactions on different system levels and to discover the principles guiding them. It therefore promises to provide the conceptual framework needed to meet the challenges outlined above.

Interestingly, there is still no unambiguous, generally accepted definition of what modern systems biology is (Kirschner 2005). Different perceptions and descriptions exist in parallel. Whereas some scientists emphasize the necessity of integrating heterogonous data into common concepts, others point to the importance of considering higher levels of organisms or underline the need of mathematical modeling and computation. Whereas some systems biologists strive to make biology more similar to physics and engineering, others do not think that the standards of physical sciences are appropriate to measure the achievements of systems biology and doubt whether they will ever be applicable to "dirty, unruly living systems" (Calvert and Fujimura 2011, 55).

The following quotations are illustrative of the conceptual pluralism of systems biology. For instance, Denis Noble, one of the early protagonists and promoters of systems biology, described its character and goals quite literarily. For him, "[s]ystems biology [...] is about putting together rather than taking apart, integration rather than reduction. It requires that we develop ways of thinking about integration that are as rigorous as our reductionist programs, but different. [...] It means changing our philosophy, in the full sense of the term" (Noble 2006, 176). Hans Westerhoff and his colleagues, pioneers of modern systems biology as well, underline that systems biology "addresses the missing links between molecules and physiology" (Bruggeman and Westerhoff 2007, 45), and that it aims "to understand how biological function absent from macromolecules in isolation, arises when they are components of their system" (Westerhoff et al. 2009, 7).

In other cases, emphasis is put on systems biology's capacity for better manipulating and controlling biological systems. Systems biology therefore studies such systems "by systematically perturbing them (biologically, genetically, or chemically); monitoring the gene, protein, and informational pathway responses; integrating these data; and ultimately, formulating mathematical models that describe the structure of the system and its response to individual perturbations" (Ideker et al. 2001, 343). Irina Borodina and Jens Nielsen (2005) stress that systems biology relies on the integration of experimental and computational approaches to achieve comprehension and prediction of complex cellular functions. Its characteristic is an iterative process of model building, comparison with new sets of experimental data, improvement of the model to account for new features, and so on (Kitano 2002).

For Olaf Wolkenhauer, one of the scientists who paved the way for modern systems biology in Germany, understanding is the major motive for applying systems approaches. For him, systems biology "aims at a system level understanding of genetic or metabolic pathways by investigating interrelationships (organization or structure) and interactions (dynamics or behavior) of genes, proteins and metabolites" (Wolkenhauer 2001, 258). Supplemented is this colorful bouquet of perceptions and descriptions by more functionalist definitions pointing to systems biology as a "research endeavor that aims at providing the scientific foundation for successful synthetic biology" (Breitling 2010, 1).

What all definitional efforts have in common is that they try to make sense of the vast amounts of data on biological processes by focusing on the mechanics behind the emergence of functionality (Westerhoff and Alberghina 2005). However, definitions of systems biology are still heterogeneous and rather illustrations of exemplary cases and phenomena than unifying framings of the core elements of this new approach to biological objects and processes. Nevertheless, they do not necessarily contradict each other but describe different aspects and perspectives of systems biology. This has been true for the foundation period of systems biology as well as for today.[1] Therefore, one could rightly state that "the current number of definitions for the term systems biology is close to the number of its practitioners" (Medina 2013, 1035). Hence, systems biology is still elusive today (Cowley 2004). To better describe and define its goals and approaches, its potentials and limitations, its prerequisites, and perspectives would not only help to hedge it into a definitional frame, but also to ask more precise questions in order to explore its possible scientific, ethical, and social implications.

Despite these problems of self-definition and a common understanding, systems biology has made an astonishing career since it entered the center stage of advanced life sciences: theoretical and experimental activities assembling under its umbrella are exploding. Numerous working groups worldwide are engaged in biological systems research and thousands of editorials, research papers, reviews, and books have been published. Whereas in early 2010 an Internet search for "systems biology" yielded about 2 million hits (Kohl et al. 2010, 25) it was nearly 5.5 million by the end of 2014. Not only academic interest grows but also interest in possible applications of systems approaches in the life sciences, and consequently the number of public and private funding programs supporting the advancement of the field.

1.1.2 Systems Biology in Context: Contexts of Systems Biology[2]

From this short introduction to systems biology, three tentative conclusions could be drawn. First, systems biology does not originate in a singular foundational event or central scientific question, but it developed out of molecular biology and is, at least partially, driven by technological development, but even more by the need to find new ways to integrate the massive amount of molecular data and to develop a systems-oriented, holistic perspective on biological entities in order to overcome the limits of exaggerated reductionism. Second, there seems to be no common

[1] This statement refers to early 2014, when the empirical phase of our project was completed.

[2] This title of this section is referring to the title of the book, *Science in Context – Contexts of Science*" (Wissenschaft im Kontext – Kontexte der Wissenschaft) (1993) edited by Wolfgang Bonß, Rainer Hohlfeld, and Regine Kollek, in which the editors outline a contextual theory of scientific knowledge. For theories of context also see Bonß et al. (1994) or Kovala (2014).

agenda, understanding, or unified definition of systems biology among actors of different disciplines or backgrounds participating in the establishment of this new approach. Third, systems biology develops in complex and diverse environments in which the values, agendas, goals, requirements, and interests of many different groups are at stake—scientists included. This leads us to the following research questions:

- How do these environments and factors influence the current and future development of systems biology?
- How do they shape its concepts and practices, and its possible future implications?
- How do they influence the political and public perception of systems biology?

In order to answer these questions, a research strategy is needed that is able to guide the analysis of such complex historical, cultural, scientific, and social settings in which multiple and heterogonous factors are at play. To explore these settings, we choose the *concept of context*. It allows us to look at systems biology as a contextualized scientific development whose understanding is not possible by reading scientific reports, programmatic papers, or statements alone, but requires more intimate knowledge about cultural, practice-related, and social or societal factors and conditions that bear an impact on it. Such environments, settings, and conditions can be regarded as *contexts*, which—and this is our core hypothesis—shape systems biology in a particular way and are in return dialectically molded by it as well. Hence, if one wants to explore the implications for science and society of such a new development, which is still in its infancy and has not yielded much concrete outcome to date that already affects medicine or agriculture more concretely, a context-sensitive approach and analysis is an appropriate way to meet this challenge. Furthermore, a contextual understanding of systems biology and its development will most likely also enable governments and stakeholders to design contextually relevant responses to problems and hindrances that may hamper its development (Waylen et al. 2010).

The term context[3] is widespread in daily language, but has also become increasingly important in disciplines including linguistics, anthropology, psychology, the computer sciences, or the science of artificial intelligence. It has also attracted the attention of sociologists and philosophers of science, who have explored various aspects of context, contextuality, and contextualization. It is not possible to review the extended body of literature devoted to the concept of context here, but a sketchy introduction makes clear why it makes sense to analyze systems biology as a contextualized phenomenon. The origins of the concept of context are diverse and heterogenous; most prominently they are located in linguistics and conversation theory. In general, the concept reflects the conviction that actions, utterances, or expressions can, at least in important aspects, only be understood in relation to their cultural, linguistic, philosophical, social, economic, or ethical environment, inter alia (Price 2008). This holds true not only for everyday verbal or physical expressions,

[3] The word *context* is derived from the Latin verb *texere* (to weave) and is also related to the Latin verb *contexere*, which carries the meaning of "to weave together", "to interweave", "to join together", or "to compose".

but also for those happening in professional environments such as science or politics. To reveal the meaning of actions, operations, sentences, expressions, claims, or statements thus requires additional information about the locality or situation in which they were expressed. In a first approximation, a context therefore can be described as "[t]he circumstances that form the setting for an event, statement, or idea, and in terms of which it can be fully understood"[4] or, more generally, "[t]hat which surrounds, and gives meaning to, something else"[5] or "the situation in which something happens: the group of conditions that exist where and when something happens."[6]

Among the first to think more systematically about the phenomenon of context was the anthropologist Bronislaw Malinowski. He underlined, for instance, the cultural context of discourse as essential to meaning (1923). In his work he connected linguistic structures with social contexts and called for a supplement of pure linguistics by ethnographic descriptions. In this work, he coined the term "context of situation" which later on became an important part of linguistic theory. He also pointed out that language is not only an expression of thoughts but also a form of action. To speak therefore means to act practically (see also Austin 1965). Such speech acts can also be perceived as social actions that in return change the context in which they originate (Sbisà 2002). The understanding of (everyday, scientific or political, i.a.) expressions hence requires a lot of background knowledge that enables participants of a speaking situation to identify and to interpret a speech act. Consequently, the cognitive as well as the social dimension of the speech act are important to generate meaning and to understand utterances.

Since the time of Malinowski, the notion of context played an important role in linguistics (see, e.g., Goodwin and Duranti 1992; Halliday and Hasan 1985; Tracy 1998). Research coming from other disciplines, especially psychology and sociology, brought additional dimensions of context into the debate. The influential sociologist Erving Goffman, for instance, studied (amongst other areas of social life) different forms of social communication and interaction as well as the social organization (framing) of experience. He pointed out that the meaning of an action—be it a verbal or physical (bodily, corporeal) one—cannot be deciphered without knowledge of the *immediate* situation of this action (Goffman 1983), in addition to knowledge about its broader social and cognitive context. Although language itself was not in the focus of his interest, he considered it as a fundamental part of human social interaction and therefore must be paid attention to in sociological analysis (Burns 1992, 301). For Goffman "form and meaning of the social and interpersonal contexts" were important because they "provide presuppositions for the decoding of

[4] *Oxford Dictionaries*, http://www.oxforddictionaries.com/definition/english/context. Accessed December 15, 2014. The example given in the dictionary reads as follows: "the proposals need to be considered in the context of new European directives."

[5] *The free dictionary*, http://encyclopedia2.thefreedictionary.com/context. Accessed December 15, 2014.

[6] *Merriam-Webster Dictionary*, http://www.merriam-webster.com/dictionary/context. Accessed December 15, 2014.

meaning" (Schiffrin 1994, 103). In his later work Goffman elaborated the concept of "frame" and "frame analysis" (Goffman 1974), which "can be read as an unpacking of the 'global and undifferentiated' idea of context" (Scheff 2005, 284). "Framing" relates to the construction or depiction of a fact or an issue from a particular perspective; they are then framed in a system of presuppositions, selected facts, beliefs, and other relevant factors that mutually support and consolidate each other. Although Goffman was well aware of the crucial importance of context (Scheff 2005, 372), he was also critical about the concept and stated later on that "[T]raditionally no analysis was provided of what it is in contexts that makes them determinative of the significance of utterances, or any statement concerning the classes of contexts that would thus emerge—all of which if explicated, would allow us to say something other than merely that the context matters" (Goffman 1981, 67).

Despite this and other criticisms concerning structure, types, systematics, and properties of context, the concept nevertheless has proven not only quite successful, but also indispensable to many different disciplines as diverse as social and cultural anthropology (Dilley 2002), the computer and information sciences (Jones 2004, Floridi 2012), artificial intelligence (Serafini and Bouquet 2004), molecular biology (Kollek 1990, Cardinale and Arkin 2012), medicine (Walzer Leavitt 1990, Beskow and Burke 2010), or ethics (Musschenga 2005). One of the reasons for its ubiquitous use is that context analysis essentially has the potential to show "how the microscopic world of words and gestures is linked to the largest social structures" (Scheff 2005, 384). This is what makes context analysis so attractive: it is hoped that it will enable us to understand the meaning and relevance of isolated phenomena which are part of systems biology in the broader cultural and societal context. Contextualizing systems biology therefore represents itself as a "holistic" research perspective in the social and cultural study of science.

There have been numerous attempts to clarify what context means in more detail, how it is structured or can be systematized (comp. i.a. Austin 1965; Schegloff 1992; Scheff 2005; Sbisà 2002), although no general theory of context has been developed yet. However, in this study it is not our aim to explore the concept of context itself in more detail or contribute to context theory. Rather, we use context as a loosely defined concept and as a heuristic device in order to gain systematic insights into the complex cultural, cognitive, and social settings systems biology is situated in and by which it is shaped. Especially helpful for our analysis is a body of work coming from the sociology of science and the sociology of scientific knowledge that has already brought forward numerous studies of the contextual nature of science and its development and of scientific knowledge. They are discussed in more detail in the following chapters.

An important question for our analysis is which contexts have to be considered as relevant in influencing or even determining the development (and hence understanding) of systems biology. Because it is not feasible to identify and name all possible or even relevant contexts we had to take a pragmatic choice. We decided to take a closer look essentially at three broadly defined contexts, which are not only plausible but also amenable to empirical analysis:

- Culture, because it may reveal how core concepts and images in systems biology are shaped by cultural beliefs and images on the one hand, and embodied sociocultural knowledge.

- Scientific practice, because it may provide us with information about how systems biology is molded and structured by specific framework conditions, such as interdisciplinarity, the requirement for ICT infrastructures, or funding.
- Society, because it can inform us about how systems biology is perceived and understood in social contexts such as science policy, or the media, and about how possible implications are evaluated.

By applying such a context-oriented perspective we want to explore how systems biology has and will come into reality in these different contexts and is shaped by them. Again, we certainly are not able to cover all aspects belonging to one specific context; this is why our analysis is necessarily selective. We nevertheless are convinced that such an approach shows the plurality, dynamics, productivity, and possible implications of systems biology much better and in a richer way than focusing on single dimensions or aspects only.

1.1.2.1 Cultural Contexts: Perception of Concepts

With regard to cultural influences on systems biology, we decided to focus on the conception and construction of core concepts in systems biology such as life, holism, reductionism, model, or system. Before going into more detail, we first want to clarify what we mean when we talk about culture or the cultural context. For our purpose, a general account of culture should suffice. In its widest sense, the term culture[7] denotes everything that has been created by humans in a formative way. In short, a culture is a complex whole that includes knowledge, belief, art, morals, law, custom, and any other capabilities and habits acquired by humans as members of society (Tylor 1920). More specifically, culture can be described as an integrated system of learned behavior patterns that are characteristic of the members of a society and are not a result of biological inheritance.[8] Societies and the distinct ways that people live and impart their experiences differ. Culture is, therefore, neither unique nor static. Cultures change, and so does the conception of culture that prevails at a certain time in a specific environment.

As science is part of culture, it does—like culture—undergo changes that affect not only its general development but also its questions, objects, theories, concepts, and practices. With regard to the modern life sciences and other fields of scientific activity, this has been analyzed and demonstrated by many different scholars.[9]

[7] The term *culture* is derived from the Latin word *cultura*, which means treating, processing, adapting, or agriculture.

[8] In cultural and social anthropology, numerous definitions of culture exist (Kroeber and Kluckhohn 1952). In addition to the different approaches (e.g., Benedict 1934; Gillin and Gillin 1948; Geertz 1973; Harris 1983), the definitions all put emphasis on social habits or, in other words, social standards within a society. A current overview can be found by Hansen (2011).

[9] To name just a few of many relevant publications pertinent to the issue: Latour and Woolgar (); Pickering (ed) (1992b); Clarke and Parsons (eds) (1997); Nelkin and Lindee (1995); Goodman et al. (eds) (2003).

Saying that science is culturally embedded and shaped therefore is more or less a truism today. Coming back to systems biology, we decided to concentrate our analysis on some of its core concepts—such as holism—in order to elucidate how they are shaped and tainted by cultural contexts. In general, a concept is an idea of what something is or how it works or something conceived in the mind, like a thought or a notion.[10] In another reading a concept (or conception[11]) is an abstract idea, mental representation, or mental symbol that exists in the mind. In science, concepts may be strongly formalized (such as in mathematics), expressed as logically connected sentences, or more or less informal descriptions. As we have argued before, core concepts in systems biology do not seem to be very stringently defined but rather resemble tentative descriptions. They capture our analytical interest because in such descriptions scientists often refer to cultural images that can be identified and analyzed, for instance, by metaphor analysis (for details see Chap. 2).

One of the most interesting, but perhaps also most controversial concepts that come along with systems thinking in biology is holism. It is based on the idea that biological entities such as cells or organisms should be viewed as wholes and not as collections of parts, and that "[S]ystemic relations arising at complicated stages of integration may produce new and unpredictable characteristics of the system" (Andersen 2001, 153). It therefore may be seen in contrast to reductionism[12] that has become an integral part of modern sciences such as molecular biology. Holism is not a new concept; it has many predecessors not only in the life sciences but also in philosophy and the humanities. It is closely linked to the concept of complex systems, especially living systems, which can give rise to new or emergent behavior or properties that cannot be deduced from the properties of the elements alone (von Bertalanffy 1968). Holism in science, therefore, is the idea that natural systems such as cells or organisms show properties which are more than the sum of their parts. However, von Bertalanffy's approach to holism, known as classical systems theory, did not survive in the scientific research agenda because the technical tools necessary to dissect cells or organisms in detail, as well as powerful computers for dynamic modeling were not available. According to Robert Rosen,[13] it is therefore "not without irony" that modern systems-theoretic ideas return to "the holistic, functionally oriented view of organisms entertained by biologists prior to the emergence of biochemistry and molecular biology" (Rosen 1968, 34). Classical holism was insofar

[10] *Merriam-Webster Dictionary*, http://www.merriam-webster.com/dictionary/concept. Accessed December 17, 2014.

[11] The terms *concept* and *conception* are sometimes used interchangeably. However, a conception may also be more encompassing and detailed than a concept with regard to factors considered and theoretical reflections.

[12] Reductionism is another important concept in systems biology; it is described in more detail in Chap. 2.

[13] Robert Rosen is a theoretical biologist and biophysicist and cofounder of relational biology. *Relational biology* was developed in reaction to the current reductionist approaches to science by molecular biologists. It maintains that organisms, and indeed all systems, have a distinct quality called organization which is not part of the language of reductionism, as, for example, in molecular biology (Baianu 2006).

"stillborn theory" (Gatherer 2010, 1). However, the term survived, designating various kinds of anti-reductionisms. Section 2.5 provides a detailed account of the history of modern systems biology and holism.

Although many scientists involved in systems biology research, especially those coming from mathematics or computer sciences, rarely speak of holism or holistic approaches in their publications, modern biology experiences today the renaissance of a more holistic perspective on life that was typical of the seventeenth and eighteenth centuries (Foster 2007). One of its prominent advocates is Leroy Hood who considerably contributed to the technological foundation of the Omics sciences and who is a cofounder of the *Institute for Systems Biology* in Seattle. For him, systems biology is primarily characterized by holism and integration (Hood and Flores 2012, 613) and therefore "attempts to study biological systems in a holistic rather than an atomistic manner" (Hood et al. 2008, 239). Other scientists describe the current development towards a more holistic approach as a move away from reductionist molecular approaches focusing on the role of single genes or proteins towards networks and interactions between individual components of networks (Kuster et al. 2011, 1037). For them, holism could become "a guide in looking for a new mode of the combination of analytical and synthetic reasoning in biology" (Křeček 2010, 157). Especially scientists exploring questions related to medical applications of systems biology see the advantages of using a holistic, network-based approach to human disease because it can move medicine from a field rooted in semi-empiric reductionism to one that recognizes the importance of molecular networks (Loscalzo and Barabasi 2011, 622). In their opinion, reductionist approaches will fail because they do not account for the complexity of biological systems and the principles governing the interaction of parts evade their methodology. Contrasting the reductionism of (old) molecular biology with the presumed holism of (new) systems biology is a popular theme in review articles and editorials. Some of them stress that molecular biology and systems biology are interdependent and complementary ways to study and understand very complex phenomena (Fang and Casadevall 2011, 1401). Other scientists are more critical and point out that holism has not yet been operationalized appropriately. For instance, Filippo Conti and colleagues maintain that "[t]he word holistic […] has a too strong esoteric connotation and is decidedly too vague." For them, it can—at least in this form—not fruitfully be used in science. The main avenue for systems biology must therefore be "to give an operational meaning to the holistic perspective" (Conti et al. 2007, 161). However, despite this skepticism, holism does not equal bad science. Studies from the history of biology show that many holistic theories were developed as a scientific and rational response to empirical problems. "They drew on the cultural and scientific resources that were available to researchers at specific times and places and significantly shaped the agenda of different disciplines ranging from theoretical biology and morphology to psychology, neurology, and psychiatry" (Laubichler 2000, 289).

This short excursion into holism shows that this concept is partially taken seriously, but is also used as a mere rhetorical means without explaining more exactly what is meant by it and how it should be applied. It is also remarkable that there are efforts to find a conception of holism which can be connected to previous concepts or findings in molecular biology, but on the other hand, it seems to be important for some scientists to distance themselves from what they call "esoteric" conceptions of holism. Such conceptions possibly are a cultural source for images of wholes and wholeness, which is less respected in science and therefore needs to be denounced. Beyond this observation, which would have to be investigated more closely, however, it is not clear yet what systems biologists think about holism and how they conceive the ideas behind it. The same may apply to other notions typical to systems biology such as system or reductionism. But if core concepts are still fuzzy, it may be difficult to operationalize them and to propose experimental strategies suitable to answer pertinent questions. However, such fuzziness may as well serve as a source for cultural meanings connected to and embedded in the concepts. This is the starting point of our analysis. In order to gain a deeper and richer understanding of concepts in systems biology, we explore how different researchers active in systems biology use the terms and handle the challenges they pose (Chap. 2), and, in addition, how they perceive the past and the future of their discipline (Chap. 3).

1.1.2.2 Context of Scientific Practice: Doing Systems Biology

A second feature of systems biology that we were interested in is the changing way of doing science which obviously comes along with systems approaches and in turn has an impact on biology and other disciplines involved. What is at stake here can be called the practice of science. Practice describes human action in society. In the social sciences, especially in sociology and anthropology it is a widely used concept that broadly refers to agency, basically meaning the human capacity to act (Ortner 1984, Giddens 1984).

Important contributions to a sociological theory of practice came from Harold Garfinkel (1967) who introduced the idea of routine grounds of social action, from Anthony Giddens (1984) who defined that social structures are produced and reproduced by human interaction, or from Pierre Bourdieu's (1977) notion of field and habitus (cf. Reckwitz 2003; Bongaerts 2007). The theories and concepts orbiting social practice put either more emphasis on individual agency (Giddens) or social structure (Bourdieu). Hence, social practice can be described as a dialectic between social structure and human agency working back and forth in a dynamic relationship.

Therefore, social practices are always embedded situationally; usually they are a bunch of actions interwoven with habits and routines (Schatzki 2002, 71), implicit knowledge about the given situation (Reckwitz 2004, 320), and the practices that would be appropriate (Schatzki 2002, 78). The French sociologist Pierre Bourdieu (1977) developed, among other concepts, the notion of field. It describes a structured social space with its own rules, schemes of domination, legitimate opinions, and so on that is defined as a specific field's habitus. Fields therefore constitute specific environments by which the respective habitus and activities are structured and shaped: they can also be conceived as contexts. Fields include art, education, politics, law, and economy, but the notion can also be applied to science, which has its own rules, hierarchies, styles of reasoning, justification processes, and so on.

Other theoreticians including Andrew Pickering (1992a), Michael Lynch (1993), Joseph Rouse (1996, 2002), or Karin Knorr-Cetina (1999) put more emphasis on material practices directed towards the reorganization of things. From those and other studies it became clear that in scientific practice (as well as in clinical practice, for instance) many different factors play a role, and are not directly linked to epistemic questions. They are nevertheless indispensable for doing scientific research and are deeply engraved in what is usually perceived as context-free, generalized scientific knowledge. Knowledge in systems biology may therefore rely, for instance, on the specific ways data are produced, stored, and managed or on other local circumstances for the production of order in masses of data about a certain scientific object, that is, their interpretation. Furthermore, despite its heterogeneity, most conceptions of systems biology agree on the fact that it is a multi- and interdisciplinary endeavor (Huang and Wikswo 2006). Consequently, the turn towards a systems-oriented approach in biology will most likely not be successful without extended interdisciplinary cooperation. What kind of impact will the different disciplinary backgrounds as well as related styles of thought and practice of the participating scientist have on the development of system biology, its agenda, and performance?

Far-reaching changes are expected first and foremost for the disciplines contributing to systems approaches in the life sciences and benefiting from it, including biology, physiology, medicine, ecology, mathematical modeling, and so on. For instance, systems biology comprises and merges techniques and practical approaches belonging to different disciplines such as the ones coming from the Omics on the one hand and mathematical modeling coming from the physical and computer sciences. These changes and the practical and intellectual challenges associated with it will not only affect practices and concepts of biology and other disciplines working together in interdisciplinary systems-oriented projects, but also their social constitution as an organization. As outlined before, systems biology is—at least today—an interdisciplinary enterprise with ambiguous academic identity and open disciplinary boundaries. In interdisciplinary cooperation new social interactions are created and organizational structures established and funding strategies explored. These interactions involve participants with diverse motivations, interests, and agendas that could make communication, work, and progress in this field far more challenging compared to traditional disciplines. Calvert and Fujimura (2011) noted and studied epistemic tensions and issues emerging in the interdisciplinary context of systems biology. Ethnographic studies based on participating observations in laboratories and research institutions illustrate such difficulties (Knorr-Cetina 1981). With regard to systems biology, Miles McLeod and Nancy J. Nersessian (2013) found, for instance, that biologists are confronted with numerous cognitive and technical constraints in laboratories of integrated systems biology. Compared to other fields, they also inhabit largely unstructured task environments that make the production of models far from straightforward (McLeod and Nersessian 2013, 35). Hence, social dynamics and good interaction will be decisive factors for the design, conduct, and accomplishment of successful projects. Thus, "proper 'social engineering' will have greater role in scientific project planning and management in the future" (Kitano et al. 2011, 323).

The expected changes for the life sciences are reflected in different statements of systems scientists. Some are convinced that modern systems biology is already contributing to a "radical transformation of molecular life sciences and biomedicine" (Medina 2013, 1034). Others, critical of the shortcomings of reductionist molecular biology, even expect that "mastering the new non-linear, holistic, complex way of thinking [...] will, undoubtedly, enable one to make less errors doing research" (Skurvydas 2005, 7), obviously referring to the envisioned potential of a more holistic systems biology to overcome the limits, fallacies, and blind spots of reductionism. However, today, we only have superficial ideas about what kind of challenges the establishment of systems approaches and associated experimental or modeling practices, the intensive use of information and communication technology (ICT) infrastructure, and interdisciplinary collaboration really pose to scientists working in the field. We also know very little about how these alterations in practice retroact on the constitution of the participating disciplines. Will they enhance or decrease the methodological and epistemic differences among the scientific communities involved?[14] What is urgently needed therefore is a much more detailed analysis of "systems biology in practice" or "systems biology in action" which has the potential to provide some answers to the questions posed above. Admittedly, such an analysis would certainly require further theoretical differentiation of the notion of practice and the development of adequate mixes of methods suited to examine such practices. Despite these shortcomings and challenges we were motivated by this quite empty space in current social science research on systems biology to conduct an extensive case study on systems cancer research. By carefully analyzing and describing the conception and realization of an ICT infrastructure in the domain of systems-oriented and interdisciplinary cancer research we aimed to get a deep insight into the practice of systems biology and one of its fields of application. This allowed us to explore the contextual factors of systems biology's practice that have an immediate impact on the researchers, their interactions, and their views on ICT supporting systems science (see Chap. 4).

1.1.2.3 Societal Context: Coproducing Science

As do all sciences, systems biology did and does not develop in a vacuum. Many societal factors act upon it when it starts to operate on natural things in the social world. Science produces new knowledge, products, and procedures that can have far-reaching consequences. Society will react to such developments in multiple ways, for instance, by implementing new processes that aim at balancing the interests between different actors and translate them into political options; by new bioethical principles, best practice rules or regulations, or by establishing assessment, oversight, and monitoring procedures. Such instruments belong to the realm of law, governance, and

[14] For the idea that practice may create epistemic differences among communities see Brown and Duguid (2001) albeit they discussed this question in the context of a firm.

bioethics, which have been called "Humantechniken" (anthropotechniques) by Helga Nowotny and Giuseppe Testa (2009). They help to adapt technology to human needs, thus stabilizing the new configurations of natural and social order. Hence, a close relationship exists between the reordering of nature and reordering of society (Jasanoff 2004a), an intertwined process, which has also been called a "co-production" of modern societies (Jasanoff (ed) 2004b). This relationship, which shapes science as well as society, has to be considered when new scientific developments and their implications are to be analyzed and assessed.

In wealthy Western societies support for science has been readily available when new visions emerged or new approaches appeared to be promising in terms of scientific, economic, or medical benefit. Systems biology has been welcomed as such an approach, especially because molecular genetics did not live up to promises delivered earlier and its research agenda no longer opened up new perspectives for knowledge generation (Rheinberger 2008). But with the establishment of systems biology fundamental changes in the life sciences and beyond its narrow academic and societal context loom at the horizon. They may be as or even more far-reaching than the ones that came along with molecular biology after its inception more than half a century ago. First of all, they will affect our cultural understanding of life. Up to now, a mechanistic conception of the organism (including the functioning of the human body) prevailed in the life sciences. With the advent of systems-oriented approaches and a new interest in a holistic understanding of biological entities, such a conception may change. However, this depends very much on how holism is interpreted and translated into research strategies. If, for instance, the view gains the upper hand, that organisms are phenotypically and behaviorally more than the sum of their parts and open to environmental influences this could result in different research strategies compared to the current paradigm of molecular biology, which is still dominated by a gene-centric, deterministic view (Keller 2002). Furthermore, it could also culminate in different socioscientific perceptions of living entities and, hence, in the development of alternative strategies in applied sciences as well as in different fields of action in society.

In addition to resulting in new insights into the complexity and dynamics of biological processes or mechanisms of emergence, systems biology also promises to be fruitful for practical applications. It is expected that progress and important achievements in different applied fields will be reached. In particular, expectations in medicine are high, with emphasis on cancer, heart, vascular, and infectious diseases (Hood et al. 2004; Wolkenhauer et al. 2009; Medina 2013). In this context, researchers interested in systems medicine often stress the shortcomings of contemporary reductionist views of human diseases. Although it is generally acknowledged that this approach has served society well for many years, systems biology applied to human disease is hoped to offer a unique approach for developing individualized (personalized) treatment strategies (Loscalzo and Barabasi 2011, 619). This vision "of a personalized systems medicine" is propagated extensively by many different actors from systems medicine (Hood et al. 2012). Even though systems medicine is the most prominent vision of application in the scientific as well as public perception of systems biology, there are other fields of applied sciences where a profound and positive impact of emerging systems biology is expected. For instance, far-reaching implications have been stated for agriculture, energy production, or environmental protection.[15]

Whether visions and expectations will become reality is a matter of debate and future development. It also has to be kept in mind that visions and expectations, too, can change social reality, exert social power, and generate implications in the social and cultural sphere (Grunwald 2004, 2009). Potential implications of systems biology regarding social, legal, or ethical topics have already been pointed out by scientists active in systems biology as well as social analysts of science (O'Malley et al. 2007; Drell 2007; Federoff and Gostin 2009) even though the implications are altogether positively anticipated. A holistic approach in medicine, for example, promising greater precision in diagnosis, opportunity for earlier intervention, risk-based prevention, individualization of care, and optimization of the patient–clinician interface and, hence, benefit to patients and society, may also lead to changes regarding the social organization of clinical sciences, issues of informed consent, allocation of resources, and social justice, which may not only be beneficial. Furthermore, privacy issues, for example, connected to the collection, storage, and processing of large amounts of personal genomic data for overarching health surveillance in the context of personalized medicine, must be taken into account, especially when data handling involves commercial cloud services (Dove et al. 2014).

These are just a few examples that demonstrate the interdependencies between science and society. They have always been complex and sometimes conflict-laden, especially when risks, social challenges, or ethical problems arise from new knowledge and/or its application. The societal context is populated with different stakeholders, experts, companies, organizations, governmental bodies, funding agencies, interest groups, and so on which participate in the discourse. Two sectors of this complex formation are especially important and therefore of interest for our analysis: science policy, and the public who perceive scientific development via the media. Science policy is one of the most important actors for establishing new research programs and financing research. In Germany, for example, since 2001 systems biology has been continuously funded via research programs set up by governmental bodies such as the Federal Ministry of Education and Research. The programs, however, favor large-scale project-oriented research, such as modeling liver cells or understanding aging processes. The aim of such projects is not only doing subject-related investigations, but also to set up infrastructures, explore new methodologies, and focus on medical or other applications. Thus, they have two effects: they are a precondition for the establishment of interdisciplinary research, and for stabilizing it and giving it some continuity. On the other hand, they also create discontinuity and uncertainty as funding is limited to the time frame of the projects. In addition to issuing research programs, funding organizations also discuss, deliberate, and commission statements regarding possible ethical and societal

[15] See, for example, Trewavas (2006), and *Institute for Systems Biology*, https://www.systemsbiology.org/about-systems-biology. Accessed Oct 28, 2014.

implications, and regulation. Such issues, however, are also discussed by scientists, the media, public interest groups, nongovernmental organizations, and industry. These discussions show how the different societal actors perceive systems biology and the science policy related to it. Up to now, little research has been done on how the development of systems biology relates to science policy, and how this is seen from the perspective of different societal actors, scientists included. Chapter 5 of this book represents an analysis of science policy on systems biology from these different perspectives.

From what has been said before it follows that the media are of vital importance for science. They communicate scientific achievements to the public, reflect on them, report on the public's discussion, and transfer them back into science. Hence they are an important part of the societal context of science. Quite often it is left unquestioned that the perception of a particular scientific development by the media and the public does not vary between different countries, especially if they are similar in language, geographic location, ethnic composition, and other sociocultural factors. We show that this assumption needs to be corrected. In Chap. 6 we present the results of a comparative media analysis that focuses on the images of systems biology communicated to the public in Germany and Austria, and which identified significant differences between the two countries. Here again we demonstrate that the context of systems biology shapes its perception and evolution, and that contextualizing this new scientific development is essential for analyzing and understanding its epistemic and cultural presuppositions as well as its implications for science and society.

1.2 An Avenue for Understanding Systems Approaches in the Life Sciences

1.2.1 A Social Science Perspective on Concepts, Practices, and Discourses

The first sections of this chapter showed that science, culture, and society are deeply intertwined in specific ways that have to be taken into account if one wants to understand scientific developments. This finding is in accordance with current social and cultural studies of science which mark "a loosely amalgamated, multidisciplinary research field drawing from history, anthropology, feminist theory, sociology, and philosophy of science." Its participants emphasize that scientific practices are historically situated, meaningful patterns of interaction with the world. Cultural studies offer an "interpretive, critical engagement with scientific practices, rather than an explanation of their outcome" (Rouse 2001, 1). In order to understand why and how systems biology did emerge, where it came from, and in which direction in may develop, as well as to gain some ground for estimation of possible implications, we therefore have to consider the contextual nature of this new approach and study it from different disciplinary, theoretical, and methodological perspectives.

Coming from such different perspectives such as linguistics (MD), social and cultural anthropology (IP), educational science and biology (AB), biology, and theory and sociology of scientific knowledge (RK) and, altogether, the social and cultural studies of science, we are interested in a variety of problems and questions regarding systems biology. They concern its emergence, development, and characteristics on the one hand, and its relation to cultural, scientific, and societal contexts. Such a context-sensitive analysis not only requires different analytical perspectives in order to grasp the variety of levels and dimensions involved in the formation of systems biology, but also different methodical approaches. For instance, scientists explain their goals, concepts, procedures, and results by the means of spoken or written language. By doing this they make transparent why they are interested in their research subject, what their goals are, what kind of hurdles they see, and what visions they have. However, they also use specific terms, notions, or concepts as well as rhetorical means, metaphors, and linguistic structures in order to describe what they think and want to communicate to colleagues or other actors within or outside science. For those involved in the social and cultural analysis of science, language, and discourse are therefore extremely important sources of information. They not only reveal how scientific objects are perceived and conceptualized, but also how scientists think about their work and their relationship to other sectors of culture and society. As already outlined in Sect. 1.1.2, speech acts can also be perceived as social actions that involve cognitive as well as social dimensions, which can be analyzed by different methodical approaches and means.

By exploring systems biology, we analyzed documents as diverse as scientific publications, policy papers, and newspaper articles. But first of all, we listened to individuals and experts active in systems biology as well as in science policy, and involved in the public communication and perception of systems biology. Therefore, semi-structured interviews and focus groups were our most important data source. By applying methods and approaches from the social sciences, but also from linguistics to these data, we provide interpretations of our findings and draw conclusions that will help us to understand better the origin, the current development, and possible future course of systems biology.

The methods we used are mainly derived from the linguist repertoire (analysis of linguistic structure of narratives about science, metaphor analysis) and from those of the qualitative social sciences (discourse analysis, context analysis, media analysis, case studies), complemented by reconstructive attempts coming from the history and philosophy of science. The methodical approaches are explained in detail in the coming chapters of this book. However, metaphor analysis plays a prominent role in this canon, especially with regard to the analysis of important concepts in systems biology (Chap. 2) and with regard to media analysis (Chap. 5). It denotes a newer approach in qualitative social science research based on the work of Lakoff and Johnson who showed that common phrases in everyday language are based on metaphorically structured concepts of thought, which determine how we act and feel (Lakoff and Johnson 1980). Metaphor analysis is an interdisciplinary project, where results and insights from many different disciplines converge, coming especially from linguistics and qualitative social science research. It implicates new chances to build bridges between different disciplines, but, even more important, to gain new insights about science, its concepts and its perceptions by different groups in society (Schmitt 1997). Another methodical approach located in the interface between linguistics and the social sciences is narrative analysis which was applied in order to analyze the stories about the history and future of systems biology encountered in reviews and scientific articles from the field of systems biology and in expert interviews with systems biologists. It was conceptually combined with other social science approaches, which share an interest in how pasts, presents, and futures are constructed, imagined, and combined via narrative representations and practices. Their aim is to trace converging, mutually enhancing or conflicting patterns of interpretation by which individuals of a social group such as a scientific discipline create legitimate pasts, presents, and futures (Chap. 3). Practices, especially the ones related to the interdisciplinary interplay between molecular biology, computer sciences, and biomedical research, were analyzed by the instrument of case study (Chap. 4). It is derived from the methodical repertoire of the social sciences, especially ethnography. The case—which can be persons, events, decisions, periods, projects, or institutions—is empirically inquired within its real-life context by using multiple sources and one or more methods (Thomas 2011). By carefully describing an individual case, such a study can give deep insight into a defined field thereby helping to identify its characteristic features, allowing us to formulate questions or postulate hypotheses that can then be explored further in a wider field in order to get more generalizable results. Finally, policy analysis was applied in order to explore how various actors from different societal groups and fields, including science, perceived and evaluated science policy related to systems biology. Originating in the political sciences, the term *policy analysis* describes social science approaches that aim at systematically dissecting and analyzing specific policy fields. Focusing on the discourse of science policy in science, media, industry, nongovernmental organizations, and public interest groups, questions related to systems biology's funding mechanisms and constraints, and to its possible implications and regulations were explored, using documents pertinent to science policy and funding as well as expert interviews with different actors as a data basis (Chap. 5).

1.2.2 Goals, Hypothesis, and Content of the Book

The study presented here explores the development of systems biology as a new approach to life. Its main aim is to contribute to a better understanding of the presuppositions and possible implications of the current shift in molecular biology towards a systems-oriented perspective for science and society. In conducting the research that provided the basis for this publication, we were not only interested in systems biology's capacity to give rise to a different or better understanding of complex biological entities such as cells or organisms, but also in the cognitive, social, and policy framings and contexts, in which this scientific approach has emerged and develops, and by which it will be shaped. Our starting point was the hypothesis that such an understanding requires an intimate knowledge of cultural, practice-related, and social or societal factors and conditions which frame and influence it, and such factors and conditions can be regarded as specific contexts that shape systems biology in a particular way and are in return molded by it as well.

The results of this study are laid down in this book. It is a collective monograph that has been conceptualized, continuously discussed, and written as a common interdisciplinary endeavor. Therefore, all authors assume responsibility for the written work, although differently with regard to individual chapters. The overall concept of the project was developed by Regine Kollek together with Martin Döring and the empirical work was done by Martin Döring (Chaps. 2 and 3), Imme Petersen (Chaps. 4 and 5), and Anne Brüninghaus (Chaps. 5 and 6). The first versions of the written texts were drafted by Regine Kollek (Chaps. 1 and 7), Martin Döring (Chaps. 2 and 3), Imme Petersen (Chaps. 4 and 5) and Anne Brüninghaus (Chaps. 5 and 6). They were discussed within the group several times and revised in an iterative process. Authorship order of the different chapters corresponds to the extent of the individual contribution; however, it must be clear that the book is based on collective work and authorship.

It consists of seven chapters. In addition to this first chapter, which introduces the subject of systems biology, the conceptual framework and the contents of this book, Chap. 2 investigates the framing of basic epistemic concepts of life, system, reductionism, holism, and model by scientists working in systems biology. Based on a corpus of written evidence and interviews conducted with system biologists in Germany, we analyze the metaphorical frameworks underlying their conceptualization to tackle implicit meanings and the practical relevance ascribed to them. It will become apparent that—to some extent—different professional backgrounds bear an impact on the framing of different concepts and that heterogeneous interpretations prevail. The results underline the need for theoretical clarification of basic epistemic concepts in systems biology and the implementation of a science philosophy curriculum as a basic ingredient of university education. Both aspects are important to avoid methodological and theoretical fallacies that restrict the innovative potential of systems biology.

Chapter 3 takes a closer look at the stories about the history and future of systems biology as encountered in articles stemming from the scientific field of systems biology and expert interviews with systems biologists. Our approach to explore such stories as a source for imagination about the past and the future of a discipline is based on a narrative analysis and combined with insights from other social sciences approaches which have in common that they share an interest in how pasts, presents, and futures are constructed, imagined, and combined via narrative representations and practices. Their aim is to trace converging, mutually enhancing, or conflicting patterns of interpretation by which individuals of a social group such as a scientific discipline create legitimate pasts, presents, and futures. With regard to systems biology, a combined method of document analysis and expert interviews was used to merge approved historical visions as found in research papers and more colloquial constructions of systems biology's past, present, and future. The narrative structures that developed pasts, presents, and futures laid open the conceptual foundations they rely on and from which further investigations into the present, pasts, or futures can start. Beyond this, it became clear that systems biology might profit from reflexivity with regard to further development of a disciplinary identity and to exploration of new theoretical and methodological issues at stake.

Chapter 4 is dedicated to the scientific practice of systems biology. Its starting point is the fact that information and communication technology (ICT) infrastructures are needed to facilitate the management, access, and sharing of the plethora of data on biological structures and processes on which systems biology is based. Although such infrastructures are essential for research and collaboration, they are often not regarded as being part of knowledge production. In contrast to this, we hypothesize that ICT infrastructures are not mere service facilities to support research activities, but enable and restrict doing systems research at the same time. Based on a case study in systems research on cancer, we argue that the understanding and modeling of biological systems is profoundly shaped by ICT technologies and their underlying conceptualizations. Furthermore, from the perspective of sociological actor network theory, our analysis also showed that such ICT infrastructures will become new powerful actors for knowledge production within the knowledge-producing community of systems biology. Individual scientists and research institutions often neglect the challenges related to standardization, integration, and management of data that complicates and sometimes impedes innovation and translation of new developments into practice. This implies that standardization and integration in systems biology are as important as data generation.

Chapter 5 examines the science policy of systems biology and perspectives thereof. It is based on interviews with actors from different fields such as science, politics, media, and economy. The scientists' perception of how science policy conceptualizes and assesses systems biology is contrasted with that of societal actors. Our analysis shows that the discourses in these fields are interconnected; they influence each other and mutually interact. Results coming from this analysis relate to the identity of systems biology as a new science, to the similarity of the scientific and public images of what systems biology is, and to the sustainability of funding. Although participation of the general public in the discussion is seen as important by politics and the media, it does not seem to be important for scientists. This prompts the question whether public participation in science in general, and in systems biology in particular, is ascribed an appropriate role.

Media are a principal mean for communicating science and its achievements to the public, for the public's discussion of science, and for transferring public opinions and perspectives back into science. In Chap. 6 the public perception of systems biology in Germany and Austria in the print media is analyzed, both quantitatively and qualitatively based on a sample of print media. This analysis focuses on the images of systems biology communicated to the public, including its application, on research funding, and on regulation. In addition, it takes into account national differences between Germany and Austria. Images are derived from an analysis of metaphors found in the print media that enables us to describe the frames and concepts upon which they are based. As we compare the public images of systems biology in Germany and Austria, we find some significant differences between both countries in the predominant metaphorical frames. The public image is well reflected in these metaphors, and we suggest that they have an important role for systems biology as an emerging science.

Instead of merely summarizing the content of the book, Chap. 7 rather aims to reveal basic assumptions and constitutive conditions of systems biology, and to embed the results of our study in a broader scientific and societal context. It first carves out some presuppositions of contemporary science in general, analyzes the past and present presuppositions of systems biology, and reflects them with regard to different paradigms and to past and future developments. After that it turns to the practice of systems biology by focusing especially on its dependence on ICT, and the epistemic impact this has on the structure and content of the data on which systems research relies. It then addresses the question of whether systems biology is an approach or a discipline and offers a new and refreshing answer to this lasting controversy. This is followed by a critical reflection on science policy pertinent to systems biology and how it is perceived by researchers active in the field. The final section takes up the initial question of this book of how the cultural and societal contexts as well as the demands and rules of scientific practice influence and shape this new development in the life sciences. It wraps up our findings and draws some conclusions with regard to the further development of systems biology.

The chapters reporting on the contents and results of our work are supplemented with a glossary that explains relevant terms and key concepts, mainly from the social sciences, from linguistics, and from philosophy of science, which have been used in the book.

References

Andersen H (2001) The history of reductionism versus holistic approaches to scientific research. Endeavour, 25(4):153–156.

Austin JL (1965) How to do things with words. Oxford University Press, Oxford

Baianu IC (2006) Robert Rosen's work and complex systems biology. Axiomathes 16(1–2): 25–34

Benedict R (1934) Patterns of culture. Houghton Mifflin, New York, NY

Beskow LM, Burke W (2010) Offering individual genetic research results: context matters. Sci Transl Med 2(38):38cm20. doi:10.1126/scitranslmed.3000952

Bongaerts G (2007) Soziale Praxis und Verhalten – Ueberlegungen zum Practice Turn in Social Theory. Z Soziol 36:246–260

Bonß W, Hohlf1eld R, Kollek R (eds) (1993) Wissenschaft als Kontext – Kontexte der Wissenschaft. Junius Verlag, Hamburg

Bonß W, Hohlfeld R, Kollek R (1994) Vorüberlegungen zu einem kontextualistischen Modell der Wissenschaftsentwicklung. Dtsch Z Philos 42(3):439–454

Borodina I, Nielsen J (2005) Fromgenomesto in silico cells via metabolic networks. Curr Opin Biotechnol 16:350–355

Bourdieu P (1977) Outline of a theory of practice. Cambridge University Press, Cambridge

Breitling R (2010) What is systems biology? Front Physiol 1(Article 9):1–5

Brown JS, Duguid P (2001) Knowledge and organization: a social-practice perspective. Organ Sci 12(2):198–213

Bruggeman FJ, Westerhoff HV (2007) The nature of systems biology. Trends Microbiol 15(1): 45–50
Burns T (1992) Erving Goffman. Routledge, London
Calvert J, Fujimura JH (2011) Calculating life? Duelling discourses in interdisciplinary systems biology. Stud Hist Philos Biol Biomed Sci 42(2):155–163
Cardinale S, Arkin AP (2012) Contextualizing context for synthetic biology – identifying causes of failure of synthetic biological systems. Biotechnol J 7:856–866
Clarke A, Parsons E (eds) (1997) Culture, kinship and genes. Towards cross-cultural genetics. St. Martins's, New York, NY
Conti F, Valerio MC, Zbilut JP, Giuliani A (2007) Will systems biology offer new holistic paradigms to life sciences? Syst Synth Biol 1(4):161–165
Cowley AW (2004) The elusive field of systems biology. Physiol Genomics 16:285–286
Dilley RM (2002) The problem of context in social and cultural anthropology. Lang Commun 22:437–456
Dove ES, Joly Y, Tassé AM, (2015) Public Population Project in Genomics and Society (P3G) International Steering Committee, International Cancer Genome Consortium (ICGC) Ethics and Policy Committee, Knoppers BM, Public Population Project in Genomics and Society P3G International Steering Committee, International Cancer Genome Consortium ICGC Ethics and Policy Committee (2014) Genomic cloud computing: legal and ethical points to consider. Eur J Hum Genet 23:1271–1278. http://www.nature.com/ejhgljournal/vaoplncurrent/full/ejhg2004196a.html
Drell D (2007) The relation of ethics and science: a commentary on "The study of socioethical issues in systems biology". Am J Bioeth 7(4):79–80
Fang FC, Casadevall A (2011) Reductionistic and holistic science. Infect Immun 79(4):1401–1404
Federoff HJ, Gostin LO (2009) Evolving from reductionism to holism: is there a future for systems medicine? JAMA 302(9):994–996
Floridi L (2012) Turing's three philosophical lessons and the philosophy of information. Phil Trans R Soc A 370:3536–3542
Foster MW (2007) Everything old is new again, including systems biology. AJOB 7(4):85–86
Garfinkel H (1967) Studies in ethnomethodology. Polity Press, Malden, MA
Gatherer D (2010) So what do we really mean when we say that systems biology is holistic? BMC Syst Biol 4(22):1–12
Geertz C (1973) The interpretation of cultures. Basic Books, New York, NY
Giddens A (1984) The constitution of society. Outline of the theory of structuration. University of California Press, Berkeley, CA
Gillin JL, Gillin JP (1948) Cultural sociology. An introduction to sociology. Macmillan, New York, NY
Goffman E (1974) Frame analysis: an essay on the organization of experience. Harper and Row, London
Goffman E (1981) Forms of talk. University of Pennsylvania Press, Philadelphia, PA
Goffman E (1983) Felicity's condition. Am J Sociol 89(1):1–53
Gomez-Cabrero D, Abugessaisa I, Maier D, Teschendorff A, Merkenschlager M, Gisel A, Ballestar E, Bongcam-Rudloff E, Conesa A, Tegnér J (2014) Data integration in the era of omics: current and future challenges. BMC Syst Biol 8(Suppl 2):I1
Goodman AH, Heath D, Lindee MS (eds) (2003) Genetic nature/culture. Anthropology and science beyond the two-culture divide. University of California Press, Berkeley, CA
Goodwin C, Duranti A (1992) Rethinking context: an introduction. In: Duranti A, Goodwin C (eds) Rethinking context: language as an interactive phenomenon. Cambridge University Press, Cambridge, pp 191–227
Green S, Wolkenhauer O (2012) Integration in action. EMBO Rep 13(9):769–771
Grunwald A (2004) Vision Assessment as a new element of the Technology Futures Analysis Toolbox. In: Proceedings of the EU-US scientific seminar: new technology foresight, forecasting & assessment methods. Seville, 13–14 May 2004

Grunwald A (2009) Vision assessment supporting the governance of knowledge – the case of futuristic nanotechnology. In: Bechmann G, Gorokhov V, Stehr N (eds) The social integration of science. Institutional and epistemological aspects of the transformation of knowledge in modern society. Edition Sigma, Berlin, pp 147–168
Halliday MAK, Hasan R (1985) Language, context and text: aspects of language in a social semiotic perspective. Oxford University Press, Oxford
Hansen KP (2011) Kultur und Kulturwissenschaft, 4th edn. UTB, Tübingen
Harris M (1983) Cultural anthropology. Harper & Row, New York, NY
Hood L, Balling R, Auffray C (2012) Revolutionizing medicine in the 21st century through systems approaches. Biotechnol J 7(8):992–1001
Hood L, Flores M (2012) A personal view on systems medicine and the emergence of proactive P4 medicine: predictive, preventive, personalized and participatory. N Biotechnol 29(6):613–624
Hood L, Heath JR, Phelps ME, Lin B (2004) Systems biology and new technologies enable predictive and preventative medicine. Science 306(5696):640–643
Hood L, Rowen L, Galas DJ, Aitchison JD (2008) Systems biology at the Institute for Systems Biology. Brief Funct Genomic Proteomic 7(4):239–248
Huang S, Wikswo J (2006) Dimensions of systems biology. Rev Physiol Biochem Pharmacol 157:81–104
Ideker T, Galitski T, Hood L (2001) A new approach to decoding life: systems biology. Annu Rev Genomics Hum Genet 2:343–372
Jasanoff S (2004a) Ordering knowledge, ordering society. In: Jasanoff S (ed) States of knowledge: the co-production of science and the social order. Routledge, London, pp 13–45
Jasanoff S (ed) (2004b) States of knowledge: the co-production of science and the social order. Routledge, London
Jones R (2004) The problem of context in computer mediated communication. In: LeVine P, Scollon R (eds) Discourse and technology: multimodal discourse analysis. Georgetown University Press, Washington, DC, pp 20–33
Keller EF (2002) Making sense of life. Explaining biological development with models, metaphors, and machines. Harvard University Press, Cambridge
Kirschner MW (2005) The meaning of systems biology. Cell 121(4):503–504
Kitano H (2000) Perspectives on systems biology. New Generat Comput 18:199–216
Kitano H (2001) Foundations of systems biology. MIT Press, Cambridge
Kitano H (2002) Systems biology: a brief overview. Science 295(5560):1662–1664
Kitano H, Ghosh S, Matsuoka Y (2011) Social engineering for virtual 'big science' in systems biology. Nat Chem Biol 7(6):323–326
Knorr-Cetina K (1981) The manufacture of knowledge. an essay on the constructivist and contextual nature of science. Pergamon Press, Oxford
Knorr-Cetina K (1999) Epistemic cultures: how the sciences make knowledge. Harvard University Press, Cambridge
Kohl P, Crampin EJ, Quinn TA, Noble D (2010) Systems biology: an approach. Clin Pharmacol Ther 88(1):25–33
Kollek R (1990) The limits of experimental knowledge. A feminist perspective on the ecological risks of genetic engineering. Issues Reprod Gene Eng 3(2):125–135
Kovala U (2014) Theories of context, theorizing context. JLT 8(1):158–177
Křeček J (2010) Holism and life manifestations: molecular and space-time biology. Physiol Res 59:157–163
Kroeber AL, Kluckhohn C (1952) Culture – a critical review of concepts and definitions. Vintage Books, New York, NY
Kuster DW, Merkus D, van der Velden J, Verhoeven AJ, Duncker DJ (2011) 'Integrative Physiology 2.0': integration of systems biology into physiology and its application to cardiovascular homeostasis. J Physiol 589(Pt 5):1037–1045
Lakoff G, Johnson M (1980) Metaphors we live by. University of Chicago Press, Chicago, IL
Latour B, Woolgar S (1979) Laboratory life: the construction of scientific facts. Sage, Beverly Hils, CA

Laubichler MD (2000) The organism is dead. Long live the organism! Perspect Sci 8(3):286–315
Loscalzo J, Barabasi AL (2011) Systems biology and the future of medicine. Wiley Interdiscip Rev Syst Biol Med 3(6):619–627
Lynch M (1993) Scientific practice and ordinary action. Ethnomethodology and social studies of science. Cambridge University Press, Cambridge
MacLeod M, Nersessian NJ (2013) The creative industry of integrative systems biology. Mind Soc 12:35–48
Malinowski B (1923) The problem of meaning in primitive languages. In: Ogden CK, Richards IA (eds) The meaning of meaning. Routledge & Kegan Paul, London, pp 146–152
Medina MÁ (2013) Systems biology for molecular life sciences and its impact in biomedicine. Cell Mol Life Sci 70(6):1035–1053
Musschenga AW (2005) Empirical ethics, context-sensitivity, and contextualism. J Med Philos 30(5):467–490
Nelkin D, Lindee MS (1995) The DNA mystique. The gene as a cultural Icon. Freeman Comp, New York, NY
Noble D (2006) The music of life: biology beyond the genome. Oxford University Press, Oxford
Nowotny H, Testa G (2009) Die gläsernen Gene. Die Erfindung des Individuums im molekularen Zeitalter. Frankfurt a. M., Suhrkamp
O'Malley MA, Brigandt I, Love AC, Crawford JW, Gilbert JA, Knight R, Mitchell SD, Rohwer F (2014) Multilevel research strategies and biological systems. Philos Sci 81(5):811–828
O'Malley MA, Calvert J, Dupré J (2007) The study of socioethical issues in systems biology. American Journal of Bioethics, 7(4):67–78
O'Malley MA, Dupré J (2005) Fundamental issues in systems biology. Bioessays 27:1270–1276
Ortner S (1984) Theory in anthropology since the sixties. Comp Stud Soc Hist 26(1):126–166
Pesce F, Pathan S, Schena FP (2013) From -omics to personalized medicine in nephrology: integration is the key. Nephrol Dial Transplant 2013(1):24–28
Pickering A (1992a) From science as knowledge to science as practice. In: Pickering A (ed) Science as practice and culture. University of Chicago Press, Chicago, IL, pp 1–28
Pickering A (ed) (1992b) Science as practice and culture. Chicago Press, Chicago, IL
Price AW (2008) Contextuality in Practical Reason. Oxford University Press.
Rheinberger H (2008) What happened to molecular biology? Bio Soc 3:303–310
Reckwitz A (2003) Grundelemente einer Theorie sozialer Praktiken: eine sozialtheoretische Perspektive. Z Soziol 32(4):282–301
Reckwitz A (2004) Die Reproduktion und die Subversion sozialer Praktiken. Zugleich ein Kommentar zu Pierre Bourdieu und Judith Butler. In: Hörning K, Reuter J (eds) Doing culture - neue Positionen zum Verhältnis von Kultur und sozialer Praxis. Transcript, Bielefeld, pp 40–54
Rosen R (1968) A means toward a new holism. Science 161(3836):34–35
Rouse J (1996) Engaging science. How to understand its practices philosophically. Cornell University Press, Ithaca, NY
Rouse J (2001) Cultural studies of science. Division I faculty publications. Paper 15. http://wesscholar.wesleyan.edu/div1facpubs/15. Accessed 15 Dec 2014
Rouse J (2002) How scientific practices matter. Reclaiming philosophical naturalism. University of Chicago Press, Chicago, IL
Sbisà M (2002) Speech acts in context. Language & Communication 22: 421–436
Schatzki T (2002) The site of the social. A philosophical account of the constitution of social life and change. Pennsylvania State University Press, Philadelphia, PA
Scheff TJ (2005) The Structure of Context: Deciphering "Frame Analysis". Sociological Theory, 23(4):368–385
Schegloff EA (1992) In Another Context. In: A. Duranti and C. Goodwin (eds.), Rethinking Context: Language as an Interactive Phenomenon. Cambridge: Cambridge University Press, pp 193–227
Schiffrin D (1994) Approaches to Discourse. Oxford: Blackwell

Schmitt R (1997) Metaphernanalyse als sozialwissenschaftliche Methode: mit einigen Bemerkungen zur theoretischen "Fundierung" psychosozialen Handelns. Psychol Gesellschaftskritik 21(1):57–86

Serafini L, Bouquet P (2004) Comparing formal theories of context in AI. Artif Intell 155:41–67

Skurvydas A (2005) New methodology in biomedical science: methodological errors in classical science. Medicina (Kaunas) 41(1):7–16

Thomas G (2011) A typology for the case study in social science following a review of definition, discourse and structure. Qual Inq 17(6):511–521

Tracy K (1998) Analyzing Context: Framing the Discussion. Research on Language and Social Interaction, 31(1):1–28

Trewavas A (2006) A brief history of systems biology. "Every object that biology studies is a system of systems." Francois Jacob (1974). Plant Cell 18(10):2420–2430

Tylor E (1920) [1871] Primitive culture, vol 1. New York, NY: J.P. Putnam's Sons

von Bertalanffy L (1968) General system theory. Foundations, development, applications. George Braziller, New York, NY

Walzer LJ (1990) Medicine in context: a review essay of the history of medicine. Am Hist Rev 95(5):1471–1484

Waylen KA, Fischer A, McGowan PJK, Thirgood SJ, Milner-Gulland EJ (2010) Effect of local cultural context on the success of community-based conservation interventions. Conserv Biol 24:1119–1129

Westerhoff HV, Alberghina L (2005) Systems biology: did we know it all along? In: Alberghina L, Westerhoff HV (eds) Systems biology: definitions and perspectives. Springer, Berlin, pp 3–9

Westerhoff HV, Kolodkin A, Conradie R, Wilkinson SJ, Bruggeman FJ, Krab K, van Schuppen JH, Hardin H, Bakker BM, Moné MJ, Rybakova KN, Eijken M, van Leeuwen HJ, Snoep JL (2009) Systems biology towards life in silico: mathematics of the control of living cells. J Math Biol 58(1–2):7–34

Wolkenhauer O (2001) Systems biology: the reincarnation of systems theory applied in biology? Brief Bioinform 2(3):258–270

Wolkenhauer O, Fell D, De Meyts P, Blüthgen N, Herzel H, Le Novère N, Höfer T, Schürrle K, van Leeuwen I (2009) SysBioMed report: advancing systems biology for medical applications. IET Syst Biol 3(3):131–136

Chapter 2
Basic Concepts of Systems Biology as Seen Through Systems Biologists' Eyes: Metaphorical Imagination and Epistemic Presuppositions

Martin Döring, Regine Kollek, Anne Brüninghaus, and Imme Petersen

Abstract After the successful structural analysis of the human and other organisms' genomes the last decade witnessed a fundamental shift in the area of research in molecular biology: the move into Omics. It produced a plethora of data that require methodological and conceptual approaches to systematize, integrate, and interpret data which go beyond a linear understanding of biological processes and systems. The promise of the rapidly developing field of systems biology is to extend—if not overcome—the methodological and theoretical limits set by previous research undertaken in molecular biology. Taking this contemporary development seriously, this chapter investigates the framing of basic epistemic concepts (life, system, reductionism, holism, and model) by scientists working in systems biology. Based on a corpus of written evidence and interviews conducted with system biologists in Germany, we analyze the metaphorical frameworks underlying their conceptualization to tackle implicit meanings and the practical relevance ascribed to them. It becomes apparent that (to some extent) different professional backgrounds bear an impact on the framing of different concepts and heterogeneous interpretation prevails. The results underline the need for theoretical clarification of basic epistemic concepts in systems biology and the implementation of a science philosophy curriculum as a basic ingredient of university education. Both aspects are important to avoid methodological and theoretical fallacies that restrict the innovative potential of systems biology.

Keywords Systems biology • Basic concepts • Conceptual analysis • Metaphor

---

M. Döring (✉) • R. Kollek • A. Brüninghaus • I. Petersen
Research Centre for Biotechnology, Society and Environment (FSP BIOGUM), University of Hamburg, Hamburg, Germany
e-mail: doering@metaphorik.de

M. Döring et al., *Contextualizing Systems Biology*,
DOI 10.1007/978-3-319-17106-7_2

Science and any branch of it are practically based upon a certain number of relevant or basic concepts. These concepts may take different forms in different disciplines, but within a specific discipline they are commonly shared and exhibit an imaginative mindset upon which basic methods and interpretations rest. This is also the case in the present context of systems biology. Generally conceived as a successor of molecular biology and heavily influenced by information and communication technology (ICT), it conceptually merged these two research fields promising not only new insights into the workings of biological systems but also a variety of innovations ranging from new medical or pharmacological applications to the development of new methods in biology (Sect. 5.1). The ICT-driven approach in systems biology represents a technologically induced and data-driven remathematization of biology that aims at a deeper understanding and better prediction of molecular processes at, between, and above all levels of biological organization. It is thus not surprising that the ambitious enterprise of systems biology has been welcomed at the wake of the new millennium as an improved approach of addressing and dealing with biological complexity.

The question, however, is what basic concepts systems biology exactly relies upon and what their content and practical use is. Both aspects are dealt with in this chapter by using an empirical approach based on a combination of discursive and linguistic approaches that provide insight into the conceptualization of important concepts used and applied in systems biology. The chapter addresses thus the following questions:

- What are basic concepts in systems biology?
- How are they framed and semantically conceptualized by scientists working in the field of systems biology?
- Are there any significant con- or divergences between scientists holding different professional backgrounds? If so, why and how do the concepts analyzed di- or converge?
- What are the possible implications contained in these basic concepts?

For this to be done, we start with a short overview of how epistemic concepts in biology have been analyzed to date (Sect. 2.1.1). Against this background, we outline the theoretical aspects of our language-oriented approach, which is mainly based on the analysis of metaphors and metaphorical concepts that are conceived of as basic mechanisms to produce, maintain, and share meaning (Sect. 2.1.2). This section is followed by methodological considerations about the data gathered and their analysis before we turn to the systematic investigation and interpretation of the concepts and their metaphorical expression encountered in expert interviews. Paradigmatic examples taken from the interviews are used to illustrate the metaphorical concepts that permeate and semantically structure the basic concepts analyzed. These are life (Sect. 2.2), system (Sect. 2.3), reductionism (Sect. 2.4), holism (Sect. 2.5), and model (Sect. 2.6). Once they have been analyzed, the final section of this chapter (Sect. 2.7) summarizes the findings and provides an assessment of the basic concepts as seen through systems biologists' eyes. It analyzes the aspects of intangible creativity nestling in the metaphorical conceptualizations of the basic concepts and partly compares them with everyday practices as alluded to in the interviews. Let us now turn to an impressionistic overview of research undertaken in the philosophy of biology and systems biology. Here, we outline the theoretical and methodological aspects of our analysis which are then applied throughout this chapter.

2.1 Basic Concepts and Implications

Questions about the epistemic dimensions of biology have been raised by a variety of scientists stemming from disciplines such as the philosophy or sociology of science, science and technology studies, the history of science, and other disciplines. They mostly converge in the fact that they understand scientific knowledge as an experiential, socially, and culturally generated construct. Even though there are different theoretical and methodological approaches within these different branches, science studies or the social study of science in general investigate the sociocultural and historical contingencies underlying the production, maintenance, and dispersion of scientific knowledge. The new philosophy of biology represents within this framework a subdiscipline that emerged at the start of the 1970s (Byron 2007) as a reaction to the traditional philosophy of science which was grounded in logical positivism and mainly addressed in physics. This led to fruitful debates and interactions of scientists and philosophers introducing a critical and above all reflexive perspective on the scientific enterprise of biology. Questions addressed in the philosophy of biology revolve around the structure and content of concepts or particular kinds of explanation that are, again and again, combined with questions addressing methodological or practical aspects. At the beginning, the philosophy of biology focused on evolutionary biology and biological systematics, but the philosophical grounding of molecular and experimental biology also received considerable attention. This was due to the advent of genetics and molecular biology, and a consequence of debates on whether biological disciplines could be reduced to molecular biology.[1] The further development of biology and the emergence of the life and biomedical sciences, including the previously mentioned Omics approaches, contributed to the development of a field that was—because of its tremendous methodical and technical advances—leading to ever better insights into genetic structures and mechanisms. However, it also provoked far- reaching ethical, social, and legal questions.[2] As a consequence, more and more scholars started to address and investigate such aspects innate in modern biology and medicine.

---

[1] We can only provide an impressionistic overview of the developments here. For further information see Sattler (1986), Sober (1993), Sterelny and Griffiths (1999), Hull and Ruse (2007) and Ayala and Arp (2009).

[2] Examples of this extensive body of work are Marteau and Richards (1996); Tutton and Corrigan (2004); or Forgó et al. (2010).

In this context, the recent advent of the interdisciplinary endeavor of systems biology represents an attempt to reintroduce a holistic perspective.[3] Based on a new technologically driven guise it mainly attracted attention from some researchers working in the overlapping areas such as the philosophy of biology, science and technology studies, and the history of science. Interest in the area of science studies was rather moderate even though different aspects have been investigated since 2005 by a rather small group of researchers. O'Malley et al. (2007) were among the first who addressed the issue of systems biology and they provided a fruitful conceptual overview of its socioethical issues. Dupré and O'Malley (2007) investigated questions about metagenomics and the impact of this discipline on reshaping biological categories and ontologies. O'Malley and Dupré (2005, 1250) also investigated fundamental issues in systems biology studying the identification of systems and the different causalities that operate at different levels of organization. Besides these more or less philosophical approaches Calvert and Fujimura (2011) studied epistemic problems and issues emerging in the interdisciplinary context of systems biology whereas Kastenhofer (2013) applied to this newly emerging approach the concept of epistemic cultures[4] (Knorr Cetina 1999) in order to analyze differences between systems and synthetic biology. Calvert (2007) also examined questions of patenting and problematized aspects of data-driven research (Calvert and Joly 2011) and De Backer et al. (2010) explored the conceptual and disciplinary borders between molecular systems biology. Ofran (2008) analyzed the emergentist's and reductionist's views underlying systems and molecular biology whereas Fujimura (2005) referred, though not systematically, to the relevance of metaphors in the conceptual language of post-genomic research and systems biology. Fox Keller (2002) provided a diachronic and conceptual analysis of synthetic and systems biology. She emphasized the importance of metaphors and models for making sense of observational and experimental data in biology providing a large overview and a deep insight into the development of biological thinking. The only paper explicitly investigating a basic concept is O'Malley's and Soyer's (2012) paper on integration in molecular systems biology: they and Green and Wolkenhauer (2012) depict the different meanings of integration and show how it has been discussed from scientific and philosophical points of view (O'Malley and Soyer 2012, 58; McLeod and Nersessian 2014).

Against this background, this chapter aims at an empirical analysis of life, system, reductionism, holism, and model as basic concepts in systems biology. Such an analysis has, to our knowledge, not been carried out to date. Emphasis is put on a grounded approach which means that data were gathered during qualitative expert interviews. These interviews were transcribed and thoroughly analyzed according to the converging requirements as outlined in grounded theory (Charmaz 2006;

---

[3] The notion of holism is further contextualized and explained in Sect. 2.5.

[4] The notion of epistemic cultures refers to the analysis of how scientific disciplines create knowledge. The concept refers to the idea that different disciplines possess intermingled scientific processes and social rationales which determine the way they do science and bear an impact on what kind of knowledge is created.

Clarke 2005; Corbin and Strauss 2008) and in the systematic analysis of metaphor (Schmitt 2000, 2005, 2011; Döring 2014). It is, furthermore, important to bear in mind that two fundamental assumptions underlie this chapter: first, that metaphors are a ubiquitous phenomenon in scientific language and thinking and, second, that the analysis of these linguistic images provides a valuable insight into the unconscious and implicit dimensions underlying the scientific theories, models, and concepts implicated in both (Paton 1996). A systematic analysis of metaphor thus holds the potential to unravel implicit foundations and connected presuppositions informing scientific thinking and acting because scientific language, like all sorts of human language and discourse, is permeated by metaphor. The systematic analysis of metaphor hence offers a constructive way to make transparent the semantic content of basic concepts in systems biology and it helps to better understand the styles of thought (Fleck 2011) in systems biology. These subconscious styles of thought and their structuration are of particular interest here as they bear an impact on how research in systems biology is done and by what kind of basic concept it is informed. It is, however, necessary to explain our analytical concept of metaphor as an essential ingredient in the workings of science in more detail. We therefore turn in the next section to a more general outline of its functions in language, thought, and action before we briefly describe the methods we applied to our data.

2.1.1 Science, Tacit Knowledge, and Linguistic Imagery

Nowadays it is a truism to state that metaphors and other kinds of linguistic imagery pervade and are creatively applied in scientific thinking (Brown 2008; Katherndahl 2014). Numerous scholars working in science and technology studies, in philosophy of science, and the social study of science have undertaken research on the constitutive role of metaphors and metonymies for scientific thinking, the development of concepts, theories, and methods of science. Important publications have paved the way for further research on the use of metaphor in biotechnological and biomedical science.[5] Especially this scientific field took up speed in the context of the human genome project at the turn of the century when attention was redrawn to the constitutive role of metaphors for science.

Numerous articles have since then been published on all sorts of aspects revolving around a large variety of topics ranging from the role of linguistic imagery in scientific thought or everyday practices via the pervasive role of metaphor in policy and regulatory discourses to the media-metaphorical framings of biotechnological developments and innovations. Especially the latter, often running under the heading of "public understanding of science", received considerable attention and this is why it is simply impossible to review all theoretical and methodological approaches

[5] See for example Black (1962), Gentner and Jeziorski (1993), Hesse (1970, 2005), Fox Keller (1992, 2002), Haraway (2004), Kay (1997, 2000), Knorr Cetina (1981), Maasen and Weingart (2000) and Nerlich et al. (2009).

applied to analyses. A closer look at the main bulk of research undertaken, however, indicates it in many cases lacks a systematic methodological approach and theoretical rigor about what analytical notion or theoretical concept of metaphor had been applied to the data gathered. It has, furthermore, been taken for granted that metaphor is an essential ingredient in constructing scientific meaning whereas its philosophical implications in terms of constructivism and objectivism are often omitted. It is thus not astonishing that important questions have not been investigated, such as, for instance: is there an existing reality or is the notion of reality just a construct science lives by? In this study, we take an experientialist position based on the assumption that our observations of biological phenomena are deeply shaped by social, cultural, and many other factors which thus contribute to form the knowledge and practices with which human beings wend their way through the world.

The philosophical implications of constructivism and objectivism have again and again been addressed by philosophers and psychologists such as Kant (1993), Giam Battista Vico (1990), Ernst Cassirer (1923), Karl Bühler (1934), and Nelson Goodman (1968), but it was mainly the historian and philosopher Michael Polanyi (1958) who started to identify and outline the shortcomings of so-called objectivism or logical empiricism in the 1950s. Polanyi himself was an experienced scientist and did not reject the notion of an existing reality. Adhering to an experiential notion of reality (Polanyi 1966), he was convinced that scientists in their daily life develop scientific theories and concepts on the basis of ideas about a hidden reality underneath the phenomena perceived. Following this conviction, ideas or visions are determined by imagination and intuition (Polanyi 1958) which define a so-called tacit knowledge. This tacit knowledge is based on experiences gathered from all kinds of encounters that are not communicable and provide a conceptual background which informs scientific thought and action (Polanyi 1966). This concept has much in common with early developmental psychology as outlined by Jean Piaget (1954) and Lew Vigotsky (2012) as Polanyi's work develops a comparable interactive and dynamic understanding of knowledge based on physical and social experiences and the humanly embodied conceptual framework. It is thus an experience-based approach that highlights the body in the mind (Johnson 1987). This experienced approach has (though not consciously) been revitalized by the philosopher Mark Johnson (1987, 2007) and the linguist George Lakoff (1987) in series of coedited and single-authored monographs (Johnson 1987, 2007; Lakoff and Johnson 1980, 1999; Lakoff 1987). Similar to Polanyi, Johnson and Lakoff underline the relevance of the human imaginative capacities and an embodied experience to construct what they call mental representations. They too build on the premise that imaginative capacities and an embodied experience are the basic ingredients for a subconscious form of tacit knowledge which materializes in spoken language and provides an interpretative access to so-called conceptual metaphors used in scientific reasoning. The distinction of linguistic and conceptual metaphors is of importance here because the latter could be understood as configurations underlying and structuring what Polanyi calls tacit knowledge. This means that the analysis of metaphor holds the possibility to reveal and analyze the cognitive patterns and processes used to reason about a scientific problem or concept. In this view, science and scientific reasoning should be understood as embodied processes of metaphorical reasoning that transforms knowledge into so-called experientially gained structures. This view obviously holds philosophical implications for notions such as objectivity and truth as they are not awaiting discovery (Putnam 1981, 1991, 1993). Thus, scientific facts are consequently a product of embodied and experiential metaphorical reasoning that provides a fragmentary but nevertheless important perspective towards reality.

Lakoff's and Johnson's experiential approach might hold serious epistemological limitations because metaphor as one of these experientially gained structures has long been conceived as a mere linguistic decorum and rhetorical device that contributes to the confusion of categorical distinctions between words and reality. Seen from this perspective of mere language games, the outlook is rather purely constructivist. Here, the experiential approach as created by Polanyi and put forward by Lakoff and Johnson offers a productive perspective because metaphor holds a great conceptual "[...] power to evoke images and complex ideas" (Chawla 2001, 115). It could be understood as a dynamic device or embodied mechanism anchored in experience that enables scientists to interpret and analyze their scientific problem under review in productive, imaginative, creative, and new ways. Scientific work and thought experiments basically involve imagination, and imagination is thus an endeavor based on accumulated and embodied experiences canalized and structured by scientific training. The resulting mental representations are of vital interest because they are metaphorical in character and hold the potential to provide access to tacit or unconscious or submerged knowledge at work in scientists' minds and practices. The aim consequently consists in uncovering and assessing the metaphorical forms of scientific reasoning and knowledge as reflected in the language and its imaginary use by scientists. In the next section, we take a closer look at the conceptual theory of metaphor and introduce the analytical tools that are used later in the analysis of basic concepts encountered in systems biology.

2.1.2 A Theoretical Slant on Metaphor

One central thesis of this chapter is that metaphorical reasoning lies at the heart of what scientists do in their everyday lives. Scientists rely on metaphors and metaphorical thinking when they communicate about and design experiments, formulate theories, develop models, make discoveries, and think about and apply basic concepts. Especially the latter are of specific interest as they provide an unconscious social and embodied knowledge against which scientific endeavors are carried out. Following the classical perspective, metaphor was long regarded as a purely rhetorical phenomenon acting on the level of words and linked only to poetic discourse or the aesthetic creativity of writers. It was therefore not considered as referring to a linguistically describable reality and rather relegated to the artistic use of language. However, according to most linguistic research on metaphor, it can no longer be regarded as a mere aesthetic figure in poetic discourse but must be understood as a ubiquitous phenomenon and constitutive element of cognition in everyday life that pervades and structures all kinds of discourse. This becomes apparent if we look at the following linguistic examples taken from a corpus of spoken language:

1. My assumption about the omnipresence of metaphor in science completely *collapsed* because my supervisor has questioned [...].
2. The results taken from her research *undermine*, at least to some extent, the argument that cancer could be, ah yes that it could be understood as genetically determined [...].

Both examples literally do not make sense but it is nevertheless astonishing that, as soon as one reads them, one understands their meaning immediately. The process of understanding these linguistic metaphors is in fact quite easy for most people who read these sentences and what is even more striking about them is the fact that the structures marked in the examples do not really appear to the reader or listener as metaphorical. Their metaphorical content only becomes obvious at second glance and provides an underlying image of arguments as objects or buildings that can collapse or be undermined. What is involved in the two examples quoted above is actually the very basic ingredient of metaphor, the metaphorical mapping. This means that a concrete domain of discourse (a building), called the *source domain*, projects its information and connected associations on an abstract domain of discourse (conceptualization of cancer), called the *target domain*. Thus, abstract entities, assumptions, or arguments could metaphorically be framed as buildings or objects which entail that they can collapse or be undermined even though they do not do such things in reality. What is interesting about the examples is not only the mapping process as a mechanism that conceptualizes abstract domains of knowledge but the intangible background knowledge implicated in it. The metaphorical transfer and its "[...] implication complexes [...]" (Black 1993, 28) enable a wide range of possible associations that open up avenues for creative thinking and acting. It is thus possible to talk about the foundation of an argument or, more creatively, to understand a theory as a building that lacks a roof. Another example taken from molecular biology might be illustrative and has been taken up as an example inter alia by Brown (2008, 25–26) because it shows how everyday language and practice might have once entered the realm of science. Brown states that one of the most active fields in molecular biology nowadays is devoted to research on how proteins change their shape and constitution in a solution. This branch of research investigates how proteins active in biological systems rearrange their chain lengths to maintain their characteristic shape. This active process was called *folding* due to a comparison or analogy between the process taking place within a protein and folding practices in the human world. In brief, the folding practices encountered in the human world were metaphorically mapped on the more abstract biological process taking place on the molecular level. As Brown (2008, 25) states, the metaphorical projection of the everyday concept of folding on a molecular biological process provoked a variety of questions that exerted an impact on further experimental arrangements and research undertaken and, at the same time, added an extra shade of meaning to the semantics of the verb "folding". What becomes clear is that the metaphorically used language creatively connects everyday discourse with scientific discourse. Metaphors thus "[...] play

[…] an essential role in establishing links between scientific language and [experiences taken from; the authors] the world" (Kuhn 1993, 539).

Thus far, we have encountered two analytically important characteristics of metaphor, namely that metaphor is—according to its etymology—a cognitive mapping mechanism which carries meaning from one domain of knowledge to another and structures the semantics of the target domain by using social, cultural, and bodily experiences.[6] A closer look, however, uncovers other important elements (Jäkel 1997, 40–42) which are relevant for our analysis. One is that metaphors, as already alluded to, hold a creative potential for thinking and acting due to the background knowledge implicated in them. To understand this process in more detail, it is important to see that the use of a linguistic imagery semantically highlights certain aspects while it hides others. Framing the way chemicals take to pass through a cell membrane metaphorically as a *channel* emphasizes the functionality of such a structure for transporting ions (see Brown 2008, 100–120). It is interesting that the noun "channel" was taken from everyday experiences because other words such as "corridor" or "tunnel" would have been available as well. The use of the word "channel", however, seemed to fit best due to the original idea of water or fluids in which ions can flow. With regard to extending the image, the metaphorical use of "channel" offers further opportunities to reframe or explore its implications. Thus, the introduction of certain membrane proteins to be conceptualized as sluices might hold the potential to block the transfer of ions and unintended reactions between molecules on the molecular waterway could be metaphorically framed as shipping accidents in the channel. The implications inherent in the channel image thus offer a variety of ways to postulate and explore creatively the functioning of the cell and possible ways to understand processes running on the molecular level as outlined in the shipping accident metaphor.

What is astonishing is that some of these images are self-explanatory whereas others such as the shipping accident metaphor require a certain amount of reflection to be understood. It follows that some images are more accessible than others in the sense that they conform more to experimental results than others. Furthermore it seems that the aspects of accuracy and comprehensibility are directly related to the important aspect of conventionality. What becomes thus apparent in the previous examples is the fact that the channel metaphor is rather easy to understand and it is nowadays an integral part in research. Lakoff and Johnson (1980) stated that scientific discourses are replete with and based on so-called *conventional metaphors* which semantically structure their content. This neatly links up with the previous example because the channel metaphor shows that a certain domain of discourse is structured by it even though empirical research indicates whole domains of discourse are based on a restricted set of conventionalized metaphors. Other good examples are the pervasive text and script metaphors used in the press coverage during the sequencing of the human genome (Nerlich et al. 2002; Nerlich and Clarke

[6] The notion of embodiment refers to the fact that the semantic content of metaphors is inter alia motivated by the human biological body.

2003; Döring 2005). They became a conventionalized image to convey the chemical structure of the DNA is information that could be read, understood, and—to use another metaphor—be rewritten in research. Another illustrative is the noun *cell* which denotes a small and functional biological unit that can be perceived through the microscope. Although the word nowadays rather appears to be a conventional noun, in the nineteenth century it was used first in the metaphorical sense because the elements perceived through the microscope in a monastery by a monk structurally resembled his cell. The initial metaphorical mapping has hence disappeared during the last decades as it underwent a process of standardization that finally changed the metaphor into a standalone word. Generally speaking, there are thus two kinds of metaphors: the novel ones that can be encountered at any time but also the *conventional metaphors* that often go unnoticed. Conventional metaphors, especially, develop underlying systems or models that deserve further attention due to their structuring force, inbuilt implications, and connected associations. This means certain domains of discourse are based on an underlying system of *conceptual metaphors* that materialize in the form of a variety of linguistic instantiations. Conceptual metaphors are thus to be conceived as cognitive meaning structures that help to make an abstract domain accessible. This aspect becomes visible in the conceptual metaphors used to frame mental activity. Jäkel (1997, 184–188) has shown that mental activity has been depicted in conventional metaphors such as IDEAS ARE OBJECTS, THINKING IS WORKING ON PROBLEM-OBJECTS WITH THE MIND-TOOL, or FORMING IDEAS IS SHAPING RAW MATERIAL whereas doing science has been metaphorically framed in terms of SCIENCE IS A JOURNEY or as SCIENCE IS THE STRUGGLE FOR THE SURVIVAL OF THE FITTEST. These conceptual metaphors develop coherent models or so-called cognitive models that represent experiential simplifications of an even more complex reality and at the same time provide a semantic structure which pervades scientific thought and practice.

In this brief overview of the conceptual theory of metaphor we have identified some basic characteristics of and assumptions on metaphor and types that are of vital importance for our analysis of basic concepts in systems biology. These are:

- Metaphors are based on a cognitive mapping process in which more concrete experiences are projected upon an abstract domain to make it semantically and cognitively accessible. The analysis of these mapping processes provides insight into the experientially informed processes of meaning making while it also opens up the possibility of analyzing and assessing possible implications transferred to the target domain.
- Metaphor is a ubiquitous phenomenon that pervades scientific discourses too. They are not an element that could be relegated to the realm of artistic discourses or poetics. Of special interest are the conventional metaphorical concepts because they subliminally shape a domain of discourse and often pass unnoticed.
- Metaphors possess a focusing function. They highlight certain semantic aspects of the discursive domain while hiding others. This offers the opportunity to analyze how a certain domain of discourse is framed and at the same time opens up the possibility to question current framings and to develop alternative ones.

- Metaphors are creative mechanisms for the production and shaping of meaning. This meaning and semantic productivity cannot be reduced to the propositions of the words involved. This aspect refers to the important aspect of malleability because metaphors hold the power to change or restructure ingrained thought patterns. In science, they clearly possess a heuristic function.
- Conceptual metaphors form so-called cognitive models. These models provide an experiential and simplified structure that semantically conceptualizes a whole domain of discourse. Cognitive models could be understood as cultural models of thought that determine the worldview of a social group or scientific discipline.

Having outlined the relevant and sometimes overlapping characteristics of metaphor here does not mean this list is exhaustive. It, however, provides a practical overview of the analytical aspects and assumptions of the conceptual theory of metaphor as first outlined by Lakoff and Johnson (1980). This plays an important role in the current context inasmuch as systems biology introduced a change in at least some biological concepts that is "accompanied by a change in some of the relevant metaphors in the corresponding parts of the network of similarities" (Kuhn 1993, 539). Our aim, consequently, consists in unraveling, analyzing, and critically assessing these metaphorical networks that inform the basic concepts of systems biology. But before we turn to our empirical analysis it is necessary to outline our methodological approach for the analysis of linguistic imagery. This is done in the next section.

2.1.3 Tracking and Analyzing Metaphor in Scientific Discourse

As we have outlined in the previous sections, metaphorical language and thought are deeply rooted in physical, social, and cultural experiences and play an essential role in science. How can data on these processes be raised and analyzed and what kind of method should be applied to do justice to the data raised? The methodological approach we chose to take represents a combination of linguistic (Jäkel 1997, 141–146; Döring 2005; Steen et al. 2010) and discourse analytical approaches[7] (Semino 2008; Döring 2014). These have been informed by recent attempts to analyze metaphor from a social science perspective systematically (Maasen and Weingart 2000; Kruse et al. 2012; Schmitt 2010, 2011, 2014).

The question we had was whether there are metaphors to be found in the scientific discourse on systems biology. For this to be partly answered, we started analyzing scientific reviews and edited volumes on systems biology to get a preliminary insight

[7] Discourse analytical approaches (as referred to here) investigate from an empirical point of view the language used to describe and frame a problem, situation, or prevailing topic under question. The analysis of different linguistic structures in the language reappearing helps us to better understand the social ascription and contesting of meanings.

into the current state of the art while at the same time a provisional analysis of metaphor was carried out. This procedure resulted in the insight that metaphors are at work in the discourse and led us at the same time to papers written inter alia by Ouzounis and Mazière (2006) and Bruggeman et al. (2005). The authors of the former made explicit reference to metaphor in their title "Maps, Books and Other Metaphors for Systems Biology" (Ouzounis and Mazière 2006, 6) however, the latter outlined that they "think that it is important to reveal the philosophy of notions such as life or cell to broaden the—sometimes—too narrow scope of systems biologists" (Bruggeman et al. 2005, 395). Both papers have in common that they emphasize the relevance of a reflexive perspective to analyze the philosophical and cultural embeddedness of systems biology. Having Polanyi's notion of tacit knowledge in mind, these statements provided by scientists working in the area of systems biology motivated us to start a systematic literature research to get a better understanding of systems biology. We thus gathered different kinds of written evidence such as conference proceedings, edited volumes, textbooks, scientific articles, and reviews to get a deeper insight into current topics and debates of systems biology. The literature search on scientific reviews was undertaken with the help of the search tool *Pubmed-Pubmed-Reminer* which offers a variety of search options to combine keywords and fine tune the search according to different parameters such as the date of publication, relevant journals, main scientists working in the field, and the main topics addressed. This helped us to set up a database of written sources that was then read and tentatively pre-structured. This procedure provided a first structure of the field of systems biology and also showed that basic concepts such as life, system, reductionism, holism, and model were quite often used. This led us to conclude that these concepts are of vital interest to the scientific community. Astonishingly, they are extensively used in the systems biology literature, but definitions or thorough discussions of them are more or less lacking. We come back to this point later on.

With this provisional result in mind and using the concept boundary object[8] (Griesemer and Star 1989; Bowker and Star 2000) as a heuristic device we decided to concentrate on such basic concepts as the main object of research. The aim was to address and assess their embeddedness and metaphorical structure (Ouzounis and Mazière 2006; Bruggeman et al. 2005) from an empirical point of view. For this to be done, we established a dataset of scientists working in systems biology in Germany whom we had identified during our literature research. We furthermore extended this list by undertaking an extensive search on the Internet that provided us with information about the contact details and, more important, with the main fields of research, the current professional status, and important publications of the respective scientists. To guarantee an adequate social distribution, representative scientists from different career levels in Germany were chosen, some of them coming from other countries than Germany. Twenty-five semi-structured interviews

[8] Boundary objects are socially constructed entities or things around which scientists or other social actors unite and which enable communication and coordinated action towards a commonly conceived goal. It is interesting though that the conceptions of such a boundary object vary considerably among the parties involved.

addressing the history and development of the discipline, the understanding of basic biological concepts, disciplinary controversies within the field, national idiosyncrasies, and the future potentials of systems biology were carried out. Interviews lasted between 1½ and 2 h, were tape-recorded and transcribed. The section addressing the conceptualization of basic concepts was cut out and analytical emphasis was put on the metaphors used by the scientists to explain their notion of the above-mentioned concepts.

The method used to analyze the metaphors started with an initial cursory reading of the transcript. The next step consisted in a close line-by-line reading trying to reveal all metaphors occurring in speech. The linguistic imagery encountered was transferred into a table where the mapping processes were analyzed. This procedure explored the degree of metaphoricity and at the same time helped us to study which source domains were used to conceptualize the target domain. Once the mappings were analyzed, all metaphors were—if possible—divided into different categories and grouped under so-called conceptual metaphors. These elements, to be understood as generic structures that pervade each concept, were studied with regard to what they highlight and hide. The final analytical step consisted in comparing whether certain kinds of conceptual metaphors could be associated with disciplinary backgrounds of interviewees and possibly refer to conventional modes of conceptualizing the basic notion under question.

The following scheme provides an overview of the different steps undertaken during our investigation and shows how these steps analytically build upon one another.

- Choose a domain of discourse in science, a discipline, or a specific research project → predefine your research object.
- Immerse yourself in the discourse by gathering different kinds of written material and iteratively read through the written evidence gathered → contextualize yourself.
- Take notes of all kinds of aspects that attract your attention and systematize them after having read through your sources → open up the field in a structured way.
- Define a tentative research question and reanalyze whether and, if so, how the research question fits → assess your research question.
- Develop a systematic database of written evidence and immerse yourself systematically in the domain of discourse → get a deep conceptual insight into the domain of discourse and develop a questionnaire for semi-structured interviews.
- Set up a table in which you note main actors in the field, their affiliation, scientific topics addressed, their current professional status, and the most important publications → become familiar with the main actors in the field.
- Choose a representative sample of scientists according to their professional status and their disciplinary backgrounds → gather a representative group of interviewees.
- Do interviews, transcribe them, and read them → iterative approximation to the interviews.

- Analyze the interviews on a line-by-line basis and gather the metaphors encountered in a table → systematically search for linguistic metaphors.
- Analyze the mapping processes in the metaphors → assess the metaphoricity of each metaphorical instantiation.
- Check whether some metaphors could be subsumed under one heading → definition of conceptual metaphors.
- Discuss aspects of highlighting and hiding → assess the possible implications of the metaphorically triggered styles of thought and propose alternatives.
- Check whether certain kinds of conceptual metaphors can be connected to disciplinary backgrounds → attribute metaphorically informed thought styles.

In sum, the approach outlined here aims at meeting the complex requirements for a detailed and methodological sound metaphor analysis of the complex issue of basic concepts in systems biology. It tries to provide a systematic examination of metaphor in scientific discourses which so far has been applied by Döring (2014) to assess and analyze metaphors in media discourses on synthetic biology. In the next section we now turn to the metaphorical framing of the notion of life by systems biologists.

2.2 The Conceptual Framing of Life in Systems Biology

As we have seen in the previous section, metaphors are a basic ingredient in scientific as well as in everyday discourses. They run through all stages of scientific thinking and acting. This also holds true for systems biology. Having read through a representative bulk of publications dealing with systems biology and having tackled the prevalent concepts of life, system, reductionism, holism, and model we now turn to the analysis of the understanding of life in systems biology. Life represents a multifaceted concept that has been described, defined, and explained in many fields such as biology, philosophy, religion, psychology, and many more. These studies are extremely interesting in themselves. However, the empirical question remains of how life is conceptualized by members of different scientific disciplines in their academic and scientific work? Is the notion of life relevant to them? If so, how do scientists working in the area of systems biology conceptualize life? An empirical investigation of these questions seems vital inasmuch as the concept of life is not plainly based on a definition or theory of life as outlined by Oparin (1924), Schrödinger (2012), Crick (1981), Monod (1970), Maynard-Smith (1986), and others. On the contrary, it is often based on scientists' associations and attitudes that in many cases are not overtly articulated. They rather reside in the unconscious and represent different kinds of tacit knowledge (Polanyi 1958, 1967), styles of thought (Fleck 2011), or cultural presumptions (Kather 2003) that are often not explicitly formulated and thought through, but nevertheless hinge on and display a certain mind-set. The task thus consists in empirically investigating conceptions and framings of life by systems biologists. Because systems biology considers itself as a highly interdisciplinary endeavor the question remains as to how the different disciplinary backgrounds and related styles of thought exert an explicit or implicit impact on the conception of life. In this section we show that the different characteristics of the concept of life play an important role, but these are merged with conceptualizations connected to scientific training and professional background. As a result, it seems appropriate to hypothesize that systems biology introduces new facets to the multifaceted concept of life.

2.2.1 *Life and Its Characteristics in Biology: An Impressionistic Overview*

Before we start to analyze the differing concepts of life in systems biology, we provide an impressionistic introduction to the different understandings of life in biology. Life represents a generic concept that is semantically difficult to grasp. It designates a phenomenon that often is explained as a property, especially as a property of organisms (see Table 2.1).

Box 2.1: Meanings of and Distinction Between Term, Notion, and Concept

In this chapter, we use specific names or terms in order to label abstract or mental constructs such as *notion* or *concept*. In scholarly discourse these constructs are not always clearly defined. Their use varies and they may have different meanings. In order to be as clear as possible in our terminology, we use the following definitions.

- A **term** is a word or compound word that in specific contexts is given a specific meaning. This may deviate from the meaning the same word may have in other contexts and in everyday language. Terminology studies the development of terms, their interrelationships, and their use.
- A **notion** in philosophy is a reflection in the mind of real objects and phenomena in their essential features and relations. Notions are usually described in terms of scope and content. Notions are often created in response to empirical observations (or experiments) of covarying trends among variables.
- A **concept** (or conception) is an abstract idea, mental representation, or mental symbol that exists in the brain. The terms concept and conception are sometimes used interchangeably. However, a conception may also be more encompassing and detailed than a concept with regard to considered factors and theoretical reflections. In metaphysics, and especially ontology, a concept is a fundamental category of existence.

When biologists try to give details about the nature of life, they refer in many cases to a set of criteria or a list of features that exemplify living organisms (see , e.g., Deamer 2010; Ganti 2003, 76–80; Mayr 1997, 20–23). Throughout the history of biology numerous efforts have been undertaken to elucidate what life is or could be (Kather 2003; Toepfer 2005) with this kind of feature-procedure ranging from Bernard's (1878) properties (organization, reproduction, development, nutrition, and vulnerability) via Crick's characteristics (reproduction, genetics, evolution, and metabolism) to Gibson et al. (2010), who understands life as exclusively based on reproduction. What becomes apparent is that biologists use these central features with the aim of exploring what life is, even though this task seems to be rather speculative and has led now and then to attempts to develop a universally shared definition. One of these endeavors was, for example, undertaken in Murphy's and O'Neill's (1997) book entitled, *What is Life? The Next Fifty Years. Speculations on the Future of Biology*. The book simply showed that it is impossible to agree and rely on a fixed set of basic features. As indicated by the subtitle, the endeavor of defining what life is via a fixed or agreed-upon set of characteristics rather represents a speculative task as these are in many cases context or temporally bound features emerging in a specific historical, social, technological, and scientific milieu, although certain features (for instance, reproduction) remain constant and can be found in almost any set of life-defining features. Having this in mind, one might conclude that investigating the notion of life is a useless venture. However, we would like to reject this conclusion inasmuch as we are interested in exactly this sociocultural contextuality. Features assigned to life are markers that meander through history and display a prevalent conception of life in a certain sociotemporal context. This sociotemporal context not only engenders a specific understanding of life but also determines questions, methods, and instruments employed in order to analyze it and to deploy its parts and processes for human goals. Different conceptions of life therefore may have different implications for science and society, and this is but one of the many reasons why it is worthwhile to re-explore them in detail once relevant framework conditions changed. Interestingly enough, some of these markers mentioned above were also encountered in the interviews we led with systems biologists in Germany. To sum up, there is a large diversity of features that have been used by different disciplines to describe and define life which reflect the richness of scientific and cultural perception of this seemingly unfathomable phenomenon. On the other hand, there is a historically generated set of so-called canonical features, which serve as indicators for life (see Table 2.1).

As collectively shared and combined markers, these canonical features fulfill the function of providing a common ground for partially defining when an entity should be considered to be alive. It is thus necessary for the analysis undertaken here to provide a thorough nonexhaustive but still representative insight into the basic characteristics of life.

The features outlined here could be conceived as discrete characteristics, but most of them are conceptually related. Therefore, some compilations or lists about basic features of life merge characteristics whereas others are divided into two or even more discrete traits overlooking that scientists more often tend to use their

Table 2.1 Defining 'life' via certain characteristics (cf. Kather 2003; Toepfer 2005)

Author	Characteristic(s)
Bernard (1878)	Organization, reproduction, development, nutrition, and vulnerability.
Oparin (1924)	Organization, metabolism, reproduction, irritability.
Crick (1981)	Reproduction, genetics, evolution, metabolism.
Monod (1970)	Teleonomy, morphogenesis, reproduction.
Maynard-Smith (1986)	Metabolism, different segments holding functions.
Gibson et al. (2010)	Reproduction

self-defined idiosyncratic features. In sum, the characteristics described above attribute basic abilities to living organisms such as to form, to develop, and to reproduce on the basis of a natural layout, which makes organisms forms of being that exist in principle independent of any kind of human or other assistance.[9]

Thus far, we have outlined in a nonexhaustive attempt the so-called traditional features of life prevalent in biology as well as in or traditional biotechnology. Systems biology, however, applies a different and perhaps more fundamental perspective on organisms, organs, cells, or even single metabolic pathways. Although not to be conceived as a uniform scientific approach, it holds its own history and has emerged in the context of different disciplines such as molecular biology, genomics, biochemistry, computer science, and engineering. The heritage from its predecessors as well as the ideas and approaches from other scientific disciplines contributed to an expansion or possible reformulation of concepts of life in the context of systems biology. We now explore and analyze the metaphorical conceptualization of life encountered in interviews with systems biologists.

2.2.2 Depicting Life as Seen Through Systems Biologists' Eyes

As we have seen in the previous section, life is perceived as a fuzzy concept comprising certain characteristics but undergoing change throughout time. In systems biology, the word *life* frequently appears in the titles of conference talks or scientific reviews, often in combination with other words such as elements, principles, basics, and so on. Furthermore, books dealing with systems biology in general as well as textbooks and articles use the notion of life in their titles or devote a considerable section or chapter to it (see, e.g., Ideker et al. 2001; Kaneko 2006; Noble 2008a; Westerhoff et al. 2009). A closer look at a representative bulk of the literature paradoxically revealed that neither characteristics nor explanations of the notion of life

[9] This holds true no matter whether a specific animal or plant has evolved naturally or by breeding or genetic engineering. The ensuing organism is alive, although it may represent a new version of its natural predecessor.

are given. Systems biologists seemingly wish to make scientific statements about life, but without describing the object of inquiry more accurately. Furthermore, in scientific practice, they focus on elucidating complex networks, processes, or (emergent) functions, but not on "life" as such, whatever that means. One could obviously argue life is a concept too complex to explain or simply not relevant in the context of systems biology. But why then is it so often used in the corresponding scientific literature? Is it only referred to for scientific marketing purposes with the aim of pushing the newly emerging approach as *the* approach that provides an answer to what life is? This does not seem to be true as a search on the ISI-Web of Knowledge and PubMed indicated the notion is constantly in use and not only employed during the starting period of systems biology. It could also be possible the notion does not apply to everyday problems and practices encountered in scientific work. This mismatch, however, attracted our attention.

Given the fact that studies devoted to the empirical conceptualization of life are still rare (see Fox-Keller 1995, 2002; Gutman 2008; Hesse 1966; Kay 2000; Bock von Wülfingen 2007; Bölker et al. 2010) and that no preliminary answers to the so-called life-question could be deduced from the scientific literature analyzed, we decided to ask the life-question during the interviews led with systems biologists. Our hypothesis was that either the life-question would simply be rejected or an interesting discussion might emerge in which metaphors are used to conceptualize and communicate the framings of life by our interviewees. We thus hypothesized that a systematic analysis of metaphor might reveal the hidden meanings nestling in the language used to depict life. Consequently, a manual for semi-structured interviews was designed in which we first asked what systems biology is, how it developed since its advent in the German context, and what its future potentials might be. This section was deliberately used to instigate a thought process that led to a self-contextualization of the interviewee. Having outlined and discussed the individual framings of systems biology's pasts, presents, and futures, the life-question was asked in the following section. The question was carefully introduced by the interviewer using a polite and cautious language which indicated that it is a complex but nevertheless relevant query. The query was informed by insights provided by prototype theory (Rosch 1973, 1978; Rosch et al. 1976). The aim consisted in initiating a thought process that psychologically reduced the complexity of the question and provided an implicit offer to start with the features outlined in the previous section: In doing so, a shared communicative grounding between interviewee and interviewer developed. Not astonishingly, most interview partners initially answered that reproduction and metabolism represent the basic features of life. However, a subcutaneous tension emerged in the course of the interview which becomes apparent in the following two representative quotes.

> This is a question we do not get often, hmm, because, well, we are just on the technical side of it, err, yes, but it's a good question one as well, as yes, we are so immersed in our daily hassle. Yeah, you know that we lose track of these, yes philosophical but relevant questions [...]. (Scientist A)
>
> German original: Das ist eine Frage, die uns nicht oft gestellt wird, hm, weil, ja, wir befinden uns auf der technischen Seite, err, aber es ist eine gute Frage, weil ja, wir sind ja

so in unserem Alltag gefangen. Ja, wir verlieren den Kontakt zu diesen, ja philosophischen, aber relevanten Fragen [...].

Life? Oh yes, big concept, loooong history and no clear answers [...] hahahaha [...] what a mess. I think that the concept does not really play an important role in our daily working life. We make a cut in our brains and just concentrate on this and this pathway [...] but the big picture, yes, I think that we should address this mess [...]. (Scientist J)

German original: Leben? Oh ja, eine großes Konzept, laaange Geschichte und keine klaren Antworten [...] hahahaha [..] was für ein Durcheinander. Ich glaube, dass das Konzept nicht wirklich eine wichtige Rolle in unserem Arbeitsalltag spielt. Wir unterteilen unsere Gehirne und konzentrieren uns lediglich auf diesen oder diesen Pathway, aber das Große und Ganze, ja, ich denke dass wir uns darum auch kümmern sollten.

The two quotes show that the life-question appears quite relevant but at the same time too big to deal with. In the first quote reference is made to daily workloads that prevent the interviewee from addressing the question of what life could be, however, the second evidently refers to his historical knowledge. Scientist J also ironically plays with the life-question but in the end concludes that the question is, at least, of interest to him. However, it is not an explicit subject for experimental or theoretical inquiry. Withstanding the tension and attempts to resolve it, the interviewer remained in these situations often silent but provided feedback channeling with the aim of keeping the thought process going. This leads, on the side of the interviewee, to a differentiation of the previously said with the help of spontaneous metaphors that were systematized and analyzed in the transcribed interview data according to the methodological procedure previously depicted. This analysis yielded the following seven conceptual metaphors framing life: LIFE IS A MACHINERY, LIFE IS A SYSTEM, LIFE IS INTERACTION AMONG SYSTEM COMPONENTS, LIFE IS A NETWORK, LIFE IS A FORCE, LIFE IS A RIDDLE, and LIFE IS A SECRET.

To start with, life has metaphorically been depicted in terms of machinery. Words metaphorically used comprise lexical items such as machine, machinery that is projected upon the domain of life, as could be seen in the following two quotes.

Life? Yes, that is tricky to explain. I would say that what we do is **understanding life as machinery**. I mean, there are all these processes which we try to understand and I think that machinery captures it quite good. (Scientist A)[10]

German original: Leben? Ja, das ich nicht einfach zu erklären. Ich würde sagen, dass das was wir machen ist dass wir **versuchen Leben als Maschine zu verstehen**. Ich meine, da sind all diese Prozesse, die wir wir versuchen sie zu zu verstehen und ich glaube, dass Maschine das ganz gut ausdrückt.

Well that's difficult [...], I would say. Well, well one might **think of life as some sort of a machine or better machinery** where different bits and pieces work together. Hm, yes, one could understand life in this way. (Scientists K)

German original: Ja, das ist schwer [...], ich ich würde sagen. Gut, gut, man könnte sich **Leben als eine Art Maschine oder besser als Maschinerie vorstellen**, in der unterschiedliche Stücke und Teile zusammen arbeiten. Hm, ja, auf diese Weise könnte man Leben verstehen.

[10] Letters in bold indicate the metaphor or metaphorical phrase.

What have been highlighted by these metaphors are clearly technical and engineering aspects. This includes a constant need for energy and at the same time relates to old images of the mitochondria as power stations of the cell, a culturally well-engrained idea in the German context. The second quote, furthermore, differentiates between machines on the one hand and then introduces the noun machinery as a generic concept. The metaphorical transfer visibly highlights images of steel, oil, gearwheels, and lubrication but also develops on a connotative level a relation to cellular and biochemical processes. In sum, the conceptual metaphor and its inherent transfer convey images of factories.

Life has also been metaphorically depicted as a system. In this case, the metaphorical transfer relies on an abstract notion used in a variety of ways ranging from economics via politics and the German waste disposal system to scientific systems. In the present context, however, the interpretative background refers to systems theory and systems biology.

> Life? Oh dear! Ok, I think **life is a system**, a fuzzy system. It is hard to explain but to me it's a structured whole. (Scientist D)
>
> German original: Leben? Oh je! Ok, Ich denke, **dass Leben ein System**, äh ein unscharfes System ist. Es, es ist wirklich schwer zu erklären, aber für mich, äh, ist es ein strukturiertes Ganzes.
>
> Life, that's a difficult notion. I see, ah [...] **life for me is a system**. Yeah, that's what it is. (Scientist E)
>
> German original: Leben, das ist ein ein schwieriger Begriff. Ich sehe ah [...] **Leben als ein System**. Ja, so könnte man es ausdrücken.

The systems metaphor, though semantically opaque, highlights aspects of structured or organized entities. These, in turn, develop out of smaller components that hold functional relations among these entities and are governed by principles in a functional way. The explanatory value of the system metaphor, however, remains small due to its imprecise meaning and open semantic content.

In addition to these first two conceptual metaphors, life is also metaphorically framed as interaction between system components. The notion of system appears in the following two cases again, but is now determined by the metaphorical use of the word "interaction". Moreover, the quotes are more precise than the previous two because they indicate who interacts with whom.

> Well, for what I now say probably a lot of people would kill me, but [...] haha [...] anyway. So in my version, **the interaction between the DNA and the proteins**, that's what I think is life. (Scientist M)
>
> German original: Gut, für das, was ich jetzt sage würden mich wahrscheinlich viele Leute umbringen, aber [...] haha [...] egal. So, meine Version von **Leben ist, ist die Interaktion zwischen der DNA und den Proteinen**, ich denke, dass das Leben ist.
>
> Yes, life is to me rather small and **rather interaction on the molecular level**, you know. That is my version of the whole thing. (Scientist F)
>
> German original: Ja, Leben ist für mich eher klein und und **eher Interaktion auf der molekularen Ebene**, verstehen sie. Das ist meine Version dieser ganzen Sache.

The metaphorical use of interaction highlights relational and mutual aspects of interdependence and cooperation. Interaction furthermore holds obvious connotations and refers to life-world experiences of social and communicative interplay and exchange. The conceptual metaphor, LIFE IS INTERACTION AMONG SYSTEM COMPONENTS, thus subcutaneously introduces a social aspect.

Furthermore, scientists in systems biology frame the notion of life using a network metaphor. The metaphorical concept, LIFE IS A NETWORK, appears to be prominent among researchers holding an IT background as the following two quotes indicate.

> The concept of life? That's a tricky question but **in my view it is rather a network**, the interaction and regulation of metabolic networks; it is this functional coupling thing that we have to deal with, we have to understand. (Scientist K)
>
> German original: Das Konzept Leben? Das ist eine schwierige Frage, aber ich sehe **es eher als Netzwerk an**, die Interaktion und Regulierung metabolischer Netzwerke; es ist dieses funktional verbindende Moment, mit dem wir uns beschäftigen, das wir verstehen sollten.
>
> That is an exciting question [...]. Well, I am quite pragmatic and I interpret life from my point of view in terms of a **network**. I mean, it is the only way I can think about it, and, yes, that is what I am interested in and how I can imagine it. (Scientist O)
>
> German orginal: Ja, das ist eine spannende Frage [...]. Also ich bin da eher pragmatisch und interpretiere aus meiner Arbeit heraus das **Konzept Leben als eine Art Netzwerk**, ok? Ich meine, ich kann das so denken und metabolische Netzwerke, ja, das ist was mich interessiert und wie ich es mir vorstellen kann.

The first interview excerpt metaphorically depicts life in terms of a network metaphor. A closer look at the example, however, refers to the implication complex of the metaphor as it could be seen in the use of the phrase "functional coupling". Here, the network metaphor is elaborated upon as connections in the system are highlighted, the integration of different levels in the system is alluded to, and the link between inside and outside is referenced. The second quote, on the contrary, displays a pragmatic and technologically driven access to the complex notion of life and at the same time outlines a quasi meta-reflection on why this metaphor has been applied: it is the work experience with IC technology that plays a vital role. The metaphor itself connects the notion of life to the semantic field of information technologies and highlights, on a connotative level, lexical items such as computers, hardware, Internets, computer programs, connections, knots, and knotting. These, although unmentioned aspects of the semantic field, resonate with each other and bear an impact on conceptualizing the notion of life using the characteristic feature of life, namely metabolism. It could thus be hypothesized that the conceptual metaphor, LIFE IS A NETWORK, theoretically merges a technologically driven vision of work experience with a biologically informed framing constitutive for systems biology.

However, in addition to such technologically driven images culturally well-established aspects of framing life emerge. The abstract metaphorical framing of LIFE IS A FORCE materializes in many interviews and often appears in conversations with interviewees trained in physics. Although not theoretically explaining the notion of life, these scientists metaphorically highlight the—to use another image—gear or impulse of life:

> We already understand complex living processes, but what is this **secret force of life** that keeps plants, humans, well all of this going? That is really, yeah, that is such a basic and interesting question and we know not much about it. (Scientist J)
>
> German original: Wir verstehen Lebensprozesse eigentlich schon ganz gut, aber was ist denn bloß diese **geheime Kraft des Lebens** die Pflanzen, uns Menschen und alles am Laufen hält? Das ist wirklich, ja, das ist eine so grundlegende und spannende Frage und wir wissen nicht wirklich viel darüber.
>
> Life, the **force of life**. That is really strange and fascinating at the same time. What agent, what kind of force keeps all these metabolic and other processes running?, Well, you can try and explain this with the law of energy conservation, but where comes this from, you understand? We have not really started yet. (Scientist A)
>
> German original: Leben, ja **die Kaft des Lebens**. Das ist schon merkwürdig und faszinierend zugleich. Welches Mittel, welche Kraft hält diese ganzen metabolischen und anderen Prozesse am Laufen, verstehen Sie? Ja, man kann das mit Energiesätzen erklären, aber woher kommt die dann. Wir sind da noch nicht mal am Anfang.

A close look at the sections in the interviews dealing with the conceptual metaphor, LIFE IS A FORCE, displays an emotional engagement. This is linguistically expressed by adjective constructions such as "[…] that is such a basic and interesting question […]" and feedback channeling such as "[…] you understand […]." However, the force-metaphor uses a generic concept that highlights the aspect of a physical and vectorial quantity which is necessary to perform work that changes the energy level of a physical system. This change in energy levels and directionality could be connected to the energy needed to provide work power for basic characteristics of life such as metabolism and reproduction. These aspects are, however, hypothetical and require further corroboration through in-depth interviews and analysis. What is interesting, however, is that the generic concept of force is used to explain the generic concept of life. Both concepts could be situated on an abstract conceptual level which might explain why the metaphorical categorization oscillates between abstract fuzziness on the one hand and human bodily experiences with forces. The conceptual metaphor thus holds an abstract concreteness based in the present case on the professional origin of the interviewee.

The penultimate conceptual metaphor we have to deal with in this section is the metaphorical concept, LIFE IS A RIDDLE. This culturally engrained concept looks back at a long history and is used in many interviews and the two quotes below are representative examples of how the metaphor is used by scientists working in the field of systems biology:

> **The concept life is a riddle to me**, you understand? What keeps replication going, ah, reproduction going on? That is so fascinating and we really have to think deeply to solve this riddle, yeah! (Scientist D)
>
> German original: Das **Konzept Leben gibt mir immer noch Rätsel auf**, verstehen sie? Was hält die Replikation, äh, die Reproduction am Laufen? Das ist so faszinierend wir sollten nach wie vor eingehend bemühen dieses Rätsel zu lösen, ja!

> It, it is still **a riddle to me** and I am sure that we will not solve it. But it is a fascinating thing, this life and, I do not know why, but it keeps me going and pesters me, this life question. (Scientist G)
>
> German original: Es, es **ist immer noch ein Rätsel für mich** und ich bin mir sicher, dass wir es nicht lösen werden. Aber es ist ein faszinierendes Ding, dieses Leben und ich weiß nicht warum, aber es hält mich am Laufen und stellt mir auch nach, diese Frage nach dem Lebensbegriff.

The conceptual metaphor of LIFE IS A RIDDLE alludes to a task or question that has logically to be solved by the process of thinking. The metaphor holds strong ties with science because of associated connotations such as scientist, to solve, to decipher, mysterious, mystery, and unresolved, and also refers to the pedagogical tasks of a riddle in terms of strategic problem-solving and education. Riddles hold a haunting if not stalking potential, as we see in the second quote where "riddle" pesters the scientist interviewed. The aspect of entertainment and pastime stemming from everyday experience with riddles or riddle magazines are not highlighted here but enable a conceptual connection between the realm of science and daily life: the abstract entity of life is conceptualized via the experienced and cultural domain of riddles.

The last metaphorical concept discussed in this section is the conceptual metaphor, LIFE IS A SECRET. This metaphor holds strong semantic ties with the analyzed concept of LIFE IS A RIDDLE. At first sight, both concepts contain connotations already encountered such as scientist, to solve, to decipher, mysterious, mystery, and unresolved, but on close inspection there are some considerable differences: secrets could reveal horrible things, are sometimes open, best-kept, and (at least in the German language) often lie in the dark. These aspects also emerged during the interviews:

> The question of what life **is or could be**? **This will remain a secret** and always stay in the dark. It might be possible that we can bring some light into darkness but that will take some time. (Scientist P)
>
> German original: Die Frage nach dem was Leben ist oder sein könnte? **Das wird ein Geheimnis bleiben** und im Dunklen bleiben. Möglicherweise kriegen wir etwas Lichts ins Dunkel, aber das wird noch dauern.
>
> The concept of life is not really interesting for us. We are working on another concrete level; this is **some sort of a secret** that will always stay in the dark. (Scientist P)
>
> German original: Das Konzept Leben ist für uns hier nicht wirklich interessant. Wir arbeiten auf einer anderen konkreteren Ebene; **das ist so eine Art Geheimnis**, das immer im Dunklen bleiben wird.

What becomes apparent in the interview extracts is that the conceptual metaphor, LIFE IS A SECRET, is often combined with linguistic light metaphors. These figurative speech patterns are based on the conceptual metaphor, LIGHT IS KNOWLEDGE, and develop a close alliance with the conceptual metaphor, LIFE IS A SECRET. The secret-metaphor again, as in the other cases, conceptually blends the abstract entity "life" with the cultural experiences revolving around the notion of secret. The quotes, however, differ considerably as the first one displays a slightly positive perspective on solving the secret of life whereas in the second quote the combination of light and secret-metaphors is used to express that the attempt to unravel the secret of life is a useless endeavor: an all-embracing concept of life seems impossible.

In summary, the metaphorical concepts analyzed demonstrate that scientists also use metaphors to conceptualize abstract scientific entities such as life. Even though it might have been problematic to ask the complex life-question, not a single interviewee rejected reflecting on it and answering it. On the contrary, the question—primarily philosophical in its character (Kather 2003; Toepfer 2005)—was in many cases conceived to be relevant and the systematic analysis of the transcripts revealed a creative and skillfull variety of ways of dealing and coping with this question, which may finally not be an explicit research subject of systems biology but nevertheless an important philosophical question for systems biologists. We now turn, in the next and final section of this subchapter to a more systematized overview of the conceptual metaphors of life analyzed. The aim consists in providing a structured overview and interpreting what kind of implications may reside in the metaphorically framed concepts.

2.2.3 Assessing Metaphorically Informed Visions of Life

The preceding analysis has shown how scientists working in the area of systems biology use metaphors to ascribe meaning to the basic notion of life. The conceptual metaphors analyzed depicted an interrelated conceptual and shared network endowing the abstract concept with meanings (see Fig. 2.1).

The interpretation of representative examples, furthermore, revealed the semantic complexities and associative networks nestling in the metaphorical concepts studied. These results offered a first insight into how and by which means a representative group of scientists working in the area of systems biology attributes meaning to the

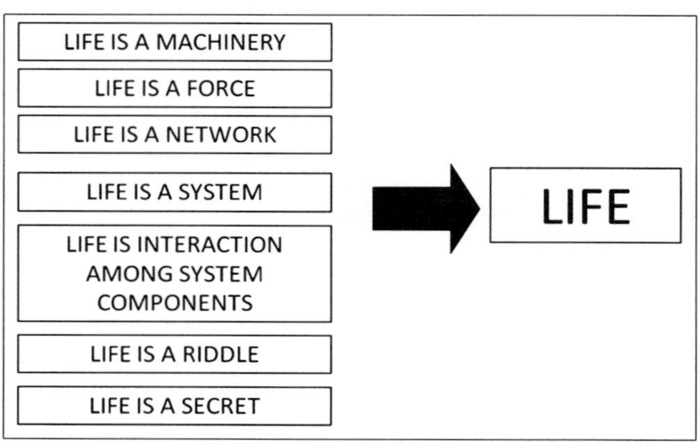

Fig. 2.1 Conceptual metaphors used to frame the notion of life

abstract notion of life in biology. The results are interesting in themselves but the question remains of what one could conclude from such a study that discloses underlying semantic networks and how such a sociocultural investigation could contribute to developing deeper insights?

First of all, we emphasize that empirical studies on the metaphorical framing of basic categories in biology as undertaken in the present context are rare. A closer look at the analyzed semantic network and its interpretation opens up the possibility for an empirically informed overview over the conceptual structure of the field. If one considers again the analyzed imagery in view of the underlying transfer processes, it shows the first three conceptual metaphors are motivated by engineering science: one is shaped by science, one stems from interpersonal experience, and two fall back on culturally established experience reports in the broadest sense of the meaning (see Fig. 2.2).

This clearly shows that in systems biology, the concept of life exhibits characteristics that are primarily technological or engineering–scientific in nature, yet there is a shift in meaning towards dynamization and complexity with the frequently encountered interaction metaphor. In view of the diverse metaphorical framings of biological relationships and their functional processes, the force metaphor highlights the gear of life and the ambiguity of the riddle or secret metaphors underlines difficulties encountered to define what life means and is. The analysis of imagery and its transfer processes therefore makes it possible to reveal a metaphorically motivated "heuristic fiction" (Black 1962, 229) with which the notion of life is explored. If one considers that in most cases the metaphors and their underlying transfer processes take place in the subconscious as elements of an implicit knowledge (Cassirer 1985, 1993; Polanyi 1966) then the possibility arises of slightly deepening the analysis in the present context. This means if we correlate the target domains with the professional backgrounds of those interviewed another interesting picture comes into view, namely that the professional backgrounds of the interviewees

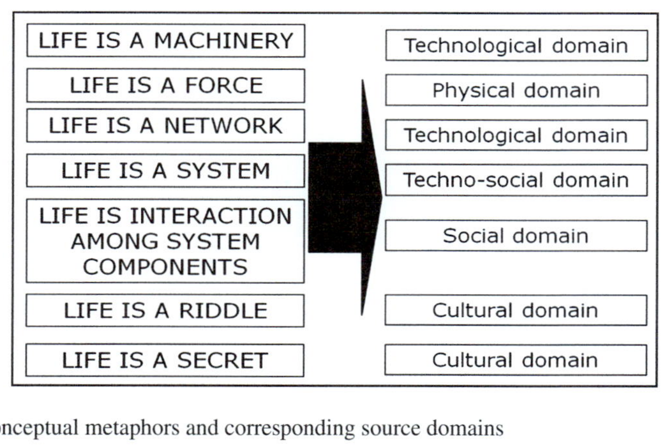

Fig. 2.2 Conceptual metaphors and corresponding source domains

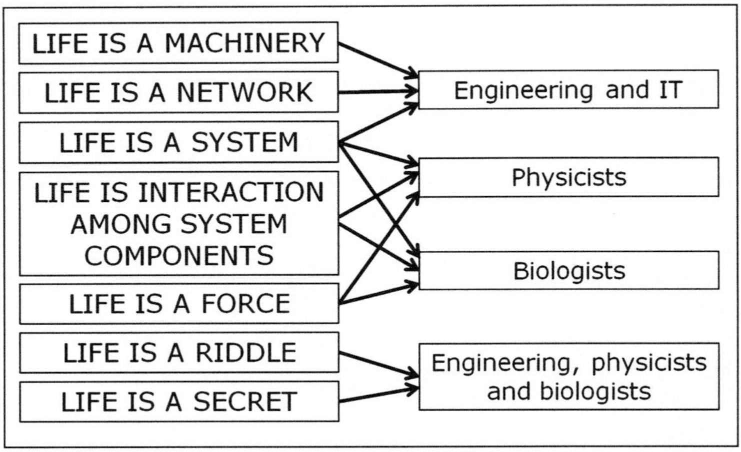

Fig. 2.3 Dispersion of conceptual metaphors among disciplinary backgrounds of scientists

influence their ways of conceptualizing the notion of life as most of those using technological source domains hold an engineering or physics background (see Fig. 2.3).

But the distinction is not as easy as that because the sociocultural target domains such as life is a riddle and life is a secret permeate all interviews. In brief, the notion of life as analyzed here is a mix of professional as well as of sociocultural experiences and knowledge, although the impact of technology-driven approaches on biology—as already outlined—becomes partly visible. Furthermore, a closer look at the transfer processes holds the potential to, though from an interpretative point of view, raise awareness about implications nestling in the transfer processes of the imagery used. Thus, the target domain of machinery clearly highlights aspects of cooperative parts, mechanical engineering, or cog wheels and the riddle metaphor alludes to implicit aspects such as playful solution, systematic deciphering and the like. Hence, the possibility arises within the framework of a critical assessment of conceptual metaphors to disclose and question these meaningful elements to a certain extent and to discuss with systems biologists various paths of technological development and their implications.

To summarize, it might have become clear in this section that metaphors play an important role in the conceptualization of abstract knowledge domains because they capture abstract circumstances with tangible representations facilitated by transfer processes. The analysis and interpretation of these processes of generating meaning provides an opportunity to reveal unconsciously constructed meanings of the notion of life and their implications for debate. The approach aims to create a form of meta-knowledge, which provides the foundation for the negotiation of evaluating technology with those working in systems biology. In this respect, an empirically grounded analysis of conceptual metaphors offers the opportunity to address implications of current, but still implicit visions of life. Even though such an analysis is still in its infancy, we now turn in the following section to an analysis of conceptual metaphors used to frame the notion of system in systems biology.

2.3 Envisioning the Notion of System in Systems Biology

The main aim of this section is to unravel the meanings attributed to the concept of system with the help of metaphor analysis. The notion of system is part of almost all sciences and each disciplinary approach developed its more or less own conceptions and applications. Consequently, system represents a multilayered concept that has been used as a heuristic tool in many disciplines ranging from sociology and economics to ecosystem or even earth system analysis. It officially gathered momentum in biology since the 1950s with the wider recognition of Ludwig von Bertalanffy's General Systems Theory (von Bertalanffy 1932, 1949) and Paul A. Weiss' *The Science of Life: The Living System—A System of Living* (Weiss 1973). These insights were taken up and put forward by theoretical biologists such as Robert Rosen (1970a, b) or Jacques Monod and Ernest Bornek (1971). The systems concept, thus, possesses a considerable theoretical history and wide range of practical applications. With regard to systems biology, the notion of system has so far not undergone a detailed theoretical clarification and empirical examination in terms of its meanings, operationalizations, and applications. Nevertheless, it has become an unquestioned and socially accepted boundary object (Griesemer and Star 1989; Bowker and Star 2000) among scientists in systems biology. However, some contemporary systems biologists such as Boogerd et al. (2007a, b), Wolkenhauer (2001, 2007a, b), and Wolkenhauer and Mesarovic (2005) already aimed—partially together with historians and philosophers of science (Drack and Wolkenhauer 2011)—for conceptual clarification and application: they address theoretical and methodological questions while providing first steps towards a philosophical examination of the notion of system in systems biology. Our task, however, is to tackle the contemporary framings of the notion of system by systems biologists. Such an empirical question concerning the system concept has to date to our knowledge rarely been addressed by system biologists or by scientists working in the area of science studies, technology assessment, and science and technology studies. To better understand how scientists working in systems biology conceive systems is important because nonarticulated conceptual differences may create misunderstandings and hamper the progress of research. We therefore try to answer the following questions. Is there some sort of a conceptual agreement on the abstract notion of system among systems biologists? Does a differentiated set of concepts exist? Furthermore, what kind of unconscious attitudes are bound to the idea of system and could they be connected to a specific professional mindset or scientific identity?

In the following analysis we show how the notion of system is conceptually framed and informed by different conceptual metaphors to disclose the otherwise intangible structures and meanings implicated in the linguistic imagery. It is, however, necessary to historically contextualize the system concept in biology to unfold its different dimensions and meanings before we turn to the empirical analysis. Consequently, the following section provides a (though nonexhaustive) diachronic insight into the notion of system and its features in systems biology. Against this backdrop, metaphors and conceptual metaphors in a representative set of interview extracts are analyzed to reveal implicit conceptualization inbuilt in the notion of system in systems biology.

2.3.1 The Systems Notion in Old and New Systems Biology: A Sketchy Overview

Generally speaking, a system could be conceived as a network of components that are interconnected and represent an interacting whole or unified entity. Systems normally demonstrate some sort of an emergent behavior holding a characteristic or property not shared by or implicated in its constituting elements. This is also a major aspect alluded to in systems biology. There seems, however, to be little awareness among systems biologists about the conceptual history of systems and theoretical impacts of so-called predecessors.[11] This is a problem because a lack in historical and conceptual awareness might lead to theoretical and methodological shortcomings that in turn bear an impact on research which currently develops many ways into biochemistry, genetics, ecology, and the like. As a result, the capacity to understand and to virtually construct complex biological systems might be affected and should hence be based on thorough theoretical, methodological, and historical knowledge about systems theory in general and in biology, especially. This obviously represents a challenge for scientists working in systems biology but holds the potential to develop historically rooted and conceptually sound models of biological systems.

It might sound counterintuitive but one has to go back to the end of the nineteenth century to understand two important concepts stemming from the seventeenth century that underpin biology and even today's systems biology. The first can be identified with René Descartes who stated that complex questions could be analyzed by reducing them to manageable pieces. Descartes' paved the way towards a reductionism that was thought to provide the relevant answers to mathematical and physical problems. Today, it still is an important principle in the sciences in general, and in biology or systems biology in particular. With regard to systems, the basic assumption of reductionism consists in the idea that characteristics of higher system levels could easily be explained by the behavior of lower biological levels (see Sect. 2.4 for more details on reductionism). This conceptual understanding was taken over by proponents of mechanistic biology which concurrently surfaced in the seventeenth century. Based on Descartes' ideas, the emergence and the development of *clockworks* enabled a mechanistic thinking of organisms or biological entities as clockwork-like. The clockwork metaphor facilitated a deterministic view that was able to draw on ideas of disassembling and reassembling and by doing so to explain the characteristics of a system via its parts (Haber 1975; Nicholson 2013). This understanding influenced many scientists such as the plant biologist Jacques Loeb (1964) whose work was based on mechanistic attitudes.

In reaction to Loeb's mechanistic ideas some concerns were articulated by a small group of theoretical and other biologists at the start of the twentieth century (Roll-Hansen 1984; Nicholson 2012). Biologists such as Woodger (2001, 31–84), Weiss (1940), and von Bertalanffy (1950a, 23; 1950b, 140; 1968, 87–89) expressed

[11] This became visible during the interviews conducted for this study.

a twofold concern with the concept of mechanistic biology (Hein 1972). They first outlined—with reference to Aristotle—that the whole is just more than its constitutive parts. This view had dominated up to the seventeenth century but vanished with the advent of experimental physics and biology. To denominate this phenomenon, the term holism was introduced by the statesman and philosopher, Jan Christiaan Smuts (1926). It included the idea that wholes such as cells or tissues, for example, hold properties which could not be understood by reference to the composition of their constituting elements. Thus, reductionism was, according to Smuts, thought to be unable to explain emerging properties of wholes based on mere information about its components.

It was Paul A. Weiss (1925) who experimentally questioned Loeb's mechanistic ideas. Weiss analyzed in his dissertation the impact of light and gravity on insect behavior and was able to show that, although all individuals displayed an identical final response, this response was achieved by unique behavioral ways. Furthermore, Roger Williams (1956) worked on biochemical individuality. He propounded molecular, physiological, and anatomic individuality in animals showing that these vary considerably in terms of chemical, hormonal, and physiological parameters. Consequently, the concept of mechanistic biology was challenged in favor of a more dynamic system-oriented conception because living cells could not be conceived of as deterministic machines but should be envisaged as adaptive and variable entities holding typical characteristics while exhibiting individual responses.

In addition to the typical characteristics, the individuality and reactivity of living systems, their hierarchical organization or structure represents an important aspect or property from a systems perspective. The theoretical biologist Joseph Henry Woodger emphasized in his book entitled, *Biological Principles* (Woodger 2001, 283–298), that higher biological entities start their life cycle from single cells and the development of complex entities follows a typical developmental order. There seem to be restricted developmental routes or constraints that are organized and controlled on a higher level. A system such as a tissue is thus constructed out of single cells and the principle underlying this development is the interaction among the constituents of the lower level that is organized and structured by a higher system level. This hierarchical organization does not follow bottom-up rules but pursues a system logic in which interactions on one level lead to emergent properties on a higher level and vice versa.

Also Paul A. Weiss (1973) aimed at unraveling important characteristics of biological systems referring to the recognition of hierarchical structures in biological systems but focused on evolutionary implications. Weiss emphasized two important aspects: he first outlined that greater variation exists at lower levels of systems and that individual metabolic pathways appear to be more ordered within a system than they would be outside a system. Especially the latter characteristic proves that molecular behaviors depend on and are coordinated by higher system levels. Comparable ideas were also expressed by Gregory Bateson (1972) who pointed to the fact that all organisms are able to adapt and deal with unpredictable environmental incidents. Later, in his book, *Steps to an Ecology of Mind*, Bateson (1972,

343–377) also refers to the gene–environment interaction. Such a systems perspective exerted a vital impact on the interpretation and understanding of evolutionary mechanisms because the interaction of an organism with a complex and variable environment was scientifically reframed as the evolutionary force of nature (Vrba and Gould 1986).

After having clarified and extensively investigated the hierarchical structure of systems, Ludwig von Bertalanffy (1950b, 1968) suggested that, among other important aspects, all complex systems are based on the common property of representing a compilation of interlinked components. This meant there are correspondences if not detailed similarities in the structure and control design of systems. Bertalanffy's outline of a General Systems Theory (von Bertalanffy 1968) gathered momentum due to its emphasis on the relevance of so-called hubs and connectors (Barabasi 2002, 63–64). These components represent the basic ingredients for a stable system structure: Hubs are thought to be connected to many connectors and these in turn are linked to only a few supplementary components (Barabasi 2002, 55–64). Bertalanffy's ideas of a general systems theory proved to be very productive because it emphasized the interconnectedness and interaction of different components: by bringing such properties of systems to the fore he aimed at explaining how they contribute to building a unified whole consisting of different levels. The interaction between different levels was also explicitly addressed by Michael Polanyi (1968) who theoretically showed that adjacent levels do restrict but not determine each other. His basic idea framed upper levels as availing entities that make constituents of lower levels perform functions or behaviors which they would not carry out on their own. Using language as an example, Polanyi (1968, 1311) showed the meaning of a sentence is an emergent property and this property restricts the use of the words to be used to express that meaning. Meaning here holds a top-down function as it bears an impact on the choice of words whereas the words themselves determine the scope of meaning to be constructed. This example can easily be transferred to biological systems and clearly explains how upward and downward causation work when mutations in the DNA appear (Polanyi 1968, 1310).

In addition to upward and downward causation, aspects of control design in systems turned out to be of vital importance. Control is carried out with the help of negative feedback and homeostasis (Cannon 1963, 98–167) which sustain a biological entity. Negative feedback is conceived to be one of the most important elements to control a system because information about the actual reactions of and performances in a system is constantly observed. Feedback controls and loops were also acknowledged by Bernard (1878) and Cannon's book, *The Wisdom of the Body* (Cannon 1963), proved to be highly influential for early proponents of systems theory such as Norbert Wiener (1948) because it anticipated basic ideas later developed in cybernetics. Wiener's book, *Cybernetics or Control and Communication in the Animal and the Machine*, cites Cannon's work (Wiener 1948, 1, 17 and 115) and conceptually owes much to it.

Inbuilt in these emerging ideas of feedback and homeostasis is the concept of stability which was thought to be an intrinsic characteristic of a biological system. Stability is conceived to be based on informational entropy which is envisaged as a driver generating a state of best stability (Beer 1965). There is, however, a problem because the subsystems' intention to aim for its own stability is in many cases overruled by interactions with higher system levels. Such aspects clearly exhibit the shortcomings of reductionist approaches that are based on invariant bottom-up behavior of internal and external system components because real-life processes seem to be more complex and interactive across a vast array of system levels. Early conceptual models of biological systems, however, took up the idea of stability and emphasized that organisms should be conceived of as open systems sustained by a recurrent stream of energy and matter (von Bertalanffy 1950a, 23; Denbigh 1951). The approach, besides its conceptual problems, paved the way towards mathematically systems-oriented *relational biology* as proposed by Robert Rosen (1970a) in his book, *Dynamical System Theory in Biology*.

At about the same time Mihaijlo Mesarovic's (1968) book, *System Theory and Biology*, appeared and his following publications such as *Mathematical Theory of General Systems* (Mesarovic and Takahara 1972), *General Systems Theory* (Mesarovic and Takahara 1975) or *Abstract Systems Theory* (Mesarovic and Takahara 1988) laid grounds for a mathematically inspired systems approach to biology. Based on the—although not new—idea that system dynamics and organizing principles of complex biological phenomena give rise to the functioning and function of cells (Wolkenhauer and Mesarovic 2005, 14), emphasis was put on the understanding of temporal aspects triggering functions of cells such as growth, differentiation, division, and apoptosis. In doing so, the need to understand the functioning of the cell from a systems perspective was stressed. The advent of bioinformatics as well as genomics and other Omics approaches driven by new ICT technologies provided biology with a plethora of genetic and genomic data and rekindled the interest in systems approaches at the end of the 1990s. Albeit their identification procedures, characterizations of main components making up cells, and first approaches to construct domain-specific ontologies provided substantive benefit for systems biology because they supplied the basic ingredients for refocusing on biological interactions, processes, and dynamics. Especially the information made available by proteomics, the listing of all proteins active in a certain state of a cell or organism and on different system levels, instigated a reconceptualization of organisms, cells, genes, and proteins as independent entities whose characteristics and relations are established and determined by their function in a whole. This conception clearly mirrors a general systems definition in which a system is conceived as a discrete number of components and the relations among them (Klir 1991). "Systems theory is then the study of organization and behavior per se and a natural conclusion therefore to consider systems biology as the application of systems theory to genomics" (Wolkenhauer 2001, 258). This concept emerged and was put forward by main proponents of the new systems biology such as Hood (2000), Kitano (2002a, b), and Wolkenhauer (2001) with the aim of developing mathematical or so-called computational models for biology.

The theoretical background to this development stems from the 1960s and is based on a conceptual transfer from physics to biology when theoretical biologists used the then contemporary systems approaches to find and analyze biological laws that govern the behavior and evolution of living entities. Analogous to the relation between physical laws and living matter, biological systems were conceived of as representing a special case of physical systems. Criticism was raised and resulted in a comprehensive discussion of systems biology by Robert Rosen (1978, 1985, 2000). Nonetheless, biologists beginning to become interested in re-emerging systems biology more than a decade ago realized there is a need to approach complex and dynamic systems in a way for which existing reductionist approaches were not suitable. Against this backdrop, the return to systems-theoretical approaches by the end of the twentieth century appears logically consistent as the plethora of data made available by advances in Omics required a conceptual rather than an empirical approach that investigates the relationship between state variables. Emphasis was in this context not put on entities themselves but on the connections between them, their functional relations, and the outcomes of these relations. These insights, along with the increase of computing capacities, initiated and promoted an interest in the mathematical modeling of biological systems. Such modeling aims at establishing rules working on different levels, thereby postulating so-called causal laws that, for example, explained functional dependencies among genes or gene products instead of describing them in terms of mere associations. Hence, the aim consisted in the description of organized and probably repeated process patterns that were envisioned to help better understand the interaction, functioning, and development of a set of biological variables on one and/or across different levels. Looking at biological processes through the system-theoretical lens thus proved to be conceptually productive and led to the first steps into mathematical modeling of biological processes. It is, however, important to bear in mind that the notion of system in biology with all its theoretical implications and recent practical transfer to mathematical modeling offers explanations that refer to the limits of the new systems biology.

Thus far, we have tried to provide a sketchy overview of the history of the systems' idea in biology and systems biology. It became apparent that the notion of system holds a long conceptual history which can be traced back in essence to antiquity and more concretely to the seventeenth century. The most important insight consists in the Aristotelian understanding that the thing is more than the sum of its constituents, a view which was abandoned with the development of Cartesian reductionism and taken up again at the start of the last century in the works of Ludwig von Bertalanffy and Paul A. Weiss. The development and application of the systems idea in biology has, as roughly depicted in this section, progressed via a variety of intermediate steps and cumulated at the end of the 1960s with the rediscovery of Ludwig von Bertalanffy's general systems theory. Bertalanffy's ideas were partly reconceptualized and mathematically put forward by Robert Rosen's and Mihaijlo Mesarovic's publications. Their mathematically inspired system approaches contributed to paving the way to what nowadays is called the "new" systems biology. Even though the works of Bertalanffy, Weiss, Rosen, and Mesarovic are rarely referred to,[12] they provided the conceptual grounding for a mathematically

[12] An exception to this rule are Westerhoff and Palsson (2004), Alberghina and Westerhoff (2005), Boogerd et al. (2007a, b), Ullah and Wolkenhauer (2007), Drack and Apfalter (2007), Drack (2009, 2013), and Drack and Wolkenhauer (2011). The authors mentioned regularly refer to the "founders" of the new systems biology.

informed understanding and modeling of systems in biology. The notion of biological systems, however, remains ambiguous and is tied to daily practices and ICT contexts. It is, hence, worth exploring how the notion of system is metaphorically conceptualized by systems biologists of different disciplines to uncover the various meanings attributed to this basic notion. This aspect is explored in the following sections.

2.3.2 *Systems Biologists Picturing the Notion of System in Systems Biology*

The previous section depicted a (though limited) historical and conceptual insight into the system notion that paved the way towards new systems biology. It became apparent that it theoretically owes a lot to the Aristotelian notion of system (wholes), to general systems theory as outlined by Ludwig von Bertalanffy and Paul A. Weiss, to cybernetics developed by Norbert Wiener, as well as to Robert Rosen and to Mihaijlo Mesarovic, who both provided important theoretical grounds for the mathematization of current systems biology. Although the system notion itself is constitutive for the discipline, its historization in new systems biology is still lacking and has also rarely received philosophical investigation. This is an astonishing fact because the concept is constitutive for the approach itself and analytically used in many ways for exploring and analyzing the functioning of systems in biology ranging from genes, cells, and organs to entire organisms. Frequent are quotes—such as the following—explaining what new systems biology is and at the same time giving an implicit idea of what a system is supposed to be.

> Systems biology is the coordinated study of biological systems by (1) investigating the components of cellular networks and their interactions, (2) applying experimental high-throughput and whole genome techniques, and (3) integrating computational methods with experimental efforts. [...] The systematic approach to biology is not new, but it has recently gained new attraction due to emerging experimental and computational methods. (Klipp et al. 2005, V)

Here we find a description of some important characteristics of systems: systems biology is depicted as a systematic study of components, their interrelation on and beyond system levels, and the experimentally grounded simulation of their interactions. Even though this textbook extract aims at introducing systems biology to students, its historical depth is reduced to just mentioning that there is a historical background which was remotivated due to emerging methods and technical innovation (Ideker et al. 2001, 345–346). One could argue that such a general outline of what systems biology is and what the systems notion means meets the needs of undergraduate students in this context. A detailed and perhaps historical introduction might simply place too much strain on undergraduates but depictions of systems in systems biology remain in many cases on a general and descriptive level.

Especially a limited outline of the historical roots often appears in the form of the following two quotes.

> Since the day of Norbert Wiener, system-level understanding has been a recurrent theme in biological science. The major reason it is gaining renewed interest today is that progress in molecular biology, particularly in genome sequencing and high-throughput measurements, enable us to collect comprehensive data sets on system performance and gain underlying information on the underlying molecules. This was not possible in the days of Wiener, when molecular biology was still an emerging discipline. (Kitano 2002a, 1662)
>
> Whereas the foundations of systems biology-at-large are generally recognized as being as far apart as of 19th century whole-organism embryology and network mathematics, there is a school of thought that systems biology of the living cell has its origin in the expansion of molecular biology to genome-wide analyses. From this perspective, the emergence of this 'new' field constitutes a 'paradigm shift' for molecular biology, which ironically has often focused on reductionist thinking. (Westerhoff and Palsson 2004, 1249)

Reference is often made to well-known scientists such as Norbert Wiener and temporal indications such as "nineteenth century" chronologically situate genealogies and disciplinary development paths. In addition to these aspects, disciplines such as molecular biology and emerging technological and methodological innovations form important narrative structures in which the notion of system is only alluded to superficially. Although these rhetoric devices refer to well-known discursive strategies of newly emerging disciplines, this is not to say that systems biologists use rather naive system conceptions. On the contrary, scientists working in the interdisciplinary field of systems biology dispose over a tacit and everyday knowledge (Polanyi 1958) of what systems are and how they should be used, but this knowledge does not appear in scientific articles or in books or is not explicitly expressed, respectively. This is why a historical anchoring and philosophical theoreticization of the systems notion in systems biology might be helpful for the reflection on presuppositions inbuilt in ideas of or about systems. This aspect has to date rarely been addressed and motivated us to explore system conceptions distributed among systems biologists.

Based on these insights, we situated the so-called systems part of our manual for the semi-structured interviews after a question addressing the conceptualization of bottom-up, top-down, and middle range approaches. This was done implicitly to allude to levels, system borders, and so on, and prepare the ground for the complicated question about what a system represents for the interview partner. The methodological procedure was triggered by the hypothesis that an organized analysis of metaphors might disclose subliminal system conceptions distributed among our interviewees whereas the aim of the question consisted in instigating a thought process in which most salient features or characteristics of systems were discursively explored. A close look at the interview transcripts corroborates the usefulness of this approach because most interviewees generally started with a typical description of the characteristics of systems such as the different layers and levels of a system, and then swiftly outlined aspects of wholeness, system borders, and interaction among system components, among others. These aspects were in many cases supplemented by focusing on current research undertaken by the interview partner and back references to technological innovation in computation technologies and methods. Once the question was conceptually grounded, a subcutaneous tension appeared as displayed in the two following interview extracts.

> You are really asking tricky questions...hmm. Well, I mean the question is really basic and I have to admit that we do not often address it because we are immersed in all these different technicalities. But now, as I start to think about it, I think that we should address this question in our seminars because I am pretty sure that, at least in our group, the meaning and the characteristics of what a system is has not been addressed. (Scientist D)
>
> German original: Sie stellen aber wirklich schwierige Fragen... Ich muss zugeben, dass die Frage wirklich grundlegend ist und wir tauchen hier immer in diese technischen Fragen ab. Aber wenn ich jetzt so darüber nachdenke, dann sollten wir schon einmal diese Frage in unserem Seminar stellen, denn ich bin mir sicher, dass zumindest in unserer Gruppe die Bedeutung und Eigenschaften von dem, was ein System ist, nicht wirklich behandelt wurden.
>
> You want to know what I think a system is? OK, we are going now in medias res, eh? Ok the whole discipline is built on this idea and I sometimes feel quite unsatisfied with the theoretical outcomes or concepts of what my colleagues think a system is. I mean, there is a diversity of system notions. Sometimes it feels like a zoo where lots of system notions are around. (Scientist F)
>
> German original: Sie wollen also von mir wissen was ein System ist? Ok, jetzt geht's aber wirklich in medias res, oder? Ok, die ganze Disziplin baut ja auf dem Begriff auf und ich bin manchmal ziemlich unbefriedigt mit theoretischen Ergebnissen oder Systemkonzepten meiner Kollegen. Ich meine, da ist eine ziemliche Diversität an Systembegriffen unterwegs. Manchmal habe ich das Gefühl, dass ich in einem Zoo bin und dort jede Menge Systembegriffe antreffe. (Scientist J)

The two quotes indicate that the question asked about the meaning of what a system is appears to be difficult. Especially in the first excerpt the phrase "tricky questions" indicates this to some extent and in the second quote the scientist interviewed refers via the phrase "in medias res" to a perceived intensity. Furthermore, both excerpts exhibit a certain amount of dissatisfaction with the conceptual framing of the systems notion in biology, as expressed by scientist J with his ironic metaphor of a zoo, whereas scientist F expresses that there is a need for further clarification. The tension, however, remained in many interviews and withstanding its resolution by the interviewer, the interviewees started giving an insight into their system notions. As a system is an abstract entity, metaphors were used to conceptualize and communicate it. This led to spontaneously used metaphors that were systematized and analyzed. The analysis of the transcribed interviews gave rise to the following five conceptual metaphors that were utilized to semantically depict what a system is. The conceptual metaphors encountered are A SYSTEM IS A WHOLE, A SYSTEM IS A STRUCTURED ENTITY, A SYSTEM IS THE RELATION OF RELATED AND INTERACTING OBJECTS, A SYSTEM IS A MACHINE, A SYSTEM IS A CYBERNETIC MACHINE, and A SYSTEM IS A BIG PICTURE.

To start with, the notion of systems was metaphorically depicted as a whole. This imagery used lexical items such as whole ("*Ganzheit*") or the big picture ("*Große und Ganze*") which are in many cases sidelined by adjectives such as entire ("*ganz*"), complete ("*komplett*"), or full ("*voll und ganz*"). These words are projected upon the abstract entity of system, as could be seen in the two following quotes.

For me, a system is some sort of a whole thing, a whole which you can deconstruct to its constituents. And here you can look for what interacts with what, well, what components interact and what comes out of it or, yes, what evolves from it. (Scientist A)

German original: **Also für mich ist ein System eine Art Ganzes, so'ne Ganzheit**, das man analytisch in seine Bestandteile auflösen kann. Und hier kann man dann schauen was mit wem, also welche Komponenten miteinander interagieren und was dabei herauskommt oder sich, ja, irgendwie entwickelt. (Scientist A)

Well, **a system that is a whole or an entity** for me that possesses borders in a way, and it's a functional unit within them. It can function and has whatever output which keeps the unit going. Yeah, this is really a rough description I would say. (Scientist K)

German original: Gut, **ein System, das ist für mich ein Ganzes oder eine Einheit**, die für mich Grenzen hat, äh eine funktionelle Einheit innerhalb dieser Grenzen. Es funktioniert und hat eine wie auch immer gearteten Output, der die Einheit am Laufen hält. Naja, das ist eine ziemlich grobe Beschreibung, würde ich sagen.

What becomes apparent in these quotes is that it seems to be quite difficult to describe the abstract entity system. What stimulated our interest is that the abstract notion of system is metaphorically conceptualized by other abstract lexical items such as whole or entity. A closer look at the linguistic structures indicates these words hold spatial implications that, roughly speaking, map out what a system is and at the same time reify it. This becomes especially apparent in the second quote where borders are mentioned on the word level and the used adjective "within" alludes to spatial structures. In sum, the conceptual metaphor, A SYSTEM IS A WHOLE, cognitively realizes systems as entities with a certain spatial extension.

Systems have also metaphorically been depicted in terms of a structured entity. An aspect which also – though implicitly – appeared in the previous quotes but is emphasized in the following three excerpts where words such as "divided into" ("unterteilt in"), "structured" ("strukturiert") or "segmented" ("segmentiert") propose an internal order.

Systems are structured entities for me, you know. They possess some sort of an internal structure comprising functional entities which stand in relationship to each other and interact. (Scientist N)

German original: **Systeme sind für mich strukturierte Einheiten**, verstehen sie das? Sie besitzen so eine Art interne Struktur, die funktionelle Einheiten umfassen und miteinander in Beziehung stehen und interagieren.

A system, phew, good question… Well I would say **that it could be understood as a whole which could be divided into functional elements**. Take for example the cell and its components which make it up. (Scientist P)

German original: Ein System, phuu, gute Frage… Gut, ich würde sagen, dass **es als ein Ganzes verstanden werden kann, das in funktionelle Elemente unterteilt werden kann**. Nehmen wir z.B. die Zelle und die unterschiedlichen Komponenten, aus denen sie besteht.

A system is a structured arrangement of components that, to my knowledge, interact and holds certain functions which emerge out of their interaction. But it's still a big question and just a definition which requires in depth thinking. (Scientist D)

German original: **Ein System ist eine Art strukturiertes Arrangement von Komponenten**, das, meinem Wissen nach, das zu Funktionen führt, die der Interaktion der Komponenten entspringen. Aber das ist immer noch eine grundlegende Frage und nur eine Definition, über die wirklich einmal intensiv nachgedacht werden sollte.

The conceptual metaphor, A SYSTEM IS A STRUCTURED ENTITY, offers another possibility to make the abstract notion of system more concrete. It holds spatial implications but implicitly refers to smaller scales and a higher degree of segmentation. Both aspects become apparent in the frequent use of the words "component" and "elements" appearing in the interview transcripts. In brief, the spatio-metaphorical downscaling provides a higher degree of segmentation which makes the system concept cognitively manageable.

The previous conceptual metaphor is further elaborated on by another one which we called, A SYSTEM IS THE RELATION OF RELATED AND INTERACTING OBJECTS. This concept goes back to the work of Mihaijlo Mesarovic who developed mathematical explanations for elucidating functional relations between associated and interacting objects in biological systems. His metaphorical concept has, although sometimes implicitly, been taken over in systems biology and points to the aspect of interaction and dynamization not covered in the previous two metaphorical concepts.

Yes, well that's quite simple because that is Mike Mesarovic's work who developed a comprehensive theory of systems. He simply said that **a system is the sum of related and interacting objects**. (Scientist E)

German original: Ja. Äh, das ist ganz einfach weil das eben Mike Mesarovics Arbeit ist, der eine umfassende Theorie von System entwickelt hat. Er hat einfach gesagt, dass ein System eine Menge von ineinander in Beziehung stehender und interagierender Objekte ist.

The system notion I adhere to is the one that emphasises the fact that **a system consists of the relations among objects**. I mean, their relation and the inherent interaction, you understand? (Scientist H)

German original: Der Systembegriff, dem ich anhänge, betont die Beziehung der Objekte untereinander. Ich meine jetzt so deren Beziehung und Interaktion, verstehen sie?

In addition to these more abstract metaphors that frame systems in terms of spatial structures, reify them in terms of an entity and apply more dynamic ideas to them, the conceptual metaphor A SYSTEM IS A MACHINE appears quite frequently. This is seen in the following two quotes.

If you like, as **system could also be understood as a machine**. Well I have now Kitano's image of an airplane in mind. There are all these subsystems consisting of their elements and components. And the whole and its subsystems work together, are interlinked and in the end the system works properly. Well, in Kitano's case, the airplane flies, if you will. (Scientist L)

German original: Wenn Sie so wollen, kann man **ein System auch als eine Maschine sehen**. Also ich meine jetzt dieses Flugzeugbild von Kitano. Das sind alle diese Subsysteme und deren Elemente und Komponenten, die zusammengesetzt sind. Und das Ganze und seine Untersystems arbeiten zusammen, greifen ineinander und am Ende arbeitet das System dann. Also gut, bei Kitano fliegt das Flugzeug dann, wenn Sie so wollen.

A system, yes, err, how could I explain this? For me, it's, **well not solely, a machine**. It's a functional unit that can work on its own but that could also be linked to other units. Well, it could also be a big and overarching entity. (Scientist F)

German original: Ein System, ja, äh, wie könnte ich das beschreiben? **Für mich ist das, also nicht ausschließlich, aber auch eine Maschine**. So eine funktionelle Einheit, die in sich selber arbeitet, aber auch mit anderen Einheiten vernetzt ist und arbeitet. Naja, ist eine große und übergreifende Einheit halt.

The machine metaphor clearly emphasizes the technical and engineering aspect of the system notion as it is outlined in the first quote with intertextual reference to Kitano's (2002a) paper. Kitano used the image of an airplane to explain what the aim of systems biology is and how a systems-oriented approach to biology could work. The plane functions here as a metaphor for a biological system constituted out of different components or subsystems which brings about a metabolism to work or a plane to fly. The machine metaphor conceptualizes the abstract domain with the help of a concrete domain and differs in this respect from prior conceptual metaphors that relied on an abstract to less abstract metaphorical mapping. It, however, holds technical and mechanistic implications that are critically assessed in the second extract. Thus, the machine metaphor seems to hold a certain explanatory potential but is also viewed critically.

This critical aspect is raised and elaborated upon in some interviews where the conception of machine is further refined into the conceptual metaphor, A SYSTEM IS A CYBERNETIC MACHINE. Here, reference is made to Norbert Wiener's cybernetics and the work done by Heinz von Foerster. Their research offered insight into the working of systems as nonlinear machines because the entities under review exposed different reactions after having received a series of the same inputs. This aspect is stressed in the following two interview sections.

> And there is an interesting story. There is, ah, an interesting American cyberneticist, Heinz von Foerster. And he coined the notion of a non-trivial machine. The non-trivial machine is a machine that – even though it gets the same input – generates different outputs. Why is this so? Well, he says that each input changes the state of the machine. **Yes, it thus is not a simple converter and biological systems are also not a simple converter**, they are not physical machines. (Scientist H)
>
> German original: Und da gibt's 'ne interessante Geschichte. Es gibt ähm einen bekannten ähm amerikanischen Kybernetiker, Heinz Foerster. Und der hat den Begriff geprägt der sogenannten nicht trivialen Maschine. Die nicht triviale Maschine ist eine Maschine, die – obwohl sie den gleichen Input bekommt – mehrmals hintereinander jedes Mal einen anderen Output produziert. Und warum ist das so? Weil er sagt, weil jeder Input, den sie kriegt, ändert den inneren Zustand der Maschine. **Ja, es ist also kein simpler Konverter und das sind biologische Systeme auch nicht**, die sind keine physikalische Maschine.
>
> You know, we have to deal with different outputs although the system is fed with the same input. I would say that this is the basics of cybernetics as outlined by Bertalanffy and von Foerster. **Biological systems are cybernetic machines** in that the hold a history and this history or experience changes the outputs even though the input is the same. Quite complicated to understand, eh? (Scientist J).
>
> German original: Wissen sie, wir haben es hier mit unterschiedlichen outputs zu tun, und dass obwohl das System mit dem gleichen Input versorgt wurde. Ich würde sagen, dass wir es hier mit den Grundlagen von Bertalanffy und von Foerster zu tun haben. **Biologische Systeme sind kybernetische Maschinen**, die eine Geschichte haben und diese Geschichte oder Erfahrungen verändern den Output auch wenn der Input gleich bleibt. Ziemlich kompliziert zu verstehen, was?

What becomes apparent in this metaphorical concept is a complex understanding of systems as nontrivial, nonlinear, and nonmechanistic entities. In fact, the machine metaphor in combination with the concept of cybernetics evokes an understanding of systems as the relation of related and interacting objects. The metaphor clearly displays some sort of cognitive dissonance because the notion of a machine holds functional and deterministic implications that cybernetic machines do not. The metaphor could thus be understood as a productive contradiction *in adjecto* or heuristic device for systems biologists as it conceptually merges a certain degree of functionality with the idea of nonlinearity.

However, along with these more technical metaphors, the visually oriented conceptual metaphor, A SYSTEM IS A BIG PICTURE, materialized in the interviews. It stresses aspects of visual perception and its relevance for the research process, and features aspects of detailed overview and insight:

> "The **system perspective really provides the big picture**, it's some sort of a vista where you can switch back and forth, from small scale to big scale, back and forth." (Scientist C)
>
> German original: Also, die **Systemperspektive führt uns wirklich zum großen Bild**, es ist so eine Art Überblick in dem man vom Ganzen ins Detail gehen kann, also einfach hin und her schalten.
>
> I mean, **the system view really is the big picture**. We can go into detail and at the same time think about the overall perspective and then see how this all evolves, the whole system. Sometimes its leads me to a new humility... because of these multifaceted interactions that make up things like petals or resistant plants such as glasswort. (Scientist G)
>
> German original: Ich finde, dass **die Systemperspektive wirklich das große und ganze Bild** ist. Wir können uns Detail gehen und zur gleichen Zeit über die übergreifende Perspektive nachdenken und schauen, wie das alles entsteht, also das System. Manchmal führt das bei mir zu einer neuen Bescheidenheit... weil, diese vielfältigen Prozesse die Blütenblätter entstehen lassen oder zu so resistenten Pflanzen wie Queller führen.

What becomes evident in the previous interview excerpts is that the conceptual metaphor, A SYSTEM IS A BIG PICTURE, is often combined with lexical items stemming from the semantic field of vision. Thus, words such as "perspective" (*Perspektive*), "systems perspective" (*Systemperspektive*), and "overview" (*Überblick*) develop a connection with the conceptual metaphor and strengthen its visual scope in terms of an improved understanding. This aspect of an improved understanding is subcutaneously endorsed by culturally well-engrained metaphorical concepts such as, UNDERSTANDING IS SEEING, and, UNDERSTANDING IS LIGHT, which relate to the visual aspect semantically inherent in the metaphorical concept, A SYSTEM IS A BIG PICTURE.

Concluding this section, we now turn to a broader picture of the conceptual metaphors encountered and analyzed in this section. The aim first consists in providing a structured overview and second in interpreting what kind of implications may reside in the metaphorical framing of the system concept.

2.3.3 Assessing Metaphorically Informed Concepts of System

As we have seen in the previous section, scientists use different metaphorical concepts to grasp and elaborate semantically upon what the abstract notion system means to them. The—sometimes—detailed analysis and interpretation revealed hidden aspects that do not appear on the word level. Although the question asked was complex and led in some cases to a short period of reflection, not a single

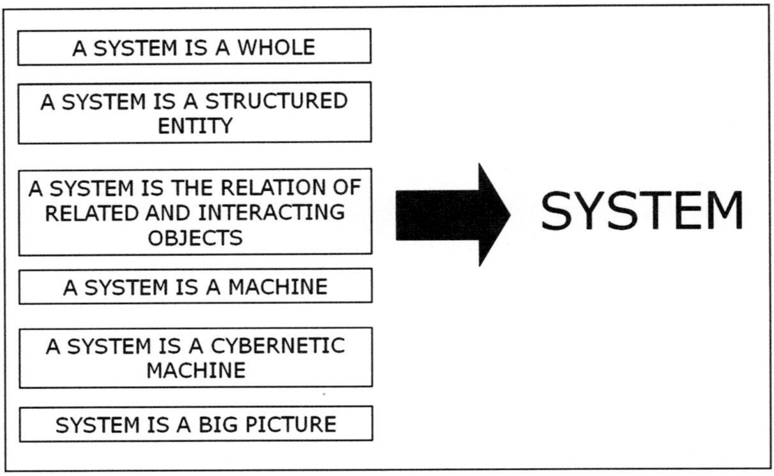

Fig. 2.4 Conceptual metaphors used to frame the notion of system

interviewee refused to answer the question. On the contrary, some scientists enjoyed exploring and depicting their understanding of what system means to them and expressed in the aftermath of the interviews that more time should be devoted to what one interviewee called "theoretical playing and exploring."

The preceding analysis, furthermore, provided insight into how the abstract notion of system was endowed with meanings, albeit different ones. These results offered a first insight into how and by which means a representative group of scientists working in the area of systems biology frames the concept of system in systems biology (see Fig. 2.4).

If one considers again the analyzed imagery in view of the underlying transfer processes, it showed that the first two conceptual metaphors were based on an implicit spatiality that contributed to reifying what a system is and applied a spatial structure to what a system could be. The aim here consisted in making the rather static system concept manageable. Meanwhile, a certain degree of dynamization was tackled in the conceptual metaphor, A SYSTEM IS THE RELATION OF RELATED AND INTERACTING OBJECTS, and in the A SYSTEM IS A CYBERNETIC MACHINE. The latter especially seemed to counteract the mechanistic implications nestling in the conceptual metaphor, A SYSTEM IS A MACHINE, by merging mechanistic aspects with a nonlinear understanding of systems. In brief, the conceptual metaphors exhibited a process of dynamization of the systems notion and this provided a bigger picture as encountered in the conceptual metaphor, A SYSTEM IS A BIG PICTURE (see Fig. 2.5).

To summarize, it should have become clear in this section that metaphors play—again—a vital role in the conceptualization of abstract knowledge domains because they capture the abstract notion of system with the help of six metaphorical concepts. Interestingly enough, the metaphorical concepts could not be connected to the scientific disciplines to which the interviewees belonged. The analysis and interpretation of the metaphorical mapping processes, furthermore, provided an

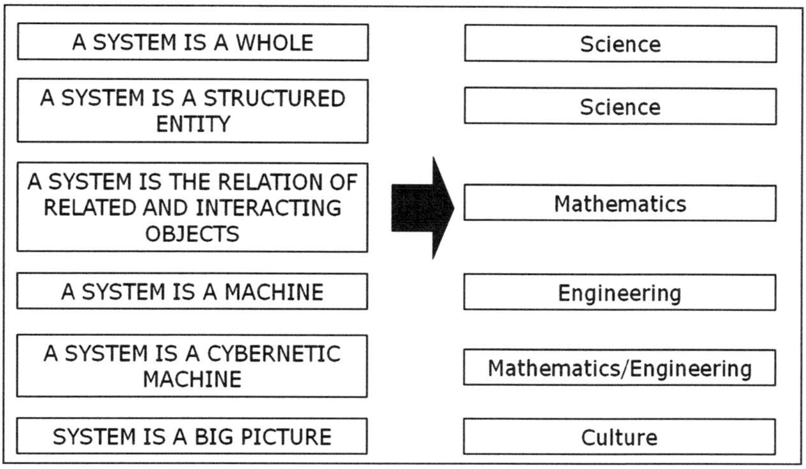

Fig. 2.5 Conceptual metaphors and source domains

opportunity to reveal the generally shared and unconsciously generated meaning constructions that seem to revolve around the aspect of "dynamizing" the understanding of biological processes with the aid of a systems concept. As the approach to systematically analyze metaphors proved to be practical and productive, we now turn to the systematic exploration of the abstract notion of reductionism.

2.4 Reimaging Reductionism in Systems Biology

The notion of reductionism seems to run counter to the logic of complexity and multilevel interaction, inbuilt in systems biology, and also to contradict the intuitions evoked by the term holism which is connected to systems biology. Reductionism refers to a concept which could roughly be characterized by the idea that the development, maintenance, and functioning of an entity can be understood and explained with reference to a basic and restricted set of underlying components. These indivisible and invisible elements such as genes (or the DNA-sequences coding for proteins or control elements, respectively) are conceived of as representing material endpoints that help to understand and explicate phenomena which go beyond them. Thus, higher levels of biological organization and their phenomena causally rely on these endpoints and represent their ostensible epiphenomena. The concept of reductionism, thus, aims at explaining complex and multifaceted phenomena of the natural world by reducing them to simpler structures of matter. It was introduced into Western thinking by René Descartes and his clockwork metaphor (Descartes 2000, 42–43 and 270–271), which we already encountered in the previous section on the concept of system in systems biology. Simply put, Descartes' idea of a clockworks is based on the belief that God, when creating the world, had a clockwork mechanism in mind (Snobelen 2012) which could be used to explain the mechanical functioning of physical, chemical, and biological systems. Reductionism, to be understood as a heuristic and philosophical standpoint, offered a rationale in which things were conceived of as being composed of a restricted set of substances (ontological reductionism), that one has to break down a system to its constituents and then to functionally reconstruct it (methodological reductionism). This procedure was thought to be a promising approach to unravel and understand the organized parts and their functionality in a comprehensive system (Pigliucci 2014). A third version of reductionism holds that concepts, laws, and theories are tied to a certain level of organization and explanations found at one level could be absorbed by theories of higher levels, or, the other way around: an explanation relevant to one level could be reduced to theories formulated for lower system levels (theoretical reductionism).

Methodical reductionism especially was adopted in molecular biology and triggered an understanding of individual components as based on their structural chemical and physical properties.[13] However, developments in biology, medicine, genomics, and proteomics indicate since the 1990s that the approach is about to arrive or already has arrived at its limits. Consequently, reductionist frameworks appear unable to explain and unravel the nature of complex phenotypes or diseases such as cancer, and efforts to explicate the complexity and indeterminacy of the human brain based on reductionist assumptions did not prove to be successful. Furthermore, certain properties inherent to biological systems could not be explained with the help of a reductionist heuristic because

> [...] proteins with identical or similar biochemical properties do not automatically also have similar biological functions. This specific protein, as found in the fruit fly, apparently catalyzes the folding of a pigment which is involved in vision, whereas the protein found in mammalian life forms seems to be involved in the regulation of the maturation of immune cells. This means that one enzyme (and the relevant gene) can influence very different biological phenomena with a different ecological relevance, depending on the genetic, cellular or phylogenetic context in which it is found [...] (Kollek 1990, 128)

and because

> [...] biological activity does not arise from the specificity of the individual molecules that are involved, as these components frequently function in many different processes. For instance, genes that affect memory formation in the fruit fly encode proteins in the cyclic AMP (cAMP) signaling pathway that are not specific to memory (van Regenmortel 2004, 1016).

Such insights instigated interest in more comprehensive and systemic approaches that materialized in the new systems biology at the end of the 1990s. Systems biology is, however, rooted in reductionist thinking which has been extremely important to molecular biology. Westerhoff and Palsson (2004, 1249) estimate that at least two reductionist roots have been important for systems biology:

> [The first] stemmed from fundamental discoveries about the nature of genetic material, structural characterization of macromolecules and later developments in recombinant and high-throughput technologies [while the second] sprung from non-equilibrium thermodynamics theory in the 1940s, the elucidation of biochemical pathways and feedback controls in unicellular organisms and the emerging recognition of networks in biology.

---

[13] It must be stated, however, that there were always currents in biology critical to an overarching methodical reductionism as, for instance, in physiology (Stange 2005) or which rejected ontological reductionism.

Reductionism and molecular biology, however, underwent considerable criticism in systems biology, even though its relevance for the development of a systems approach in biology has generally been acknowledged. The empirical question, nevertheless, remains of how contemporary scientists working in systems biology frame the notion of reductionism. Although this question has already been addressed by Calvert and Fujimura (2011) in their analysis of how scientists working in systems biology separate their discipline from molecular biology, our analysis has different aims: first, we want to elucidate how the abstract notion of reductionism is metaphorically framed by systems biologists, and second, whether different implications nestling in conceptual metaphors display a critical or positive view of reductionism. Furthermore, do these metaphorical concepts contribute to building up a professional identity or difference (Bourdieu 1976) between molecular biology and systems biology?

The overall aim therefore consists in disclosing and interpreting the intangible structure implicated in the linguistic imagery used. It is, however, necessary to provide an insight into the different theoretical backgrounds of reductionism and its current relevance in systems biology before we turn to our empirical analysis as only knowledge about its conceptual history can help to contextualize and better understand current ideas of reductionism. We therefore present in the following section a—though compressed—diachronic and synchronic insight into the notion of reductionism and its use in systems biology. Against this background, linguistic and conceptual metaphors taken from the interviews conducted are analyzed to reveal the current conceptualization of reductionism in systems biology.

2.4.1 Reductionism in Biology and Systems Biology: A Short Overview

Reductionism is a basic concept in modern sciences such as physics, chemistry, or biology since the days of Descartes and Newton, and debates revolving around it question "whether specific scientific entities, concepts or relations can replace other entities, concepts or relations. Attempts at such reductions from one area of inquiry to another have been an integral part of much modern science" (Andersen 2001, 153). Reductionism, thus, represents a historically consolidated concept that goes back to the seventeenth century where, for example, developments in mechanical philosophy used a mechanical logic to explain optical phenomena, whereas physicists at the end of the nineteenth century tried to explain the thermodynamics of ideal gases by analyzing the mechanical activities of constituting molecules.

There is thus a well-established tendency to study complex natural phenomena in relation to elements that are conceived of as constituting parts. According to this view, the world could be interpreted as a nested structure of reductive levels where the laws of higher systems levels could be reduced to the ones of lower system levels. This position is also termed *theoretical reductionism*, which aims at reducing one explanation or theory to another, simpler, but more comprehensive idea (Andersen 2001). The strongest version of this view was put forward in the 1930s by the logical positivists adhering to the ideas of linearity, causality, and the cumulative aspect of nested structures (Feigl 1981a). It remained an important approach until the late 1950s and had a considerable impact on science (Feigl 1981b). The approach is best expressed in Oppenheimer's and Putnam's (1958, 3) paper entitled, "The Unity of Science as a Working Hypothesis." The authors stress that

> [i]t is not absurd to suppose that psychological laws may eventually be explained in terms of the behavior of individual neurons of the brain; that the behavior of individual cells – including neurons – may eventually be explained in terms of their biochemical constitution; and that the behavior of molecules – including the macro-molecules that make up living cells – may eventually be explained in terms of atomic physics.

Contemporary concepts of reduction and reductionism are highly influenced by the logical empiricist Ernest Nagel. Nagel aimed at developing a formal framework for reduction in his essay "The Meaning of Reduction in the Natural Sciences" (Nagel 1960, 99) and in his book entitled, *The Structure of Science. Problems in the Logic of Scientific Explanation* (Nagel 1961). He described reduction as "the explanation of a theory or set of experimental laws established in one area of inquiry, by a theory usually though not invariably formulated for some other domain" (Nagel 1961, 338). This concept is based on a not unproblematic logical empiricist background because it is unable to explain why phenomena could not be reduced to the components or workings of lower system levels. To overcome these logical shortcomings, Nagel developed the idea of the condition of connectability and the condition of derivability with which he aimed at first allowing an assumption that connects functionally discrete entities, and second provides the basis logically to derive laws of the lower system from the higher system (Klein 2009; Peacocke 1976). Nagel's thinking is an important point of reference in philosophical discussions of reductionism that could be, according to Ayala (1974) and as already mentioned, divided into methodological, theoretical, and ontological reductionism. Although these subcategories are in reality intertwined and almost always appear in combination, they nevertheless represent, from an analytical point of view, discrete analytical concepts running through past and present philosophical analyses of scientific research and reasoning.

Ontological reductionism is based on the monist idea that all natural phenomena are composed of a minimum number of kinds of entities or substances. In essence, it is a metaphysical position claiming that all objects, properties, and processes are finally reducible to a single substance. In general, it holds that knowledge about the most basic level and the functionality of its constituting elements suffices to explain phenomena emerging at higher levels of natural entities. These phenomena are called epiphenomena and their complexity is resolved by reducing them to ever-simpler structures of matter which means that the evolution and behavior of higher levels of complexity are driven by basic laws that govern the configuration of basic elements. Change in the structure of elements and their relations is conceptualized as movement in space and the geometrical rearrangement is ruled by cause and effect (Schaffner 1993a, b).

Methodological reductionism, in addition, often builds upon ontological reductionism in that it is unconsciously implicated in the former. The concept of methodological reductionism is based on the conviction that it is a scientifically sound and sensible way to analyze any system at its lowest level. The approach consists in breaking a whole system down to its constituting elements to investigate the structures and functions of its components, and then to reconstruct it with the aim of understanding their functional interaction in the context of the whole entity (Peacocke 1985).

The French physiologist Claude Bernard may be considered one of the first scientists to apply these central principles of experimental philosophy (Böhme et al. 1977) to biology and medicine. In his *Introduction to the Study of Experimental Medicine* (Bernard 1983), he outlined his approach and research methods in the field of experimental physiology: Experimental reasoning, whose different terms we have examined in the preceding section, sets itself the same goal in all the sciences. Experimenters try to reach determinism; with the help of reasoning and of experiment they try to connect natural phenomena with their necessary conditions, or, in other words, with their immediate causes. By these means they reach the law which enables them to master phenomena (Bernard 1983, 57). Research in modern analytical biology and medicine has for the most part followed this pattern since then.

In addition to these two concepts of reductionism, theoretical or epistemological reductionism (Ayala 1974) presupposes that epistemic units such as laws or theories are tied to a certain level of organization and could be explained by implementing the rules of reduction taken from epistemic units at lower levels of the system (Nagel 1961). Epistemological reductionism clearly possesses hierarchical characteristics and obviously holds strong ties with the previous two kinds of reductionism. To summarize, there is thus not a single concept of reductionism but mainly three concepts that inform and permeate scientific discourses and practices of all sorts (Stöckler 1991). In scientific discourse, however, the difference between the analytically discrete concepts usually is neither appreciated nor consciously discussed. Rather, they are often mingled together which does not help to clarify the debate about reductionism.

Inbuilt in all concepts of reductionism outlined here are basic ideas of linear causality and predictability. They hold strong ties with a deterministic worldview in which any phenomenon of nature is tied to pre-existing causes: knowing the initial conditions and the mechanical laws triggering the behavior of the entities hence leads to predictability of the system. Both concepts—reductionism and determinism—had an impact on biology and played an important role in the rise of molecular biology which was mainly propelled by scientists trained in physics (Morange 2000, 2009) and computer sciences (Kay 2000). Consequently, genetic information was conceived to be a straight representation of the genetic code and the structure of the DNA. In this model causal linear flows trigger the transcription of genetic information from the DNA to RNA to proteins (Crick 1958). These information flows were thought to be unidirectional and were conceived of as the central dogma of molecular biology (Schaffner 2002) regardless of the fact that control genes or feedback loops were detected later on. As a result, the molecule-centered perspective of biology was coupled with a molecular-reductionist perspective which suggested that the identification of relevant molecules and their laws of interaction are the relevant units of biological analysis for understanding the functioning of biological entities (Rosenberg 1997).

Reductionism possessed and still possesses a considerable explanatory power, and it enabled scientists working in biology to explore important molecular and cellular processes. Many scientists working in molecular biology to date still rely on reductionist models (Parry and Dupré 2010; Fox Keller 2010). However, a critical point for excessive reductionism was reached when evidence was provided that gene products are not linear representations of genetic information (Falk 1986, 2010) and that their function depends on the spatiotemporal patterns of their expression and on their interactions with other genes. As a consequence, genes (and their products) today have to be conceived of as elements of complex networks on different levels, and that these levels bear an impact on their context-specific activity in cells, tissues or organs. In summary, genes have different functional purposes in an organism depending on their place and position in time. What becomes apparent is the fact that research deconstructed the belief that complex processes could be reduced to unidirectional processes or to the workings of the lower-level elements (Laubichler and Wagner 2001).

A good example for the tendency to simplify complex issues consists in the fact that the reductionist approach removes the object of investigation from its natural context (Kollek 1990; Bonß et al. 1993, 1994; Rheinberger 1997, 2009, 2010). The disciplinization of the research object for specific research purposes reduces the validity of scientific results and can lead to over-interpretation and misleading conclusions. According to this view, it appears impossible to explain processes of life by reducing them to the molecular or genetic level.

This is precisely where systems biology comes into play because it is based on the idea that biological systems are complex and interactive entities with a multitude of structural and functional entities distributed over different system levels. This approach questions the idea of a central control unit and decentralizes it, and also sheds doubts on a hierarchical mode of control and "democratizes" it even though it does not mean that scientists working in systems biology did throw out the baby with the bathwater. The systems approach leads to developing new questions (van Regenmortel 2004) and the application of novel methods. Thus, scientists working in systems biology try to detach themselves from the molecular tradition of linear causality and upward causation by generating ideas of downward causation and distributed causality and control. This change in approaching problems was instigated by the advent of innovations in IC technologies, high-throughput technologies and enhanced possibilities of simulating complex systems or biological networks with the help of mathematical models (Alm and Arkin 2003). No matter how elaborated the positions and reflections are in detail, what becomes apparent is the new conviction that the behavior of a complex system cannot be explained by the structural analysis of the systems components alone, although knowledge about these components is indispensable. But still, although a different mindset in the context of systems biology emerged, an explicit and critical reflection of reductionism and its subcategories for systems biology in general and molecular systems biology in particular is still pending.

In summary, we have tried to provide a somewhat reduced historical and synchronic overview of the main aspects and concepts of reductionism in biology and, as far as possible, systems biology. It became clear that the notion of reductionism holds a long history dating back to the days of René Descartes and Isaac Newton. Descartes' clockwork metaphor, especially, paved grounds for a mechanistic and reductionist logic that was taken up and conceptually redefined in the nineteenth and twentieth centuries by a variety of scientists and philosophers of biology and science. The three analytically discrete but in reality intertwined subconcepts of ontological, methodological, and epistemological (resp., theoretical) reductionism (Ayala 1974) were tackled in which constitutive aspects such as predictability, linear causality, upward causation, and the idea of a central control unit nestled. These aspects became constitutive elements in the rationale of molecular biology and considerably contributed to its development. Results from research and technological developments such as the advent of high-throughput technologies put superficial reductionist rationales into question but a closer look at approaches and concepts in systems biology indicates methodological reductionism cannot be relinquished and still constitutes an important research strategy, whereas epistemological reductionism has implicitly been accepted, but its challenges have not really been tackled yet by the research community. Although systems biology is thought to emphasize that biological processes are characterized by upward as well as by downward causation across system levels, distributed causality, and disseminated control, the different forms of reductionism are still at work in scientists' minds and research carried out. The notion and concept of reductionism in systems biology, however, has not received much critical inspection or in-depth reflection to date. In order to explore what is meant by reductionism in systems biology it is thus important to study its metaphorical conceptualization by systems biologists as it surely is a basic heuristic and practical ingredient in their daily scientific work. This is done in the following section where a paradigmatic set of interview excerpts displays conceptual metaphors used to conceptualize reductionism.

2.4.2 *Systems Biologists' Imaging Reductionism*

In the previous section we encountered the different dimensions of reductionism and their basic conceptual ingredients. It became apparent that the general notion of reductionism is based on three subconcepts such as ontological reductionism, methodological reductionism, and epistemic reductionism. Taking these aspects into consideration, it is remarkable that a historical, theoretical, and philosophical investigation of reductionism in systems biology has rarely been addressed.[14] Quotes, such as the following, often depict some sort of historical overview in which different concepts are generally mentioned with regard to molecular biology, but not further analyzed:

> Much of twentieth-century biology has been an attempt to reduce biological phenomena as an investigation into the inheritance of variation, such as differences in the color of pea seeds and fly eyes. From these studies, geneticists inferred the existence of genes and many

[14] For exceptions see Andersen (2001), Fang and Casadevall (2011) and Kaiser (2011).

of their properties, such as their linear arrangement along the length of a chromosome. Further analysis led to the principles that each gene controls the synthesis of one protein, that DNA contains genetic information, and that the genetic code links the sequence of DNA to the structure of proteins. Despite the enormous success of this approach, a discrete biological function can only rarely be attributed to an individual molecule in the sense that the main purpose of hemoglobin is to transport gas molecules in the bloodstream. In contrast, most biological functions arise from interactions among many components. (Hartwell et al. 1999, C47)

This paradigmatic description of twentieth-century biology clearly refers to molecular biology and displays some of the characteristics of reductionism outlined in the preceding section such as central control by genes or the attribution of discrete biological functions to a single molecule. What remains in the dark is what this means for systems biology and in what way the systems approach differs from theoretical implications of reductionism inherent in molecular biology. This aspect is historically referred to in the following quote of a paper written by Westerhoff and Palsson:

Whereas the foundations of systems biology-at-large are generally recognized as being as far apart as 19th century whole-organism embryology and network mathematics, there is a school of thought that systems biology of the living cell has its origin in the expansion of molecular biology to genome-wide analyses. From this perspective, the emergence of this 'new' field constitutes a 'paradigm shift' for molecular biology, which ironically has often focused on reductionist thinking. Systems thinking in molecular biology will likely be dominated by formal integrative analysis going forward rather than solely being driven by high-throughput technologies. (Westerhoff and Palsson 2004, 1249)

Generic reference is made to well-known historical developments to situate the genealogy of systems biology: nineteenth century whole-organism embryology is conceptually coupled with recent advancements such as network mathematics whereas reductionism is explicitly alluded to in the phrase "reductionist thinking". In fact, a closer inspection of the relation between the concepts of reductionism and holism is lacking in many papers on systems biology which often follow the rhetorical logic of a short historical introduction to be initially pursued by the specific problem under investigation. These rhetoric devices contribute to developing a narrative of systems biology as already existing and then re-emerging due to the shortcomings of the reductionist agenda inherent in molecular biology. Although this rhetoric appears to be constitutive for the discipline of systems biology, one has to bear a mind that there is a great difference between the written form of scientific papers or reviews and the knowledge of systems biologists in their scientific everyday life on the other hand. Scientists possess an implicit and pragmatic knowledge about reductionism; it is relevant for them and whether and how it should be applied to scientific problems. This kind of knowledge does not materialize in reviewed scientific papers or books but in cognitive strategies of problem-solving and scientific practices. Two interview excerpts taken from our dataset clearly indicate that reductionism is a ubiquitous phenomenon in systems biology:

You now, everyone criticizes reductionism or this reductionist agenda but honestly speaking, we have to reduce the problem to make it manageable. We can only start with the smallest units and then go up to the next level to try to understand it. I mean the small parts constitute the overall entity, which naturally has an impact back on the smaller units. (Scientist D)

German original: Wissens Sie, jeder kritisiert den Reduktionismus oder diese reduktionistische Agenda, aber ehrlich gesagt müssen wir das Problem reduzieren um es handhabbar zu machen. Wir können doch nur mit den kleinsten Einheiten anfangen und dann auf die nächst höherer Ebene gehen, um diese zu erklären. Ich meine, die kleinsten Einheiten bringen doch die übergreifende Einheit hervor, die natürlich wieder auf die kleineren Einheiten zurückwirkt.

The quote clearly displays an ontological reductionism by indicating that the "small parts constitute the overall entity" and methodological reductionism appears in the phrase "we have to reduce the problem to make it manageable." A comparable mixture of reductionist dimensions appears in the following quote where "to start with the elements" refers to ontological reductionism and, "I think that we have to be pragmatic to structure the research process," clearly refers to methodological reductionism.

I think that we need this kind of daily reductionism to solve problems. I really do not subscribe to the idea that the parts constitute the whole but I think that we have to be pragmatic to structure the research process. So we start with the elements and see what happens on the next system level, and then we try to understand this process or the interaction between the elements and the system levels. (Scientist G)

German original: Ich bin der Meinung, dass wir eine Art Reduktionismus für die Problemlösung brauchen. Ich bin wirklich kein Fan der Idee, dass die Teile das Ganze konstituieren, aber ich glaube, dass wir pragmatisch sein sollten, um den Forschungsprozess zu strukturieren. Also beginnen wir mit den Komponenten und gucken dann, was auf der nächsten Systemebene passiert. Und dann versuchen wir diesen Prozess oder die Interaktion zwischen den Komponenten und den Ebenen zu verstehen.

What is even more interesting is that in both quotes, nonreductionist and reductionist thinking surface at the same time. It looks as though systems biologists insist on methodical reductionism, which they think is essential to their work, while having problems to detach themselves from ontological reductionism.

Taking these results into consideration, we think it might be of vital interest for systems biologists to reflect philosophically upon the notion of reductionism with the aim of developing a clearer picture of what reductionism is and what role it plays in their daily work and their conceptionalization of biological research objects. As reductionism is an abstract concept, metaphors are used to concretize and to communicate it. This led to spontaneously generated metaphors that were analyzed and systematized by examining the transcribed interviews and led to four conceptual four conceptual metaphors. These are REDUCTIONISM IS AN ENTITY, REDUCTIONISM IS AN ANCESTOR, REDUCTIONISM IS A PREDECESSOR, and REDUCTIONISM IS AN ADVERSARY.

To start with, reductionism has been metaphorically framed as an entity by using the conceptual metaphor of REDUCTIONISM IS AN ENTITY. This imagery uses words such as "entity" (*eine Sache*), "scientific entity" (*wissenschaftlicher Gegenstand*), and "scientific concept" (*wissenschaftliches Konzept*) which often appear in the interview quotes. These words are projected upon the abstract entity reductionism to make it cognitively accessible and manageable:

Ehm, **reductionism, well that is some sort of entity** which is really tricky to handle. It has been and is so influential in science and even though it proved to be wrong. I do not want to throw out the baby with the bathtub because it has been important to research in biology. (Scientist A)

German original: Äh, der **Reduktionismus, das ist so eine schwierige Sache zu handhaben**. Er war und ist so einflussreich in der Wissenschaft auch wenn er sich in vielen Fällen als falsch erwies. Ich möchte nicht das Baby mit dem Bad ausschütten denn er war schon sehr wichtig für die Forschung in der Biologie.

Reductionism has been highly influential in my scientific life. **It's a scientific entity or a concept which** has been quite helpful and brought biology forward. I would say that it has been an influential concept. (Scientist M)

German original: Der Reduktionismus war in für mein wissenschaftliches Leben sehr wichtig. **Das ist ein wissenschaftlicher Gegenstand oder so ein Konzept**, das sehr hilfreich war und die Biologie wirklich vorangebracht hat. Ich würde sagen, dass es ein einflussreiches Konzept war.

What one can see in these quotes is that it is quite difficult to describe the abstract entity of reductionism, and ontological metaphors help to constitute it as a thing. Furthermore, a closer look at the quotes indicates an implicit historicization: tenses used such as "has been," "proved," and "has been and is" develop a temporal image of a past to which reductionism is implicitly relegated.

The ontological metaphor, however, is quite important as it prepares the conceptual ground for the following two conceptual metaphors which are mainly based on personifications (Jäkel 1997). Personification, to be understood as a subcategory of ontological metaphors, conceptualize an abstract entity in terms of a human being and open up the possibility to ascribe human characteristics to it. This becomes apparent in the following quotes where reductionism (and sometimes molecular biology) is metaphorically framed as an ancestor.

The concept reductionism has been around for decades and **I see it partly as an ancestor of systems biology**. It has indeed contributed so much to the development of biology and the sciences, I mean in the context of molecular biology, but it did not manage to solve the problems detected by it. Somehow a funny development. (Scientist H)

German original: Das Konzept des Reduktionismus kennen wir ja schon seit Jahrzehnten und **ich sehe es teilweise als eine Art Vorfahre der Systembiologie**. Es hat wirklich sehr viel zur Biologie und der Entwicklung der Wissenschaft beigetragen, ich meine im Zusammenhang mit der Molekularbiologie, auch wenn diese nicht die Probleme lösen konnte, die sie aufgeworfen hat. Auch irgendwie eine komische Entwicklung.

It [reductionism] **could be understood as an ancestor** that led the way to systems biology, to complexity and all these interesting questions, you know. (Scientist E)

German original: Er [der Reduktionismus] **könnte als eine Art Vorfahren verstanden werden** der den Weg zur Systembiologie bereitet hat, zur Komplexität und all diesen spannenden Fragen, wissen Sie.

The personification, REDUCTIONISM IS AN ANCESTOR, offers a further possibility to structure the concept of reductionism semantically. It develops human genealogy and situates systems biology at the end of family tree which began in the past with molecular biology.

The following personification, REDUCTIONISM IS A PREDECESSOR, elaborates on the previous concepts but introduces a more neutral aspect on the genealogical aspect because predecessors can be family members but also people not belonging to the family.

Reductionism and obviously **molecular biology are predecessors of systems biology**. I see it as such and I know that many colleagues would subscribe to this view. (Scientist I)

German original: Der Reduktionismus und natürlich auch **die Molekularbiologie sind Vorläufer der Systembiologie**. Ich sehe es zumindest so und ich weiß, dass viele Kollegen es auch so sehen.

For me, reductionism is, together with molecular biology, well, yeah, **they are predecessors of systems biology**. (Scientist N)

German original: Für mich ist der Reduktionismus, zusammen mit der Molekularbiologie, ja, also das sind Vorläufer der Systembiologie.

This personification again puts emphasis on the temporal aspect of succession and the development of scientific theories. What is even more important is the fact that both personifications, REDUCTIONISM IS AN ANCESTOR, and REDUCTIONISM IS A PREDECESSOR, construct a succession of events starting in the past and contributing to bringing about systems biology in its current state, at least in systems biology.

Personification can also be critically used because other theories or approaches can become enemies or adversaries threatening their own scientific agenda or even existence. The conceptual metaphor, REDUCTIONISM IS AN ADVERSARY, appeared not as often as the previous personifications, but it is worth noting here as it could be interpreted as some sort of a relict of former scientific struggles or enforcement techniques.

Reductionism has long been seen as an adversary. I see it today much more as a useful development which then paved the way towards new approaches such as metabolomics, network biology or systems biology. (Scientist J)

German original: **Der Reduktionismus wurde lang als Gegner angesehen**. Ich sehe es heute eher so als eine sinnvolle Entwicklung, die den Weg für Ansätze wie metabolomics, die network biology oder auch die Systembiologie freigemacht hat.

At the beginning, there was a lot dispute and **reductionism and molecular biology were conceived as an adversary**, if one can say it in this way... (Scientist B)

German original: Zu Anfang gab's schon ziemlich viel Streit **und der Reduktionismus sowie die Molekularbiologie wurden schon als Gegner verstanden**, also wenn man das so sagen kann.

What is interesting in the previous quotes is the fact that reductionism and molecular biology are metaphorically framed as adversaries of the past. The use of the past tense or the past progressive indicates the differences between systems and molecular biologists already came to an end. So, the images of reductionism and molecular biology as adversaries fade and more prominent images such as REDUCTIONISM IS AN ANCESTOR and REDUCTIONISM IS A PREDECESSOR indicate a reconciliation of both approaches.

We now turn, in concluding this section, to a short and systematized overview of the conceptual metaphors encountered and analyzed in this section.

2.4.3 *Evaluating Metaphorically Informed Images of Reductionism*

As we have seen in the preceding empirical section, the conceptual metaphors encountered and analyzed demonstrate that scientists, in fact, use a somewhat restricted set of metaphors to frame the abstract concept of reductionism. The conceptual metaphors

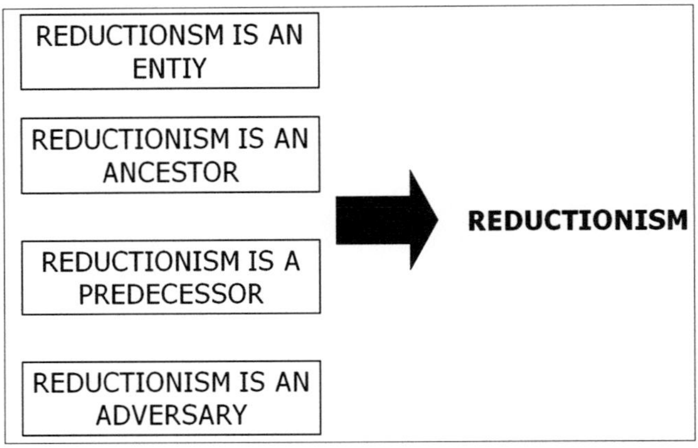

Fig. 2.6 Conceptual metaphors and personifications used to frame the notion of reductionism

REDUCTIONISM IS A THING, REDUCTIONISM IS AN ANCESTOR, REDUCTIONISM IS A PREDECESSOR, and REDUCTIONISM IS AN ADVERSARY were generally used to express and frame the relation of system biologists towards the concept of reductionism and—in many cases—molecular biology (see Fig. 2.6). This seems to indicate that a close conceptual relation between reductionism and the overarching discipline of molecular biology exists because at some stages in the interviews, reductionism almost became a metonym for molecular biology. It, furthermore, became evident that almost exclusively personifications were used as linguistic devices to frame what reductionism is. Thus, the ontological metaphor, REDUCTIONISM IS AN ENTITY, prepares grounds—metaphorically speaking—to substantiate the concept of reductionism. Based on this conceptual grounding, the two personifications, REDUCTIONISM IS AN ANCESTOR and REDUCTIONISM IS A PREDECESSOR, develop a genealogical image of advance supported by the fading personification, REDUCTIONISM IS AN ADVERSARY.

The analysis of the conceptual metaphors, moreover, showed that the ongoing quarrels between molecular and systems biology have been reassured. This was deducted from the personifications as well as from surrounding linguistic context which displayed an extensive use of verbs in the past tense and the past progressive. In brief, reductionism is no longer conceived of as an adversary but a relict of former times that has been incorporated but not overcome by systems biology.

To summarize, we have seen how metaphors and two of its special subcategories, ontological metaphors and personifications, contribute to semantically making the abstract concept of reductionism accessible. The analysis and interpretation of the metaphorical mapping processes inherent in the personifications provided an opportunity to disclose the shared meaning constructions that seem to revolve around the aspect of reconciling systems biology with reductionism, in some cases at least to be understood as a metonym for molecular biology. To complement and contrast the analysis in this section, we now turn to the analysis of imagery framing the abstract notion of holism.

2.5 The Concept of Holism in Systems Biology

As we have seen in the previous section, reductionism in systems biology is metaphorically framed as an entity and personified as an ancestor, a predecessor, and an adversary. These metaphorical notions and their delimiting implications now make it necessary to contrast reductionism with the opposite notion of holism as conceptualized by systems biologists. It is not our academic endeavor (at least not in this section) to show that systems biology is more holistic than molecular biology. It is, however, interesting that such claims, which are made by quite a number of researchers working in systems biology (see Sect. 1.1.2) often lack a theoretical consideration or explanation of what holism means for systems biologists and how they apply the concept in their scientific work. As a result, it is, first, not surprising that the reductionist–holist debate has made a limited reappearance in the context of systems biology and, second, that the notion of holism is in many cases mainly superficially used. This aspect is corroborated by the fact that 32 articles use the notion of holistic, holism, whole, or wholes in their titles and abstracts in the period from 2000 to 2014.[15] Furthermore, a close reading of the articles indicates that theoretical or conceptual considerations are often lacking while they frequently provide an ahistorical understanding of the reductionism–holism debate: it is in many cases framed as closely related to the advent of systems biology. This is astonishing as systems biology had already made contact with the concept of holism in the context of general systems theory. Ludwig von Bertalanffy, Paul A. Weiss, and Robert Rosen, to mention just a few, already addressed in the 1950s and 1960s the question of how holistic views or concepts could be integrated into systems theory and theoretical biology. Not taking these theoretical and conceptual insights into consideration (Drack and Apfalter 2007; Drack 2009) seems to have led to a certain degree of semantic confusion, conceptual inconsistency, and historical misunderstanding about the meaning of holism and from where it originates.[16] We, thus, aim at providing in the following section an insight into how systems biologists metaphorically frame the notion of holism. This investigation provides us with a synchronic insight which is then used as a backdrop for a historical contextualization and conceptual comparison with scientific precursors such as vitalism, classical holism, and modern holism. The main aim of this historically inverse procedure consists in disclosing the intangible structures implicated in the linguistic imagery and in the analysis whether and how system biologists relate to scientific predecessors. Although similar questions have already been empirically addressed by Calvert and Fujimura (2011) and Mazzochi (2012) with regard to systems biology, the scope of the present section differs from these studies as it puts emphasis on the conceptual analysis

[15] See, for example, Stange (2005), Verpoorte et al. (2005), Bennett and Monk (2008), Hood et al. (2008), Federoff and Gostin (2009) and Greek and Rice (2012).

[16] We would like to refer among others to the works of Ayala (1974), Schaffner (1969), Zucker (1981), Ruse (1988), Andersen (2001), and Allen (2005) who provided historical and philosophical insight into the different theoretical concepts and arguments underlying the debates on holism and reductionism.

of the notion of holism as used by systems biologists in their daily language. So let us now turn to study how the notion of holism is metaphorically framed in paradigmatic interview excerpts.

2.5.1 Systems Biologists' Metaphorical Frameworks of Holism

As outlined in the introduction, holism represents a theoretical concept that appears to be of interest to scientists working in systems biology even though not much theoretical activity has been devoted to clarifying what holism means for them. It is astonishing, however, that systems biology appears rather to be informed by idiosyncratic and sometimes tentative concepts of holism which lack historical, theoretical, and philosophical precision. Quotes such as the following are paradigmatic in that they represent general thoughts about holism by redepicting the holism–reductionism dichotomy and characterizing systems biology per se as holistic. What is lacking here is theoretical or conceptual elaboration on why systems biology has to be conceived as holistic.

> I would like to make the point that to obtain the evidence for the activity of traditional medicine we should not follow the reductionist approach, but go back to the holistic in vivo approach. This can be done in two different ways, one is via clinical trials. The other is through animal experiments. Besides the classic physiological observations that can be made in such in vivo experiments, e.g. blood pressure, analgesic activity, sedation, etc. nowadays we also have the possibility to measure gene expression, the proteome and the metabolome. These methods open a complete new world of possibilities, giving a much better insight in possible changes in the organism, i.e. in a holistic way. It will give us the possibilities to better understand the mode of action by comparing the changes in the transcriptome, proteome and metabolome patterns if compared to those observed with known drugs. Such an approach is now known as a systems biology approach (Verpoorte et al. 2005, 54).

Although the phrase "go back to the holistic in vivo approach" implicitly refers to historical predecessors, the quote in general appears to be a colorful conglomeration of buzzwords.

The holistic approach of systems biology is also often introduced by using visual metaphors such as holistic perspective or holistic view, as in the next quote. This all-encompassing perspective, however, neither expresses nor attracts any theoretical analysis of holism or possible implication nestling in the concept itself. This aspect becomes apparent in the following quote where the conceptual shift from genomics towards systems biology is metaphorically depicted as a revolution:

> At a ***first glance the*** present 'Western' medical approach may seem very different from holistic forms of traditional medicine. Western medicine relies on a detailed classification of diseases, empirical investigations and treatments targeting those disorders. However, the revolution in genomics that has taken place in life sciences during the past decade has provided considerable support for a ***more holistic view*** on diagnosis and treatment (Wang et al. 2005, 173).

Even though this quote stems from an article combining TCM with systems biology, it provides a typical way of vaguely outlining the value and meaning of holism. In addition to the narrative structure that often uses technical lexical elements such as "genetics," "genomics," "proteomics," and the like, the visual metaphor "first glance" opens the quote and the adjective "more" gradually characterizes the concept of holism at the end of the quote. This perspective with regard to holism, however, remains imprecise if not unclear and a closer look at the interview transcripts indicates not much reflection has been devoted to the concept of holism. This becomes apparent in the following two quotes which are spontaneous reactions to the interviewer's question of what holism means in systems biology:

> I have to admit that I did not really devote much reflection to that topic. I would say that we simply scaled the analytical levels up and tried to broaden the perspective. (Scientist F)
>
> German original: Ich habe da noch nie so wirklich drüber nachgedacht, muss ich gestehen. Wir haben einfach die Skalen oder Level erhöht und die Perspektive etwas erweitert, würde ich sagen.
>
> No, albeit this is important for our group, we have not theorized the holistic aspect. The interest, if you would like to call it holistic, emerged from the fact that we came under pressure for failing to offer explanations and were discontent with the current conceptual framework. (Scientist B)
>
> German original: Nö, das ist jetzt für unser Gruppe hier zwar wichtig, aber theoretisiert haben wir das nicht. Das Interesse, wenn man es holistisch nennen möchte, entstand eher aus der Tatsache, dass wir in Erklärungsnot kamen und auf einer konzeptuellen Ebene unzufrieden waren.

What becomes apparent in these quotes is that the concept of holism does not explicitly represent a theoretical problem: it is rather the practical side which is emphasized as the main driver to address conceptual problems implicated in research. What one witnesses here is a tentative knowledge that leads to a kind of practicotheoretical reflection in view of undertaking scientific work on a daily basis. This is an interesting fact and in the course of the interview questions on holism a theoretical reflection started in which the three conceptual metaphors, HOLISM IS AN ENTITY, HOLISM IS A BUILDING, and HOLISM IS A PERSPECTIVE, emerged.

To start with, holism was, similar to reductionism, metaphorically depicted as an entity. This designation emerged in almost all interviews and aimed at making an abstract entity cognitively manageable. This imagery is based on generic lexical items such as "entity" (*Entität*) or "thing" (*Ding*) which can be found in the following quotes.

> Holism, ok, I would say **that it is a kind of entity**, **a constructed entity** which helps to better understand or sheds light on emergent properties which develop and cannot be explained by the underlying parts. (Scientist H)
>
> German original: Holismus, ok. Ich würde sagen, **dass das so eine Art Entität ist**, **eine konstruierte Entität**, die uns ein besseres Verständnis ermöglicht und Licht ins Dunkel emergierender Eigenschaften bringt, die nicht aus den einzelnen Teilen erklärt werden können.
>
> For me holism is **some sort of a philosophical thing or better**, **a theoretical entity**, yes a theory. I think it goes back to Smuts and this Vitalist thinking, I think. (Scientist L)
>
> German original: Für mich ist **Holismus eine Art philosophisches Ding**, **also eine theoretische Entität**, ja eine Theorie. Ich glaube, dass die auf Smuts zurückgeht und vitalistisches Denken, das glaube ich.

Although references to theoretical predecessors and schools of thought rarely appear, the quotes partly outline the uneasiness in defining what holism is. It is for this reason that it is, first, metaphorically reified in terms of an entity and then, second, conceptually relegated via the adjective "theoretical" to a more abstract and theoretical level.

The reification, however, initiated in some interviews a more structured thought process in which the concept of holism was metaphorically framed in terms of a building. Thus, theories or concepts become structured and fabricated entities known from everyday life.

> Well, I do not know the **theoretical building of holism** in detail. [...] This does not mean that I know, if you will, the house as a whole but at least some rooms and perhaps one or two floors. (Scientist E)
>
> German original: Also, ich kenne das **theoretische Gebäude des Holismus** nicht en detail. [...] Das bedeutet nicht, dass ich das Haus als Ganzes kenne, aber wenigstens einige Räume und vielleicht ein oder zwei Etagen.
>
> If you wish, I would say that I do not really **know the theoretical building holism**. I know the notion has been around but it rather is not of everyday relevance for me. (Scientist G)
>
> German original: Wenn sie so wollen, dann kann ich nur sagen, **dass ich diese Theoriegebäude nicht wirklich kenne**. Es ist mir zwar begrifflich bekannt, aber es hat für mich eher keine alltägliche Relevanz.

What becomes apparent in the quotes is that the conceptual metaphor, HOLISM IS BUILDING, reifies holism as a structured but still abstract entity. This view quite roughly implies that one could home-in on such a building because buildings normally have doors to enter them and rooms to live in, but it can also remain a rather empty concept as in the last quotation. Here no inner differentiation or structure appears and it therefore comes near to the conceptual metaphor, HOLISM IS AN ENTITY. This aspect is also corroborated by the second sentence in the quote in which the theoretical building is framed as not being relevant for daily scientific work.

Finally, holism is in many cases metaphorically portrayed by the conceptual metaphor, HOLISM IS A PERSPECTIVE. This metaphor obviously emphasizes visual aspects and implicitly refers to a possibly higher standpoint from which the analytical aspect is investigated by the scientist. This can be seen in the following interview excerpts.

> The **holistic perspective on biological processes** aims at being all comprising. It tries to explain emergent properties which appear on higher levels but are determined by lower ones. (Scientist D)
>
> German original: **Die holistische Perspektive auf biologische Prozesse** versucht übergreifend zu sein. Sie versucht Eigenschaften zu erklären die sich auf höheren Ebenen abspielen und durch untere bestimmt sind.
>
> **The holistic view on biology** has been always been around but not very prominent in times of molecular biology. I mean, molecular biology has provided an important framework but I think that it is now time to broaden the scope and see what is possible. (Scientist C)
>
> German original: **Die holistische Perspektive** war eigentlich immer da, wenn auch nicht sehr prominent vertreten in Zeiten molekularbiologischer Forschung. Ich meine, die Molekularbiologie hat wichtige Konzepte entwickelt, aber ich denke dass es jetzt an der Zeit ist den Ausschnitt wieder zu vergrößern um zu schauen, was möglich ist.

Both quotes exhibit the linguistic metaphor "holistic" view. Although the first quote provides an unspecific or general example that prevails in the corpus, the second offers a relatively detailed example with regard to content. The verb "to see," furthermore, conceptually links up with the metaphorically induced visual aspects, and therefore to the conceptual metaphor, UNDERSTANDING IS SEEING (Sweetser 1990).

In sum, the interview excerpts show the concept of holism is framed by the three conceptual metaphors, HOLISM IS AN ENTITY, HOLISM IS A BUILDING, and HOLISM IS A PERSPECTIVE. The first concept clearly puts emphasis on the aspect of manageability and reification of an abstract entity whereas the other two implicitly map their inherent structure partly on the abstract entity of holism. The conceptual metaphor, HOLISM IS A BUILDING, thus offers an encultured mapping and reification by using an entity encountered in daily life that implicitly holds the potentials of conceptual differentiation in terms of doors, rooms, and windows. HOLISM IS A PERSPECTIVE, moreover, highlights visual aspects while implicitly referring to a higher viewpoint and sight with which a better overview could be gained: two important aspects for a holistic approach. In sum, it becomes evident that at least a limited theoretical knowledge about holism exists among systems biologists. This is clearly mirrored in the general metaphors used to depict what holism is and in the lack of inner differentiation that might provide entry points for further elaboration and differentiation. With these aspects in mind, we now turn to a historical and conceptual contextualization of holism. The aim here consists in providing an interpretative background against which the conceptual metaphors encountered in this section are analyzed.

2.5.2 An Incomprehensive Insight into Holism in Biology and Systems Biology

After having empirically analyzed the limited set of metaphorical framings used to ascribe meaning to the concept of holism in systems biology, it is now time to take a look at the different theoretical threads that emerged in the conceptual history of holism. This might help to better understand the current semantic void of the concept in systems biology. Generally speaking, holism is based on three interrelated theories that historically overlap and inform each other: vitalism, classical holism, and modern holism.

Vitalism (the first concept to be outlined here) is informed by the belief that a special life-force provides a necessary difference to separate living matter from inanimate entities (De Klerk 1979). The basic idea or premise of vitalism (Benton 1974; Williams 2003) is the assumption of the irreducibility of life which was conceived as being brought forth by an anti-materialist power process which could not be explained by an in-depth understanding of underlying physical, chemical or biological processes. This antimaterialist concept of a vital principle or vital force was put forward by Paul-Joseph Barthez (1806) at the end of the eighteenth century

(Canghuilhem 1994) and appeared in its most recent form in the work of Henri Bergson (1911). One of the last scientific proponents of vitalism was Hans Driesch (1914), a biologist and natural philosopher, who considerably contributed to mould developmental biology out of descriptive embryological anatomy. Driesch's vitalist concept in biology declined when Eduard Buchner (Ukrow 2004) discovered in 1897 cell-free fermentation and by doing so laid grounds for modern biochemistry as a foundation of molecular biology (Kohler 1971, 1972). Buchner's materialistic discovery more or less provided food for thought for a mechanic-materialistic understanding of life as expressed by Jacques Loeb. According to Loeb (1964, 430) living processes should and could be explained as physico-chemical processes. With regard to these and other insights provided by biological research, vitalism was abandoned due to its theoretical shortcomings and the prevalent mechanistic logic paved the way towards a variety of experimental possibilities. In brief, the mechanistic logic became the conceptual foundation of the theoretical debate in biology.[17]

In addition to vitalism a second antimechanistic and antireductionist approach entered the stage by the end of the nineteenth century. Proponents of neo-Lamarckism[18] emphasized an interactionist approach which interpreted development as the outcome of a lifelong interaction between an organism and its environment. Although neo-Darwinian ideas were gaining more and more attention in the 1920s, neo-Lamarckian concepts were prominently brought forward by Lloyd Morgan (1923) among others. His antireductionist idea of emergent evolution (Morgan 1923) became influential in debates about the dichotomy of reductionism and holism, although his theory of holism perished with the advent of a neo-Darwinism proposed by R. A. Fisher (1930) (Box 1978; Mayr and Provine 1988; Tabery 2008). What unites the turning away from vitalist and neo-Lamarckian concepts in the history of biology is the fact that approaches such as neo-Darwinism, mechanistic, and materialistic interpretations of biological functioning and development offered practicalities for doing research in the lab.

It was in this mechanistic context that Jan Smuts, a proponent of classical holism, first published his concept of holism (Smuts 1926). Based on the Greek concept of wholes, Smuts' ideas overlapped with Lloyd Morgan's theory of emergent evolution. Smuts' theory, however, was based on the concept that the universe has a tendency to form stable wholes on the basis of constituting parts. Thus, the tendency for stability was conceived to run through all levels of an entity ranging from comprising atoms to whole biological systems. His conception of life clearly differs from that of the early vitalists in that it is assumed to be triggered by a force which drives evolution and development towards upper and more complex levels of living organisms.

[17] Mechanistic biology describes the causal relationship of interacting components in a biological system that produce changes and effects in it (see Nicholson 2012). Allen (2005) investigates the importance of the context for mechanism, vitalism, and organicism in late nineteenth- and twentieth-century biology.

[18] Jean Baptiste Lamarck (1744–1829) is known for the theory of inheritance of acquired characteristics, also called Lamarckism.

This diversifying force was consequently interpreted as a materially inherent characteristic of systems and not understood in terms of a mechanistic interaction of biological parts. Smuts (1926), however, aimed for an all-comprising understanding of holism[19] and is thus not astonishing that he did not subscribe to mechanistic interpretations of biological processes and the positivist research agenda as outlined by Auguste Compte and the Viennese Circle. Their attempt to develop a unified and hierarchically ordered science was based on a mechanistic understanding of biological and other processes and in which physics was conceived of as providing the philosophical endpoint (Carnap 2011). The Viennese Circle thus proposed a layered model of reductionism that envisaged chemistry as based on physics, biology as based on chemistry, and human sciences as based on biology. This kind of step-by-step reductionism was put forward by one of the Circle's disciples, Ernest Nagel (1961) in the 1960s. It is astonishing that attempts have seldom been made to apply conceptually the Viennese reductionist framework to real cases in the biological sciences.[20] Assertions such as Crick's (1966, 10) statement that "the ultimate aim of the modern movement of biology is in fact to explain all biology in terms of physics and chemistry" have been taken for granted within the then emerging field of molecular biology and provided a comfortable and ideal philosophical background. However, reductionist conceptions clearly superseded holistic approaches even though von Bertalanffy and Rosen were publishing their work on general systems theory and the mathematization of biology.

Meanwhile, the notion of a more modern holism was taken up by scientists opposed to ontological and theoretical reductionism and skeptical of the epistemic value of methodical reductionism and its advances (Polanyi 1968; Waddington 1968, 1975; Baedke 2013). The so-called postwar holists envisaged reductionism as ill-treating complex biological and other biological phenomena while they basically rejected the idea of explaining them solely on the basis of molecular interactions (MacCay 1965). The reductionists, on the contrary, accused holists to use an, ironically speaking, include-all rationale that lacks any specific explanation why emergent

---

[19] Smuts developed the orthogenetic theory which is a biological theory based on the hypothesis that life has an innate tendency to evolve in an unlinear fashion due to some internal or external driving force (for an introduction into the theoretical foundations and the spread of orthogenetic theory between 1880 and 1926 see Ulett 2014).

[20] One of the few exceptions is represented by Kenneth F. Schaffner's early work on the reduction of biology to chemistry and physics. There he states that "[t]he outcome of this account of the development of molecular genetics—which I have characterised as being both stimulated and unified by the Watson-Crick model of DNA—is to warrant as a working hypothesis a biological principle of reduction. This principle, it seems, holds not only for genetics, but also for other biological theories. The principle can be stated as follows: given an organism composed out of chemical constituents, the present behaviour of that organism is a function of the constituents as they are characterisable in isolation plus the topological causal inter-structure of the chemical constituents (The environment must of course, in certain conditions, be specified.)" (Schaffner 1969, 346). Studying the reducibility of more complex phenomena, however, he stated later on: "It would thus seem that for the present and the foreseeable future neurobiology as well as general biology will not be fully reducible sciences. This is a position which I believe can be described as a form of as 'weak emergentism'" (Schaffner 1993a, b, 342).

properties materialize. One has, however, to bear in mind that there was not one agreed philosophical framework among those who subscribed to holism. In fact, it was rarely the case that the philosophical foundations completely converged among so-called modern holists. One good example of this variety was Erwin László (1972), a Hungarian systems theoretician and philosopher of science, who did not principally deny reductionism but questions its practicability. László used the example of car accident on specific dates to explain his conceptual problem with reductionism. Addressing the individual level of drivers such as driver's abilities or journey lengths, and so on, for the analysis of why car accidents occur might be useful but not feasible due to the complex amount of data to be gathered and conceptually coupled. Lázló argued that it might be more reasonable and practicable to use and analyze so-called middle-range data such as weather in the respective locations, the alcohol consumption of drivers during the day, and place-specific accident statistics to estimate the probability of car accidents. Obviously, László preferred another analytical level in the analysis of the car accident system which followed the idea of reducing the amount of data to be gathered and coupled for the sake of practicality. His kind of "reductionist holism" received considerable attention and others adopted the idea of ontological antireductionism.

The notion of ontological antireductionism (Nagel 1998) refers to the fact that things do not simply exist and work on the basis of their mechanical functions. This would mean, for example, that a cellist's musical performance could not solely be analyzed and understood by the physics of playing the cello. There is more to such a performance; the spiritual and musical aspects, for instance, often go unnoticed but play a vital role for the performance of a whole piece of music. Another branch, that of explanatory ontological antireductionism (Nagel 1998), provides a fuller picture as it accepts the vitality of the cellist's performance but also draws attention to the physical aspects of it and that there are emerging laws governing the presentation of a cello suite. This understanding is not based on deterministic or bottom-up ideas but rather a dynamic version of deterministic and nondeterministic thinking. In brief, the physics of performing a piece of music is as important as the processes and governing laws that emerge in the course of the presentation such as phrasing and spirituality. The latter aspects are, in principle, nonreducible elements as they emerge in practice.

Higher-level properties thus appear to be connected to low-level properties but not in a deterministic way: the laws of the higher level could not be deduced from the laws of the lower level. Comparable aspects have been addressed by the German-American physicist Walter Elsasser (1958, 1961, 1998) who coined the term of biotonic laws (Olby 1971). Biotonic laws are biological laws compatible with physical laws but they cannot be deduced from them. This conceptual framework considerably contributed to the idea that biological processes could work in terms of top-down and bottom-up causation (Drack and Apfalter 2007), a concept that runs counter to reductionist thinking and was put forward by Michael Polanyi (1968) who emphasized that knowledge about constituting parts does not fully explain properties appearing on higher system levels (Porsch 1986).[21]

[21] In modern systems biology, the idea of bottom-up and top-down causations working in parallel has mainly been propagated and substantiated by Denis Noble (Noble 2008b).

In the meantime, molecular biology proliferated into the main biological paradigm relying on a concept one could frame as "tentative reductionism" (Morange 2000, 2009). Whereas the focus was first on single gene analysis, it shifted towards the study of genomics and protein expression (proteomics) later on. The production of vast amounts of data on increasingly complex structures coming along with this development brought the reductionist–holism debate back into biology, via the medium of systems approaches. Many proponents of systems biology made general statements that it provides a more holistic approach to biology than molecular biology by generally outlining molecular biology as reductionist and deterministic (see, e.g., Li 2009; Lu et al. 2012). Systems biology was and still is thought to complement or even replace molecular biology. All these claims have considerably contributed to the framing of systems biology in terms of a fundamental paradigm shift (Seth and Thaker 2014) that will result in a biology which is done in a different and more comprehensive way. The death of molecular biology (Morange 2008) was propagated and some authors also outlined that one witnesses in biology a similar shift like the one from classical to modern physics. These claims are obviously exaggerated and a new humility surfaced in recent years in which such claims only randomly appear. Thus, not everybody working in systems biology agreed and concerns have been raised as the concept of systems biology's holism is not holistic enough (Cornish-Bowden et al. 2004; Mesarovic et al. 2004). A small group of dissidents, however, also disagrees with the concept of holism and argued for more reductionist approaches (Bose 2013; Tin and Poon 2014) and others expressed considerable skepticism towards the holistic paradigm shift induced by systems biology (Bennett and Monk 2008).

Hence, the notion of holism is in current systems biology far from being precise or clear. Although some authors have been enthusiastic about it, it has still to be explicated what holistic-oriented systems biology exactly represents. This aspect is also mirrored in the interviews in which general statements about holism appear. They show that different kinds of holism in systems biology are at work and that many proponents do not reflect their concept of holism: where theoretical reflection, however, is done, ontological explanatory antireductionism surfaces (Ahn et al. 2006; Conti et al. 2007).

To sum up, we have tried in this section to provide a limited diachronic and synchronic overview of the main concepts and characteristics of holism. They show the notion of holism holds a long, complex, and varying conceptual history in which the dichotomy of holism and reductionism proved to be productive for the differentiation of both concepts. After a short overview of vitalism which was envisaged as a theoretical predecessor for different kinds of holism we showed that at least two holisms exist: classical and modern holism. The latter was subdivided into ontological antireductionism and ontological explanatory antireductionism which theoretically differ in view of their implicit reductionism. Whereas ontological explanatory antireductionism aims at productively combining bottom-up and top-down causation to explain emergent biological phenomena and laws, ontological antireductionism theoretically remains on the level of vitalism emphasizing emergent properties without an attempt to explain them in terms of multilevel interaction. In the emerging discourse on systems biology holism can rather be conceived of as a strategically

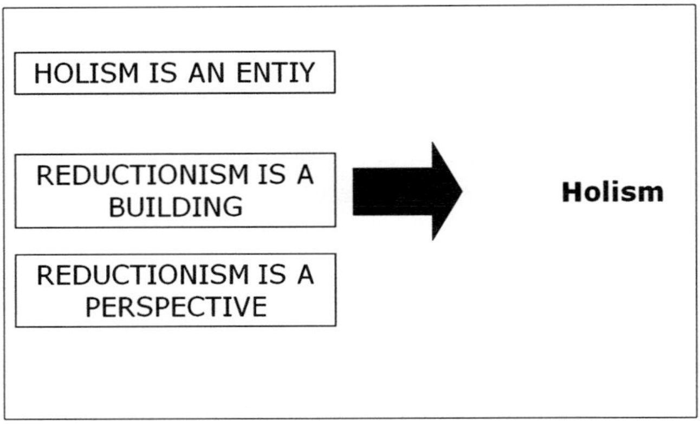

Fig. 2.7 Conceptual metaphors used to frame the notion of holism

used boundary object (Griesemer and Star 1989; Bowker and Star 2000). Although semantically vague it provides a fairly limited theoretical anchoring which, as we have seen, does not materialize in the analysis of the interview excerpts. Only a limited number of scientists working in systems biology theoretically reflect their conception of holism or reductionism; most systems biologists appear to be immersed in their daily work and devote no time to the philosophy of biology or epistemic deliberations. We now turn to the assessment of the concept of holism in systems biology.

2.5.3 Assessing the Concept of Holism in Systems Biology

The preceding section provided an overview of the conceptual history of holism in biology. It became apparent that different strands and developments of holism existed in history which finally culminated in ideas of ontological antireductionism and ontological explanatory antireductionism. Both concepts mainly materialized subcutaneously and contributed to building a somewhat vague and sometimes unconscious rationale underlying different research strands in systems biology. The conceptual and semantic imprecision of the term holism throughout its history is also mirrored in the empirical findings of the interviews with system biologists. Conceptual metaphors such as HOLISM IS AN ENTITY, HOLISM IS A BUILDING, and HOLISM IS A PERSPECTIVE demonstrated that the scientists interviewed use a quite restricted set of unspecific metaphors to frame the abstract concept of holism: these conceptually highlight aspects of reification, manageability, and perspective or sight (see Fig. 2.7).

This low metaphorical complexity and differentiation coheres with the fact that the historically generated concept of holism to date did not attract much philosophical and theoretical reflection in systems biology. This is somewhat astonishing because the main proponents of systems biology portray the approach as holistic or quasi-holistic. Even though different theoretical conceptions of holism exist in systems biology, the main bulk of research seems to subscribe unconsciously to the concept of ontological antireductionism and only a minority favors the more reductionist concept of explanatory ontological antireductionism. In brief, the concept of holism in systems biology remains unspecific and requires clarification to better understand the epistemological assumptions inherent in systems biology. After having analyzed the concepts of reductionism and holism, we now turn to the analysis of the theoretical notion of model in systems biology.

2.6 Images of Models in Systems Biology

The final section of this chapter on the understanding of basic concepts in systems biology aims at analyzing how scientists working in systems biology metaphorically conceptualize the notion of model. Models could be conceived as cognitive, epistemic, practical, and technical devices used in a variety of scientific disciplines ranging from meteorology and climate science to the social sciences to better understand a system's structure, state, behavior, and development. The term model implies, according to its theoretical outline and contextualized scientific use, different functions and applications. In brief, every scientific approach or discipline uses models as abstractions of real-world processes to understand and solve problems.

This also applies to systems biology where models are both important cognitive concepts and technical devices used for problem solving. The fact that modeling in many cases is perceived as an evidence-based technical process of depicting real-world phenomena with the help of numerics, differential equations, and computation, however, constitutes a problem because models always are also social and cultural constructions. The cultural dimensions and philosophical implications of models and modeling in systems biology hitherto did not receive much attention (Fox Keller 2002). Questions such as Wolkenhauer's (2014, 1) "Why modeling?" are rarely asked as they require a considerable effort to reconsider one's concepts used in modeling, their limiting implications for the scientific endeavor undertaken, and for its application to real-world phenomena (Boogerd et al. 2007a, b). An additional problem consists in the fact that nowadays a huge variety of different kinds of models such as stochastic models, rate equation models, multiscale models, mechanistic models, mathematical models, and the like are in use. This "zoo of models" (Wolkenhauer 2014, 3) makes it extremely difficult to maintain an overview over all types of models used and track the changes they undergo in scientific problem settings.

In addition to these more theoretical aspects, a substantial part of the scientific literature in systems biology is devoted to the practical dimensions of modeling and models. An analysis of the publications with the keywords "systems biology," "modeling," "systems biology model" on the Pubmed database conducted in May 2014 displays more than 1187 publications. This result goes hand in hand with a growing number of monographs and edited volumes devoted to modeling and models in systems biology (see, e.g., Palsson 2011; Patel and Nagl 2010; Koch et al. 2011; Ingalls 2013). Most of these publications are, however, problem-oriented and seldom address philosophical aspects of modeling. If they do so, the phrase philosophy rather refers to the structured process of defining, assessing, selecting, and functionally combining elements of biological systems without explicitly addressing underlying theoretical concepts and assumptions that inform the fabrication of a model. This perspective thus stands in sharp contrast to what is understood to be the principles and philosophy of modeling in biology (Massoud et al. 1998). The empirical question thus remains: how do scientists working in systems biology frame the notion of model and what kind of philosophical implications and tacit knowledge could be deduced from their framings?

This question might be of interest to systems biologists as it has rarely been addressed by research carried out in theoretical systems biology and studies conducted in the sociology of science (Hesse 1966) and the philosophy of science (Giere 2004; Morrison 2009; Suárez 2009). This is the reason why we took up the challenge and explored how the notion of model is framed by systems biologists. The aim of this exploration first is to analyze the metaphors used by systems biologists and second to disclose the implications inherent in them from an interpretative point of view. Yet, it is still necessary partly to contextualize the notion of model and provide an insight into the different conceptual dimensions of models. Against this background, we then turn to the analysis and interpretation of the different conceptual metaphors used by scientists in systems biology to depict what the notion of model means to them.

2.6.1 Models in Science and in Systems Biology

The use of models in biology has gained momentum in the last two decades. For systems biologists, they represent a familiar device. Currently, the development and use of models represents a daily practice in many areas of biological research such as ecosystem analysis or systems biology: they are descriptive abstractions and reduced ways of understanding a complex reality that materialize through heuristic media such as diagrams, chemical formulae, graphs, and so on. The concept of model, thus, applies to hypotheses developed by scientists in the course of scientific work carried out as well as to theories (Black 1962) to be projected upon a scientific issue. A closer inspection of the term model, however, shows a large variety of different model conceptions. There are, for example, complex numerical or mathematical models that could be implemented on computers, they can also have an epistemic function as devices for theoretical reflection, and they all converge in their potential to provide an intentionally reduced representation and explanation of dynamic processes in biology. One has, however, to bear in mind that models develop and are used in social contexts and represent outcomes of socioscientific processes that can become powerful instruments (Hastrup and Skrydstrup 2012) beyond the realm of science. Yet, for biologists, the use of models in research is still unfathomable because modeling has not been a core topic in biology. The last 40 years were mainly shaped by molecular biology and only the interpretative problems generated by genomics and other Omics approaches rekindled the interest in modeling in biology. Bearing in mind that conceptual models were already developed from the 1930s onwards (von Bertalanffy 1932, 1949; Weiss 1973; Rosen 1970a, b), the current use of mathematical and other modeling approaches in biology is developed in special contexts and for particular purposes. This problem-oriented use of modeling makes it difficult to provide an all-comprising conceptual history of the notion of model in systems biology (an exception is Krohs 2013) because there is simply an abundance of models around. Although attempts have been undertaken to investigate characteristics of models in systems biology systematically (Krohs and Callebaut 2007; Richardson and Stephan 2007; Schaffner 2007; Ullah and Wolkenhauer 2007), we think that it would increase the understanding of the model concept in systems biology if we approach it from a theoretical point of view by studying its semantic, ontological, and epistemological dimensions. Questions to be asked are the following: how do models represent processes or things? What could be learned with the help of models? How do theory and models relate to each other? What functions do they serve?

A theoretical look at the functions of models clearly indicates two aspects: they are designed to represent a selected extract from a perceived reality and, at the same time, they embody theory meaning that a model could be seen as an interpretation of empirical findings exhibiting the laws and axioms implicated in a theoretical framework. Although one might agree that scientific models are characterized by a limited perception of a phenomenon or process to be encountered in the world, the notion of phenomenon represents a generic term that scientifically refers to features of the world. Thus, the theoretical problem of scientific depictions concerns the question of what a model actually represents or what it stands for (Frigg 2006). This ontological question becomes even more difficult to be answered if we envisage models as not purely linguistically determined. We then have to clarify what kind of medium is used for the purpose of representing a scientific issue under investigation (Suárez and Solé 2006) and this goes hand in hand with the variety of representational styles by which biological entities or things could be depicted. In brief, different kinds of models are different ways of addressing and skillfully representing a certain aspect or perspective on the world. In brief, models are theoretically informed, fragmentary, and stylistic representations of a phenomenon under investigation.

Connected to the aspect of different styles of representation are different kinds of models. Scale models, for example, are basically reduced or small copies of a system to be investigated. Thus, cardboard models of housing estates are naturalistic replica or copies of such things. Scale models could be conceived of as generally restricted visions whereas idealized models contain deliberate simplifications. Such simplifications aim at constructing something less complicated to make a certain problem easier to understand. Economic theories based on the assumption of rational choice represent such idealized models of individuals whose motivation to perform any kind of action is based on calculating the profit to be generated by a certain act. Analogical models (Black 1962), moreover, are based on the idea of shared properties. Thus, the brain could metaphorically be conceived of as a computer because there are relevant similarities that could be projected upon a different domain of discourse. Thus, discussions could be framed as fights or even wars that not only share certain features but abstractly require the development of formal analogies and comparable patterns. Such analogies are important for science as they are thought-provoking heuristic devices that play a vital role in the development of theories (Gentner and Jeziorski 1993; Bailer-Jones 2003). Finally, there are phenomenological models that only represent observable properties. They refrain from analyzing underlying mechanisms that trigger a system even though they often incorporate laws and principles. The question of how models relate to reality has partly been addressed in so-called models of data (Suppes 1962). This concept simply states that the model has been derived from raw data through processes of correction, rectification, and idealization and confirms a tentative theoretical outline or approach. Such models are often constructed in systems biology and represent complicated constructions. Their development requires the application of sophistical statistical procedures that raise philosophical and methodological questions (Harris 2003). Questions such as, "What data should be included in the model?" and, "How should the functional relation between aggregated data be designed?" range among the easier questions to be addressed in models of data.

In addition to questions of what kinds of models exist and how they represent an issue under investigation, the most relevant question arises of what models actually are. Ontologically speaking, models are conceived to be physical, fictional, structural, and descriptive in nature. Physical models are thus material representations of something else such as the replication of a mammoth by a paleontologist. There are also enlived models such as knock-out mice in cancer research or certain kinds of yeast often used in experiments about heat resistance (Leonelli and Ankeny 2012). Such examples do not really provoke any difficulties with the ontological status of models as they can objectively be experienced. There are, however, nonmaterial or fictional models (Fine 1993) as well. Bohr's model of the atom only existed in his mind and exerted a considerable impact on physics. Such models appear to be purely fictional even though they—as in the case of Bohr—can exert a vital impact on a discipline (Giere 1988) and are open to modification through any kind of discussion or interaction. Fictional models cognitively reify entities and represent intangible vehicles and productive devices of research to develop or deepen scientific thinking. Although the conception of models as fictions has gathered interest in research undertaken in science and technology studies and the philosophy of science, their ontological status is to date far from clear. Important though for biology and systems biology (Lloyd 1984, 1994) are models that appear to be based on set-theoretic structures (Suppes 1960). These models are closely tied to mathematically oriented models and envisaged as a specific set and functional combination of structures. Finally, models can also take the ontological shape of descriptions. These often appear in scientifically stylized presentations, papers, and textbooks of a system under investigation. The problem here lies in the fact that the description is often confused with the model: the description becomes the model. The problem inherent in the duality between description and model is one of descriptive properties. Models do not per se possess such properties but achieve them through a kind of medium used to portray and this bears an impact on its ontological status and assessment. Different models therefore hold discrete ontological statuses that in many cases overlap. They can ontologically reside in scientists' minds and at the same time appear as drawing on a blackboard or as virtual constructs in a computer.

Finally, models also possess an epistemic and learning function as they are skillfully designed entities to acquire knowledge. Today, significant parts of research undertaken in science are based on the development of models. Thus, an important part of scientific research is carried out on (virtual) models and not on (material) reality itself. As surrogate entities, models instigate what one might call a process of model-based reasoning (Magnani and Nersessian 2002) which comprises aspects of denotation, demonstration, and interpretation (Hughes 1997). The process of learning starts by developing a representation that links the model with a targeted system to be investigated. This denotative process is followed by the demonstrative procedure in which features of the model and their relation to theoretical claims are thought through. Finally, the claims achieved will have to be projected upon the targeted system and converted into assertions about it. This interpretative procedure is intrinsic in learning about models on the one hand and in converting knowledge about the model on the targeted system. Thus, learning about the model happens in the course of its construction and manipulation which is devoted to the model's properties. It obviously depends on what kind of methodology is applied and what activities are carried out to structure the model. Hence, material models do not prompt questions that go beyond questions of experimentation. Most important, mathematical models help scientists to derive results or equations analytically. Computers and their ability to perform simulations and preliminary results are of great value at this point as they provide the opportunity to tinker with equations and test them. This means computers offer an opportunity to learn something about the model and its functioning by using simulations. Simulations could therefore be conceived of as a kind of methodology that might raise philosophical problems (Frigg and Reiss 2009) but are of enormous practical relevance because they often generate an improved understanding of dynamical models. Although the relational differences and convergences between computer simulation and experiments have not yet been resolved, current experimental setups in systems biology combine an in silico and in vivo rationale to learn more about the calibration of models (Franceschelli and Imbert 2009) and test trustworthiness. The aspect of trustworthiness addresses the question of whether equations used in the computer models adequately represent the functioning of the targeted system. In addition to these aspects, computer simulations appear to possess a considerable heuristic value as they contribute to generating theoretical improvements, amend models, and develop hypotheses. Once knowledge about a model has been amended, the knowledge generated has to be integrated into existing knowledge about the targeted system. This procedure is implicitly controlled by the scientifically informed assumption that analogies or idealizations in the model have converging counterparts in the real world.

To sum up, we have provided a tour de force on a variety of aspects connected to the notion of model. We started with the aspect of representation and the problem of what a model actually represents. It became clear that models are theoretically informed and fragmentary constructs of a certain research object. Ways or styles of representation differ according to the estimated relation between model and research object. These are, furthermore, influenced by the kind of model used and its ontological status. It, thus, matters if a model resides in scientists' minds or whether it virtually exists in a computer because experiments and falsifications carried out differ. Finally, the investigation of the epistemological dimension referred to two kinds of learning: leaning about the model itself and model-informed learning about the system or phenomenon under investigation. Although these two aspects could analytically be understood as discrete, they in reality appear as intermingled processes in scientific work and reveal the often hidden but nevertheless complex procedures at work in building and working with models. The notion of model and the philosophical aspects and practices tied to it make it now advisable to explore them in our analysis of the expert interviews. The analysis of the conceptual metaphors used not only reveals how models are conceptualized by systems biologists but also enable us—at least to some extent—to reveal and analyze theoretical aspects outlined in this section.

2.6.2 Systems Biologists Picturing Models

The previous section sketched out the different theoretical and conceptual issues inherent in models and modeling. Looking at it through the lens of philosophy of science and science and technology studies, it became clear that models have a variety of implications on different levels. They concern basic theoretical aspects such as representation, different kinds of models, and their ontological and epistemological status. Only few systems biologists and theoreticians have to date touched upon the philosophical aspects implicated in it (Boogerd et al. 2007a, b; Krohs 2004, 2013; Ullah and Wolkenhauer 2007; Wolkenhauer 2014). This is astonishing as the notion is constitutive for the discipline that aims at modeling genetic networks, cells, or even organs. There are, however, many references in papers on different kinds of models and the importance of modeling for systems biology. The general relevance of models is, for example, outlined by Kitano (2002b, 206) as follows.

> There are still issues to be resolved, but computational modeling and analysis are now able to provide useful biological insights and predictions for well understood targets such as bifurcation analysis of the cell cycle, metabolic analysis or comparative studies of robustness of biological oscillation circuits.

Bearing in mind that the quote stems from the start of the century, it outlines gaps in research on models and how they could be used in the context of systems biology. It displays narrative elements to legitimize why models are important in and for biological research by highlighting two aspects: it implicitly emphasizes the epistemological aspect of models (useful biological insights) and alludes to the process of how knowledge generated with the model could be turned into knowledge about the biological processes to be modeled. In addition to such strategic explanations and outlines, textbooks in systems biology devote whole chapters to the topic of models to be explained to students. Rhetorical questions of what models represent are answered as follows.

> What is a model? The answer to this question will differ among communities of researchers. In a broad sense, a model is an abstract representation of objects or processes that explains features of these processes [...]. A biochemical reaction network can be represented by graphical sketch showing dots for metabolites and arrows for reactions; the same network could also be described by a system of differential equations, which allows simulating and predicting the dynamic behavior of that network. (Klipp et al. 2009, 5)

What we encounter here are far more aspects about models than we have taken from the previous more strategic excerpt. First, the notion "model" is depicted as an entity standing in close connection to its scientific users (communities of researchers) and is envisaged to be differently framed by different scientists. General reference is made to phenomena (objects, processes, and features), and these abstract aspects are then explained by means of two types of models (graphical and mathematical). The descriptive measures of the models appear to stand in an isomorphic relationship to real entities or processes (arrows–reactions, equations–behavior, dots–metabolites) and their epistemic status is alluded to by indicating that the model aims at a dynamic prediction of the behavior of the target network (biochemical reaction network). What is fascinating about this excerpt of just 83 words is that it is a complex mixture of underlying philosophical assumptions of modeling which draw on means of representation, facets of isomorphism, different kinds of models, and epistemological considerations. Such aspects should not be underestimated as they permeate the varying discourses of systems biology, are basic ingredients in scientists' everyday knowledge and practices: they display, in a way, a subliminal philosophy of modeling.

Comparable aspects also materialized in the interviews conducted where system biologists were asked to explain what the term model means to them.:

> We are interested in metabolic pathways and their functioning. This means that we have to define the biological entities involved in certain reactions and functionally relate them. We do this quite often by drawing pictures or using special programs on the computer. Once we have sorted things out which means that we have decided what kind of biological units we include, discussion starts how this could be integrated in a mathematical model. (Scientist K)
>
> German original: Wir interessieren uns vor allem für Stoffwechselwege und deren Funktionen. Das bedeutet, dass wir die biologischen Substanzen, die in Reaktionen involviert sind, erst einmal herausfinden und dann funktionell verbinden müssen. Dafür fertigen wir einfach Zeichnungen an oder nutzen diese speziellen Computerprogramme. Wenn das dann klar ist, also welche biologischen Einheiten wir in das mathematische Modell einbeziehen, dann beginnt die Diskussion, wie wir das ins Modell integrieren können.

Different theoretical aspects of models are combined here: its starts with the question of what has to be represented. This process of selection goes hand in hand with the heuristic use of different models (drawing pictures, mathematical) which display different ways of representing the problem under investigation. The same holds true for the following quote in which a rather complicated way of developing and improving a model is outlined.

> The models used are more or less the same. I mean that we have to fine-tune them to the process or whatever aspect we investigate. We use quite a complicated setup as we model and then do the experiments and then go back to the model to improve it. The aim is to get a well-balanced model which helps us to better understand the processes we investigate. Failures are quite important as they help us to better understand what goes on in the system. (Scientist B)
>
> German original: Die Modelle, die wir hier nutzen, sind mehr oder minder dieselben. Ich meine, dass wir die natürlich an das anpassen müssen, was wir gerade untersuchen. Wir nutzen dafür einen ziemlich komplizierten Versuchsaufbau, da wird ein Modell entwickelt und dann Experimente durchgeführt, um dann wieder das Modell zu verbessern. Ziel ist es, dass wir ein gut ausbalanciertes Modell bekommen, das uns ein besseres Verstehen oder Verständnis von den Prozessen gibt, die wir untersuchen. Misserfolge sind übrigens ziemlich wichtig, denn die helfen uns besser zu verstehen, was im System los ist.

What we can see here is a procedure used in the area of in silico and in vivo experimentation. The aim consists in calibrating the model with the help of so-called lab models. This means different kinds of models holding distinct ontological statuses are combined with the epistemic aim to know more about the respective model and the natural entity to be analyzed. The two showcase examples visibly exhibit the manifold conceptual and amalgamated theoretical processes at work in developing models of biological processes.

In addition to the analysis of these subliminal philosophical and theoretical aspects at work in modeling biological processes, it is also of vital interest to analyze how scientists in systems biology frame the entity model itself because here further aspects of what a model represents become analytically accessible. The systematic investigation of the interviews conducted gave rise to five conceptual metaphors: A MODEL IS A CONSTRUCT, A MODEL IS A HEURISTIC DEVICE, A MODEL IS A HEURISTIC MACHINE, A MODEL IS A TOY, and A MODEL IS AN INTEGRATING ENTITY.

To start with, the notion of model has been metaphorically depicted as a construct that highlights the aspects of abstraction and the selection of relevant units to become functional parts of the model.

> […], **it [a model] is a construct of a natural process** that we aim to understand. There are lots of things going on in the area of modeling and we always check first what kind of models are around, what their characteristics are and for what purposes they have been used. At the same time we start to try to understand the constituents of the system and how they relate. (Scientist G)
>
> German original: […], **das ist ein Abstraktionsprozess von natürlichen Prozessen**, den wir verstehen möchten. Im Modellbereich passiert derzeit viel und wir schauen immer, was für Modelle gerade benutzt werden, was deren Eigenschaften sind und für welche Zwecke die benutzt wurden. Aber gleichzeitig schauen wir uns natürlich auch das System und seine Elemente an, in welcher Beziehung die stehen.

The example clearly displays the metaphorical concept, A MODEL IS A CONSTRUCT, and explains in the rest of the quote the different processes underlying the development of a model: the monitoring of model development, their area of application, and the comparison with the entity to be modeled. Such complex aspects also appear in the following quote, although here emphasis is put on the status of the model as a construct.

> **Models are constructs**. I always have to emphasize this because many students and colleagues often shift from model to reality and back without drawing a line between the natural process which we cannot tackle as a whole and the scientific construction we make of it. (Scientist M)
>
> German original: **Modelle sind Konstruktionen**. Ich muss das immer wieder betonen und viele Studenten und Kollegen wechseln zwischen Modell und Realität hin und her, ohne zwischen dem natürlichen Prozess, den wir als Ganzes eben nicht verstehen können und der wissenschaftlichen Konstruktion, zu unterscheiden.

Here, the aspect of construction is again highlighted but puts emphasis on the problem of confusing the model and the real biological process. What becomes apparent is that, as a construct, a model is partial simulation but not a representation of nature. The distinction between these two aspects is implicated in the metaphorical mapping but it still requires critical reflection and emphasis.

Building on the metaphorical concept, A MODEL IS A CONSTRUCT, the mapping inherent in the conceptual metaphor, A MODEL IS HEURISTIC DEVICE, more evidently plays with the aspect of construction albeit it emphasizes a more technical aspect.

> There is a lot happening at the moment in our group. We have some PhD students coming from informatics which bear a positive impact. I mean, **models are for me some sort of a heuristic device** which helps me to understand not only the phenomenon under investigation but also my thinking about it. (Scientist I)
>
> German original: Ins unserer Gruppe passiert gerade sehr viel. Wir haben da diese Doktoranden aus der Informatik, das wirkt sehr positiv. Ich meine, also für mich sind Modelle **eine Art heuristisches Instrument** das mir nicht nur hilft, das Phänomen zu verstehen, sondern auch mein eigenes Denken über den Forschungsgegenstand.

Models, metaphorically to be understood in terms of a heuristic device, exhibit a certain degree of theoretical reflection because they are conceived to be instrumental. Such aspects are involved in the mapping of the lexical item device on the notion of model. Furthermore, theoretical repercussions models might bear on the scientific process and scientific thinking become quite evident in the quote. This interaction is also highlighted in the following excerpt.

> The modeling is very important, I would say. I mean it is some sort of a way to think and do something about a scientific problem. A model is always a construct, but built according to a scientific logic or convention. **It is a way or a heuristic device** to better understand things in manifold ways as a model replicates but is not a one to one representation. (Scientist Q)
>
> German original: Die Modellierung ist sehr wichtig, würde ich sagen. Ich meine, das ist eine Art und Weise über ein wissenschaftliches Problem nachzudenken und was zu machen. Ein Model ist immer ein Konstrukt, das einer bestimmten wissenschaftlichen Logik und Konvention folgt. **Das ist ein Weg oder ein heuristisches Instrument**, einen Sache aus vielen Richtungen zu verstehen und es überträgt ja und stellt es nicht so dar wie es ist.

Here, we clearly witness a process of pondering what a model actually is. The concealed metaphorical concept, A MODEL IS HEURISTIC DEVICE, drives a reflection on the different aspects of models whereas at the end of the quote a variety of theoretical reflections appears. This metaphorical concept is technically elaborated upon in the following quote where models are metaphorically framed as heuristic machines.

> You know, technically speaking it [model] is **some sort of a heuristic machine** with which we come to grips with scientific problems. They often appear to me as skillfully designed constructs in which we invest a lot of time. (Scientists N)
>
> German original: Wissen sie, technisch gesprochen sind sie **eine Art heuristische Maschine** mit der wir wissenschaftliche Problem in den Griff bekommen. Für mich sind das gekonnt entwickelte Konstrukte in die viel Zeit reingeht.

The conceptual metaphor, A MODEL IS A HEURISTIC MACHINE, appears here again and maps technical aspects on the domain of models. The linguistic element "some sort of" in the quote indicates the metaphor is deliberately used to explain what the interviewee means. The technical implications nestling in the metaphorical mapping are not further discussed and models are again abstractly framed as constructs, but in this case as a skillful construct.

Such aspects are, however, critically discussed in the following interview section where the potential of models is obviously stressed but also critically assessed.

> To me, **models are very important because they are a background or heuristic machine** against which I can better understand the processes and phenomena we investigate. But this machine, well the computer, is just a man made machine. It can help us to better understand a biological system but also it helps us to improve our ways of approaching scientific problem in terms of what is relevant to be modeled and what not. (Scientists O)
>
> German original: Für mich **sind Modelle sehr wichtig weil sie eine Art Hintergrund oder heuristische Maschine sind**, mit der wir besser die Prozesse und Phänomene verstehen können, die wir untersuchen. Aber diese Maschine, also der Computer, ist eine menschgemachte Maschine. Sie kann uns helfen eine biologisches System besser zu verstehen, aber sie hilft uns auch dabei, ein wissenschaftliches Problem besser anzugehen, also in dem Sinne, was wichtig für das Modell und die Modellierung ist und was nicht.

The machine metaphor here holds an epistemological status as it instigates a critical reflection on what kind of knowledge is generated about the biological process under investigation. Furthermore, the model is depicted as computational and manmade machine that helps to revisit methodologically whether the selected components in the model are acute (improve our ways of approach a scientific problem).

In addition to the more technical aspects as expressed by the metaphorical use of machine or device, models are also metaphorically portrayed as toys. The metaphorical transfer of toy onto the target domain model semantically infuses it with ideas revolving around childhood and, more important, playing. This aspect is accentuated in the following quote.

> For some scientists, **their model is a toy** they have been playing with for a long time. My colleagues really invest a lot of time in building them and we share a considerable amount of our scientific lifetime with them. (Scientist D)
>
> German original: Also für einige Wissenschaftler ist **ein Modell ihr Spielzeug** mit dem sie lange Zeit herumgespielt haben. Meine Kollegen investieren wirklich viel Zeit in die Entwicklung und wir verbringen wirkliche eine nicht zu unterschätzende Zeit unseres wissenschaftlichen Lebens mit ihnen.

What we can take from this quote is the fact that there is, first, a perceived close relationship between the model and the scientist who works on it and, second, that playing is metaphorically used to frame the more sober process of constructing. It is interesting that the metaphorical use of playing is informed by the context in which it is carried out: a scientifically informed tinkering to develop or improve a model. Such aspects are also found in the next excerpt:

> Sometimes a **model also takes the shape of a toy**. You develop it and then you play with it to see whether it works and how it has to be amended. This takes some time and can also be a fun part – well sometimes obviously not. (Scientist L)
>
> German original: Manchmal **nimmt ein Modell auch die Gestalt eines Spielzeugs an**. Du entwickelst es und dann spielst Du mit ihm herum, um zu sehen, ob es funktioniert und wie es verbessert werden könnte. Das verbraucht viel Zeit und kann Spaß machen – manchmal aber auch nicht.

This quote still refers to the play- or joyful development of models, but underlines the act of playing as a scientifically guided procedure to improve them which can also turn into a laborious exercise of amending or revising elements of a model.

Finally, models also appear to hold a social function as an integrating device because they are, metaphorically speaking, personified entities that gather scientists around them. This could clearly be seen in the following two quotes where the conceptual metaphor, A MODEL IS AN INTEGRATING ENTITY was tackled:

> Sometime we literally gather around a model, well a sketch of it and discuss it. Deep thinking sometimes occurs and an exchange of ideas materializes that, at least in my view, tightens the bonds in the group. I mean, we all learn from this thinking together and **the model really brings us together**. (Scientist M)
>
> German original: Manchmal versammeln wir ich buchstäblich um ein Model herum, also um eine Skizze und dann diskutieren wir die. Da werden oft wichtige und essentielle Überlegungen ausgetauscht und das, so sehe ich das zumindest, trägt schon dazu bei, dass sich Beziehungen entwickeln. Ich finde, dass wir alle **von diesem gemeinsamen Durchdenken lernen und das Modell bringt uns zusammen**.
>
> Certain approaches and **models really bring people together** while they sometimes exclude others. There is this what I have called a model culture but this is perhaps too general. (Scientist P)
>
> German original: Also einige Ansätze und **Modelle bringen die Leute wirklich zusammen** und natürlich grenzen die auch aus. Da gibt es halt auch diese Modellkulturen, auch wenn das vielleicht zu generell ausgedrückt ist.

Models are thus not only technical devices but entities that are envisioned to exert an integrating function. Certain practices and routines are tied to them which contribute to establishing human interaction and bonds while, at the same time, model cultures and scientific identities are built. In sum, the metaphorical concept A MODEL IS AN INTEGRATING ENTITY not only refers to the integration of data and functions, but more important, conceptualizes the social process of bonding among scientists.

The previously outlined metaphorical concepts of A MODEL IS A CONSTRUCT, A MODEL IS A HEURISTIC DEVICE, and A MODEL IS A HEURISTIC MACHINE, clearly revealed more technical images of what models are and sometimes disclosed knowledge-theoretical aspects revolving around the questions of what one learns about the model, what about the targeted biological entity, and about one's own epistemological assumptions. Albeit these conceptions prevail in the interviews, it is interesting that models are also metaphorically envisaged as toys in A MODEL IS A TOY, and as socially integrating devices in A MODEL IS AN INTEGRATING ENTITY. The toy metaphors stress aspects of scientific and creative tinkering whereas the personification of models as integrating social agents underline the bonding effect they might exert on the scientific community. We now turn to the concluding section to provide a more systematized overview over the conceptual metaphors of model encountered here. The aim consists in summarizing the outcomes, in interpreting what kind of results nestle in the metaphorical concepts, and in analyzing whether a specific set of metaphorical concepts could be connected to scientists holding a particular scientific background or training. This might help us to develop a clearer picture about and grasp the multifaceted dimensions of the notion of model in systems biology.

2.6.3 Picturing Metaphorically Informed Images of Models

The previous section depicted five metaphorical concepts that were encountered and analyzed in the course of the study conducted. The analysis showed how the abstract entity of model was metaphorically conceptualized and demonstrated how scientists try to grasp and elaborate upon the notion of model. Having roughly outlined the philosophical aspects of models and modeling before, the metaphorical analysis provided insight into the theoretical playing and exploring of what models mean to scientists working in systems biology. The analysis, thus, gave way to an interrelated and mostly shared conceptual network that endowed the abstract entity of model with meaning and at the same time offered insight into the theoretical processes underlying the development and idea of models. Interestingly though, it was not possible to relate specific metaphorical concepts to a particular scientific training or background. All concepts were shared among the scientists interviewed and represent a rather limited insight into how people working in systems biology frame the notion of model (see Fig. 2.8).

Considering the imagery analyzed in view of the underlying metaphorical mappings, one might say that the first three metaphors elaborated upon convey the idea of models as technical constructs whereas the latter two were refined by developing a more instrumental and mechanistic perspective. Although one might be tempted to hypothesize that scientists holding an engineering background preferably use these images, the evidence provided did not corroborate this claim: the imagery was shared by almost all interviewees. The concept of model as construct, as heuristic device, and heuristic machine thus appears to be mainly structured by a more technical imagery. One has, however, to bear in mind that the conceptual mappings in the metaphors, A MODEL IS A TOY and A MODEL IS AN INTEGRATING ENTITY, highlight other and complementing aspects: creative scientific tinkering and the social dimension of models and modeling. Both counteract the subliminal technical and mechanistic

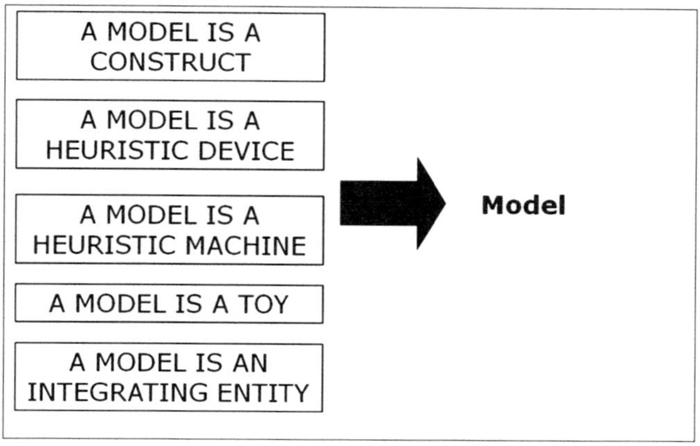

Fig. 2.8 Conceptual metaphors used by scientists to frame the notion of model

rationale of models and modeling and might deserve further scrutiny in terms of models and modeling as imaginative and social activities. To finish this chapter, we now turn to an overall summary of all basic notions analyzed in this chapter.

2.7 Concluding Remarks: Basic Concepts, Metaphors, and Scientific Imagination

The previous sections provided thought-provoking insights into the framing of basic biological concepts by systems biologists. Although the scope of the study was limited[22] and most of the time we were only able to scratch the surface of complex issues to be further investigated, the systematic analysis and contextualization of conceptual metaphors revealed a network of intangible but effective characteristics connected to five basic notions in systems biology. Thus, the analysis of the concept of life discovered a complex network of seven conceptual metaphors used to frame it (see Fig. 2.9). First of all, they made the concept of life accessible by applying a variety of technological, social, and cultural source domains to it that semantically permeate it.

A closer look at the mapping processes indicated that meaning was in many cases determined by metaphors stemming from a technological source domain. This could suggest that life might undergo a technological reframing within the field of systems biology due to the influx of scientists coming from engineering, physics, mathematics, and computer science, but the almost ubiquitous presence of

[22] Results are based on 25 interviews conducted with German scientists, on an extensive analysis of the scientific literature published on the topic of systems biology, a reading of historical precursors, and the analysis of it with the help of secondary literature that dealt with them.

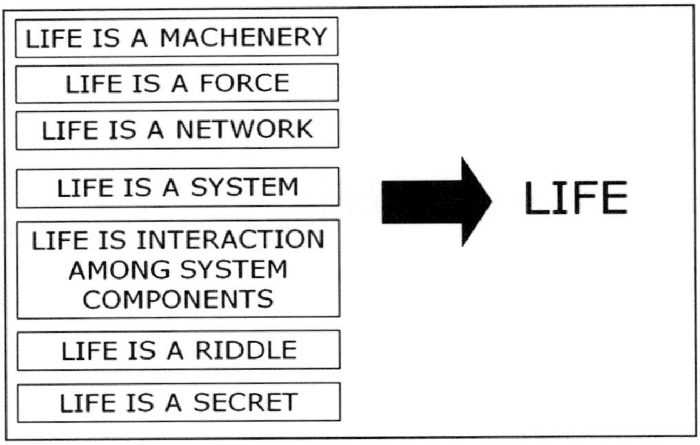

Fig. 2.9 Conceptual metaphors used to frame the notion of life

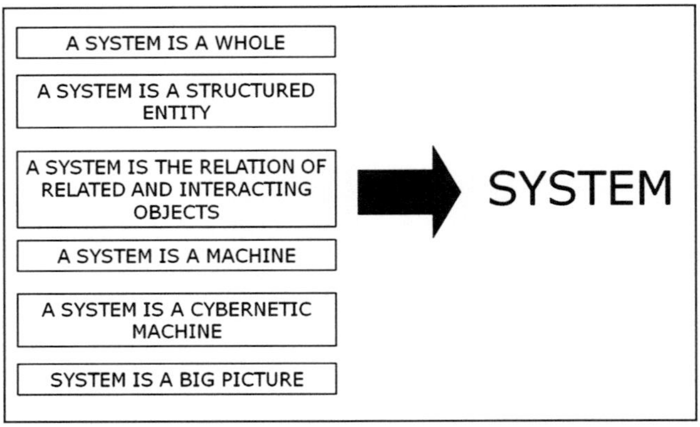

Fig. 2.10 Conceptual metaphors used to frame the notion of system

the conceptual metaphors, LIFE IS A RIDDLE and LIFE IS A SECRET, seemed to counteract this hypothesis: They give the impression that a conceptual path towards a technological understanding of life is stepped on, at least at the period of the interviews, but that the complexity of the life-issue to date remains unresolved and still represents a riddle. One could, however, hypothesize that the prevalence of technological metaphors and their implications in the long run might reshape the notion of life or hold the potential to solve the riddle or secret of life. Further discussion among systems biologists is needed as to whether these conceptualizations of life represent accurate ways of approaching the life-problem from a reflexive point of view.

With regard to the metaphors used to frame the abstract notion of system, six conceptual metaphors were revealed by the analysis (see Fig. 2.10). The transfer processes underlying the imagery reified the notion of system as an entity and applied a spatial structure to it. In using the metaphorical concepts, A SYSTEM IS A WHOLE and A SYSTEM IS A STRUCTURED ENTITY, the notion system became a cognitively manageable entity.

Furthermore, efforts of dynamization are implicated in the conceptual metaphors, A SYSTEM IS THE RELATION OF RELATED AND INTERACTING OBJECTS and A SYSTEM IS A CYBERNETIC MACHINE. Especially the latter appears to counteract the mechanistic implications inherent in the conceptual metaphor, A SYSTEM IS A MACHINE, because it merges a rather mechanistic conception of systems with a nonlinear, complex, and dynamic understanding of systems. What can be deduced from the conceptual metaphors are efforts to increase dynamization which is conceived to lead to an improved understanding of complex processes constituting biological systems. One could try to pursue this way of a metaphorically induced dynamization by developing alternative or complementing metaphors with systems biologists. Such an enterprise could build upon existing conceptual metaphors and help to reconsider reflexively the implications nestling in them to develop alternatively perspectives and approaches based on critically revised system conceptions.

The turn towards the important concept of reductionism exposed four conceptual metaphors used to frame what reductionism represents (see Fig. 2.11). Here, interesting aspects emerged in the course of the analysis which first disclosed reductionism as an entity. The conceptual metaphor, REDUCTIONISM IS AN ENTITY, made the abstract entity reductionism manageable for further clarification and discussion. Indeed, the reification appeared to be a basic process of meaning ascription by which the following metaphorical concepts, REDUCTIONISM IS AN ANCESTOR, REDUCTIONISM IS A PREDECESSOR, and REDUCTIONISM IS AN ADVERSARY were applied to create a relationship of systems biology towards the concept of reductionism. Implicated in the personifications, REDUCTIONISM IS AN ANCESTOR and REDUCTIONISM IS A PREDECESSOR, are genealogical images that subliminally depict the more holistic approach of systems biology

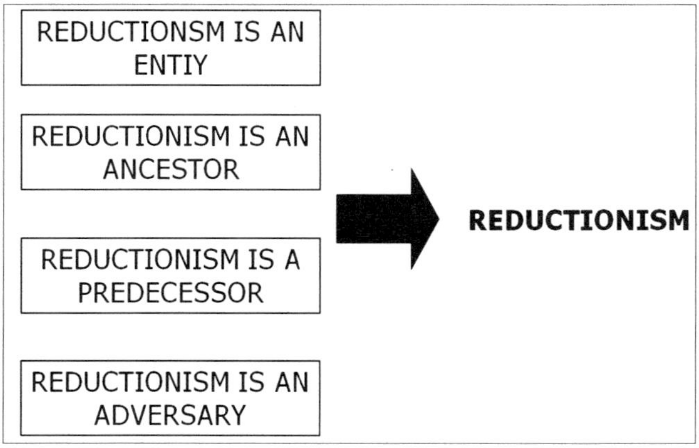

Fig. 2.11 Conceptual metaphors and personifications used to frame the notion of reductionism

as some sort of offspring. Albeit reductionism is now and then metaphorically depicted as an adversary, such an attribution was rare.

Surprisingly, theoretical awareness of the different subconcepts of reductionism was rather less marked in the interviews. The lack in awareness about ontological, methodological, and epistemological reductionism, so it seems, was concealed by the metaphorically motivated dissociation from a reductionist framework. In theoretical terms, this is problematic because approaches in systems biology are and also will in future be based on methodological reductionism. Nevertheless, it must face the challenges posed by the implicit epistemic (theoretical) reductionism found to be prevalent among systems biologists in order not to impede progress with regard to theoretical and experimental or practical progress. Therefore an open reflection of what kind of reductionism is necessary to tackle a certain scientific problem might be important to outline clearly the scope of results that can be achieved. A critical reflection of the conceptual metaphors in view of different subconcepts of reductionism might also help to divulge important theoretical aspects productively for doing research, help to examine concepts of holism in systems biology critically, and inspect the invention of a holistic tradition (Hobsbawm and Ranger 1992) or conceptual framework.

The analysis of holism displayed the three conceptual metaphors, HOLISM IS AN ENTITY, HOLISM IS A BUILDING, and HOLISM IS A PERSPECTIVE (see Fig. 2.12). These metaphors conceptually highlighted aspects of reification, manageability, perspective, and sight. A closer look at the conceptual metaphors, however, disclosed a low degree of metaphorical and semantic differentiation which was an astonishing fact because the concept of holism forms an important theoretical anchor in systems biology.

This shows that although systems biologists are thought to be holistically oriented, they in many cases undertake their research in view of unconscious reductionist frameworks. Those who theoretically reflect on holistic concepts mainly apply the concept of ontological antireductionism and work on the basis of their mechanical

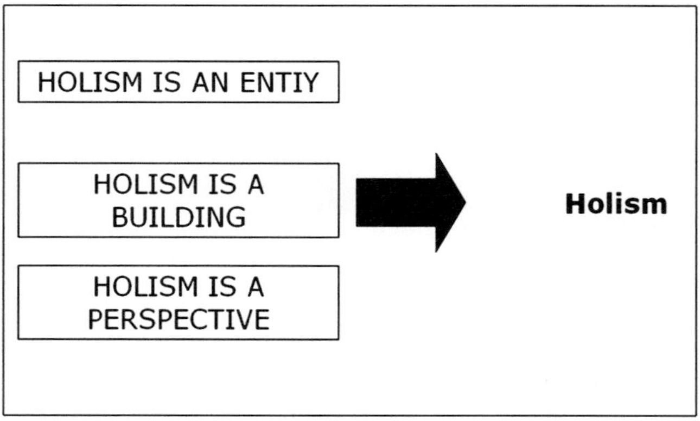

Fig. 2.12 Conceptual metaphors used to frame the notion of holism

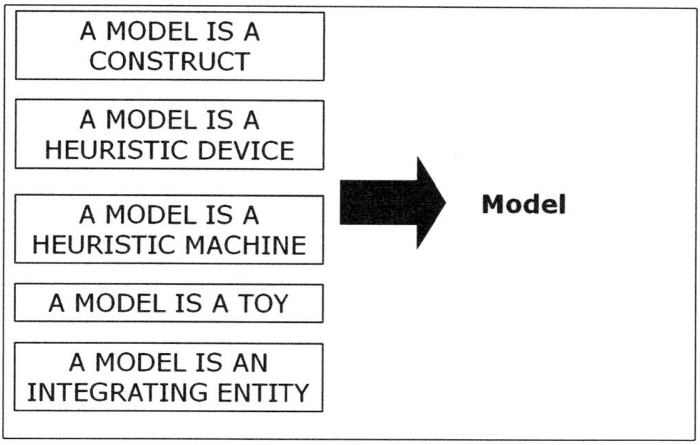

Fig. 2.13 Conceptual metaphors used to frame the notion of model

functions. However, this antimechanistic mindset was rare and encountered in the interviews only now and then together with different kinds of subconcepts of reductionism. Such a theoretical imprecision also surfaced in the metaphorical concepts and it would be important to elaborate on the different kinds of holism to define more precisely the underlying holistic rationale to be applied in research undertaken in systems biology. The metaphors could be taken here as starting point for analyzing the tacit knowledge[23] among system biologists about holism to be aligned with theoretical conceptions of holism. This would help to develop a situated and bottom-up definition of holism obviously implicitly inherent in research.

Finally, the analysis of the notion of model exposed five metaphorical concepts (see Fig. 2.13). The conceptual metaphors, A MODEL IS A CONSTRUCT, A MODEL IS A HEURISTIC DEVICE, and A MODEL IS A HEURISTIC MACHINE, elaborate that models are technical constructs and the latter two emphasize more instrumental aspects. It is, however, important to stress that in the three concepts awareness materialized that models do not represent exact replications but are abstract constructs of biological processes based on system components considered to be relevant. Although one might be tempted to hypothesize that scientists holding an engineering background preferably use these images, this does not hold true as the imagery was mainly shared by all interviewees.

Equally important are the conceptual metaphors, A MODEL IS A TOY and A MODEL IS AN INTEGRATING ENTITY, because they highlight aspects of scientific tinkering and the social dimension of models and modeling: they might deserve further scrutiny in terms of models and modeling as nonrationalist imaginative and social activities.

The number of these metaphorical concepts of systems biologists is by far not comprehensive and others such as interaction, dynamism, emergence, integration

[23] Tacit knowledge is a kind of knowledge that is difficult to be verbalized or communicated to others. It is gained through experience and not formalized.

(Green and Wolkenhauer 2012; O'Malley and Soyer 2012), and basic principles (Wolkenhauer and Green 2013) await further and much deeper linguistic, sociological, and philosophical analysis. However, the investigation of the figurative language used in speech in this chapter revealed how scientists semantically frame basic biological concepts and explain it to the scientifically informed interviewer.[24] The systematization of these linguistic images as conceptual metaphors proved to be methodologically useful and productive as they uncovered a wide range of interpretations and exposed to a certain degree the dynamics of knowledge (Maasen and Weingart 2000) inherent in the basic biological concepts discussed. It should have become clear that the conceptual metaphors "create what we would call cultural cosmologies or meaning-worlds that, once built, for better or for worse become the 'homes' in which we reason and act [...]" (Harrington 1995, 359). They exhibit the poetics of scientists' minds (Bono 1999; Gibbs 1994) and disclose a culture in mind (Shore 1996) that generates the mindset to manufacture scientific knowledge (Knorr Cetina 1981) and the development of epistemic cultures (Knorr Cetina 1999). Hence, the conceptual metaphors encountered in the previous sections partially depicted the cultural, social, and philosophical grounding or metaphysics of scientific work undertaken in systems biology.

In addition to the previous aspect which mainly addresses the cognitive dimension in the conceptual theory of metaphor, images in speech are also ways of doing things with words (Austin 1962). This means the process of ascribing meanings to abstract entities entails different kinds of action or practices connected to or inherent in it. Thus, metaphors could be understood as "[...] a field of embodied, materially interwoven practices centrally organized around shared practical understandings" (Schatzki 2001, 3). Consequently, possible kinds of actions or practices (Bourdieu 1976; Pickering 1995; Schatzki 1996; Schatzki et al. 2001) could interpretatively be deduced from the conceptual metaphors informing practical understandings. Hence, metaphorically framing, for example, a model as a toy, possibly leads literally to playing around with a model on a computer to improve it scientifically whereas conceptualizing reductionism as an enemy brings about defensive attitudes towards certain experimental procedures in research or even the stigmatization or exclusion of colleagues subscribing to it. Michael Polanyi's (1967) tacit dimension comes into play here as it outlines the idea of knowing without being able to express this knowledge. In brief, he refers to knowledge that triggers action and practices where the individual is only partly able to explicate it verbally. Here metaphor as a meaning generating function of language gets into the game again. Although Polanyi is not explicit on the relevance of metaphor (Polanyi 1977, 66–81), for his concept of tacit knowledge, we would like to indicate that metaphors represent a way to verbalize the inexpressible partly. This means that the diversity of conceptual metaphors

[24] One has to keep in mind that the interview-setting represents in some way an unbalanced situation as the interviewer was not scientifically trained but conducted extensive research on the scientific and social aspects of systems biology. The interviews should thus be seen as sophisticated encounters in which scholars explain their framings of basic biological concepts in an interdisciplinary setting.

encountered in the preceding sections represent incorporated (Johnson 1987) tacit knowledge which is expressed and held together by daily behavioral routines of scientific work. Thus, metaphors could be understood as dialectical entities because daily practices and thinking hold them together. Therefore conceptual metaphors, as taken from the interviews, could also be understood as socially stabilized constructs constituted by acting and thinking at the same time: they represent in our study the diversity of dialectical knowledge systems underlying basic biological concepts in systems biology. These conceptual frameworks play an important role as they are "subconscious forms of understanding [manifest] in the metaphorical reasoning, as reflected in the language used in reasoning and communicating about science" (Brown 2008, 11).

Such a theoretical perspective on metaphor holds important philosophical and practical implications for science, policy, and the public. As embodied structures (Lakoff 1987; Johnson 1987, 1993; Lakoff and Johnson 1999), conceptual metaphors challenge the concept of scientific work as a logically driven and disembodied enterprise. Indeed, the mapping processes encountered in the previous sections display a wide range of bodily experiences and sociocultural reservoirs used to make abstract concepts meaningful and apply them to scientific problems. It is thus not surprising that both reductionism and holism are metaphorically conceptualized as entities or buildings because the metaphors give them concrete shape. But what follows from these insights?

First those basic biological concepts are perceived "through the lenses of embodied and social experience" (Brown 2008, 195) and second a rejection of objectivism (Putnam 1988) in favor of an experientialist perspective. This means in the context of this study that the basic biological concepts analyzed are grounded in and structured by conceptual metaphors. Such a position might sound too relativistic but many technological developments ranging from airplanes to medical therapies clearly show that metaphorically informed scientific knowledge is indeed able to solve problems and develop useful technologies or drugs. John Ziman (2000, 6) eloquently summarizes that imagery "is the vital link between the social and epistemic dimensions of science." This is an important point as Ziman (2000) refers to the productive analysis of metaphors that could be applied in metaphor assessment (Döring 2013; Jäkel 1997; Mambrey and Tepper 2000; Katherndahl 2014). Critical metaphor assessment helps to broach the subject of implications and assumptions nestling in basic biological concepts in order to empirically analyze and critically reflect on them. This could, for example, be seen in the previous sections of metaphors framing reductionism and holism where we took stock of conceptual metaphors to tackle how the biological concepts were framed. Complemented with a philosophical analysis, the assessment of conceptual metaphors revealed the subconscious rationales at work on system biologists' minds. Such an approach offers space for critical reflection and theoretical development. The complex problems addressed by the interdisciplinary enterprise of systems biology could obviously profit from such an assessment as it possesses the means to unravel hidden rationales driving research and the metaphorically generated professional backgrounds of the disciplines involved in research. As seen in the sections on life and systems, disciplinary framings were based upon occupation-related conceptual metaphors. An integration of these professional perspectives and attitudes could be brought forward once metaphors are shared, critically inspected, and reframed (Liebert 1995). Such a process holds the potential to contribute to a better mutual understanding, to initiate integrative problem understanding, to explore cooperative ways to approach scientific problems more creatively and to develop new or improved methods and theories for systems biology. The question, however, remains of how one could approach the theoretical diversity encountered in the previous sections from a reflexive point of view.

One way to instigate a reflective process would consist in providing scientists during workshops with the philosophical grounding of their scientific thinking and discuss the theoretical and practical implications nestling in it. This could create awareness about the impact of metaphorically coined worldviews on doing scientific research. Furthermore, courses on the philosophy of science and the relevance of language as a framing device in the curriculum of systems biology appear to be quite important as they could provide tools to develop a reflexive understanding of the epistemological and theoretical foundations innate in doing systems biology. These elements would theoretically and practically complement the worldwide mushrooming curricula of systems biology and productively match their existing interdisciplinary scope from a reflexive point of view. In fact, the theory of conceptual metaphor could pave the way towards an integrative or even experientialist teaching practice and open up interdisciplinary avenues of doing science. Such curricula would do justice to the complex issues addressed by systems biology from a very practical point of view because the analysis of metaphorically generated basic concepts in systems biology could quite easily be combined with a historical perspective. In doing so, new generations of systems biologists would learn about the diachronic and synchronic elements of their epistemologies and that enables them to reflect on their philosophical grounding of doing science. Thus, a course on different concepts of life, their philosophical sources together with the metaphorical framings of students, might help them to better understand and contextualize the origins of their thinking. To conceptualize life as machinery or a network holds certain implications but could also be seen as creative ways to rethink or improve one's scientific thinking. Hence, a curriculum that includes embodied experiences of students offers the opportunity comprehensively to educate and prepare them for the complex endeavor of systems biology.

In summary, we have tried in this chapter synchronically and diachronically to contextualize five important biological concepts permeating systems biology. The analysis provided a rich framework of different meanings metaphorically ascribed to the respective concepts under review. Philosophically grounded in experientialism (Johnson 1987) that emphasizes "that we know the world only in terms of perceptions, categorizations, and reasoning, both conscious and unconscious, grounded in our bodily capacities and life experiences" (Brown 2008, 187), metaphor was conceived as an analogical anchor in reality. And this anchor truly deserves more attention with regard to the analysis of basic biological concepts.

References

Ahn A, Tewari M, Poon C-S, Phillips R (2006) The limits of reductionism in medicine: could systems biology offer an alternative? PLoS Med 3(6):e208

Alberghina L, Westerhoff H (eds) (2005) Systems biology. Definitions and perspectives. Springer, New York, NY

Allen G (2005) Mechanism, vitalism and organicism in late nineteenth and twentieth-century biology: the importance of historical context. Stud Hist Philos Sci 36(2):261–283

Alm E, Arkin A (2003) Biological networks. Curr Opin Struct Biol 13(2):193–202

Andersen H (2001) The history of reductionism versus holistic approaches to scientific research. Endeavour 25(4):153–156

Austin I (1962) How to do things with words. The William James Lectures delivered at Harvard University 1955. Clarendon, Oxford

Ayala F (1974) Studies in the philosophy of biology: reduction and related problems. University of California Press, Berkeley, CA

Ayala F, Arp R (eds) (2009) Contemporary debates in philosophy of biology. Wiley-Blackwell, Oxford

Baedke J (2013) The epigenetic landscape in the course of time: Conrad Hal Waddington's methodological impact on the life sciences. Stud Hist Philos Biol Biomed Sci 44:756–773

Bailer-Jones D (2003) When scientific models represent. Int Stud Philos Sci 17:59–74

Barabasi A (2002) Linked: the new science of networks. Perseus Publishing, Cambridge

Barthez P-J (1806) Nouveaux éléments de la science de l'homme. Goujon et Brunot, Paris

Bateson G (1972) Steps to an ecology of mind. Ballantine, New York, NY

Beer S (1965) The world, the flesh and the metal. The prerogatives of systems. Nature 205:223–231

Bennett M, Monk N (2008) The flowering of systems approaches in plant and crop biology. New Phytol 179:567–568

Benton E (1974) Vitalism in nineteenth-century scientific thought: a typology and reassessment. Stud Hist Philos Sci A 5(1):17–48

Bergson H (1911) Creative evolution. Henry Holt and Company, New York, NY

Bernard C (1878) Lecons sur les phénomènes de la vie communs aux animaux et aux végétaux. J.-B. Baillière et Fils, Paris

Bernard C (1983) An introduction to the study of experimental medicine. Dover Publications Inc, New York, NY

Black M (1962) Models and metaphors. Cornell University Press, New York, NY

Black M (1993) More about metaphor. In: Ortony A (ed) Metaphor and thought, 2nd edn. Cambridge University Press, Cambridge, pp 19–41

Bock von Wülfingen B (2007) Genetisierung der Zeugung. Eine Diskurs- und Metaphernanalyse reproduktionsgenetischer Zukünfte. Transcript, Bielefeld

Bölker M, Gutman M, Hesse W (eds) (2008) Menschenbilder und Metaphern im Informationszeitalter. Lit, Münster

Bölker M, Gutmann M, Hesse W (eds) (2010) Information und Menschenbild. Springer, Berlin

Böhme G, van den Daele W, Krohn W (1977) Experimentelle Philosophie. Ursprünge autonomer Wissenschaftsentwicklung. Suhrkamp, Frankfurt am Main

Bono J (1999) A new Ithaca: toward a poetics of science. 2B J Ideas 14:63–73

Bonß W, Hohlfeld R, Kollek R (1993) Kontextualität – ein neues Paradigma der Wissenschaftsanalyse? In: Bonß W, Hohlfeld R, Kollek R (eds) Wissenschaft als Kontext – Kontexte der Wissenschaft. Junius Verlag, Hamburg, pp 171–191

Bonß W, Hohlfeld R, Kollek R (1994) Vorüberlegungen zu einem kontextualistischen Modell der Wissenschaftsentwicklung. Deutsche Zeitschrift für Philosophie 42(3):439–454

Boogerd F, Bruggeman F, Hofmeyr J-H, Westerhoff H (2007a) Towards philosophical foundations of systems biology: introduction. In: Boogerd F, Bruggeman F, Hofmeyr J-H, Westerhoff H (eds) Systems biology: philosophical foundations. Elsevier, Amsterdam, pp 3–20

Boogerd F, Bruggeman F, Hofmeyr J-H, Westerhoff H (2007b) Afterthoughts as foundations for systems biology. In: Boogerd F, Bruggeman F, Hofmeyr J-H, Westerhoff H (eds) Systems biology: philosophical foundations. Elsevier, Amsterdam, pp 321–336
Bose B (2013) Systems biology: a biologist's viewpoint. Progress Biophys Mol Biol 113(3): 358–368
Bourdieu P (1976) Entwurf einer Theorie der Praxisauf der Grundlage der kabylischen Gesellschaft. Suhrkamp, Frankfurt am Main
Bowker G, Star SL (2000) Sorting things out: classification and its consequences. MIT Press, Cambridge, MA
Box JF (1978) R.A. Fisher: the life of a scientist. John Wiley and Sons, New York, NY
Brown T (2008) Making truth: metaphor in science. University of Illinois Press, Champaign, IL
Bühler K (1934) Sprachtheorie. Die Darstellungsfunktion der Sprache. UTB, Stuttgart
Byron J (2007) Whence philosophy of biology? Br J Philos Sci 58:409–422
Calvert J (2007) Patenting genomic objects: genes, genomes, function and information. Sci Cult 16(2):207–223
Calvert J, Fujimura J (2011) Calculating life? Dueling discourses in interdisciplinary systems biology. Stud Hist Philos Biol Biomed Sci 42(2):155–163
Calvert J, Joly P (2011) How did the gene become a chemical compound? Shifting ontologies of the gene and the patenting of DNA. Soc Sci Inf 50(2):157–177
Canghuilhem G (1994) A vital rationalist: selected writings. Zone Books, New York, NY
Cannon W (1963) The wisdom of the body. Norton, New York, NY
Carnap R (2011) The unity of science. Routledge, London
Cassirer E (1923) Philosophie der symbolischen Formen. vol 1. Bruno Cassirer, Berlin
Cassirer E (1985) Symbol, Technik, Sprache. Aufsätze aus den Jahren, 1997–1933. Meiner, Hamburg
Cassirer E (1993) Erkenntnis, Begriff, Kultur. Meiner, Hamburg
Charmaz K (2006) Constructing grounded theory: a practical guide through qualitative analysis. Sage, London
Chawla S (2001) Linguistic and philosophical roots of our environmental crisis. In: Fill A, Mühlhäusler P (eds) The ecolinguistics reader. Language, ecology and environment. Continuum, London, pp 115–123
Clarke A (2005) Situational analysis: grounded theory after the postmodern turn. Sage, London
Conti F, Valerio M, Zbilut J, Giuliani A (2007) Will systems biology offer new holistic paradigms to life sciences? Syst Synth Biol 1:161–165
Corbin J, Strauss A (2008) Basics of qualitative research: techniques and procedures for developing grounded theory. Sage, London
Cornish-Bowden A, Cardenas M, Letelier J, Soto-Andrade J, Abarzua F (2004) Understanding the parts in terms of the whole. Biol Cell 96:713–717
Crick F (1958) Central dogma of molecular biology. Nature 227:561–563
Crick F (1966) Of molecules and men. University of Washington Press, Seattle, WA
Crick F (1981) Life itself: its origin and nature. Simon and Schuster, London
De Backer P, De Waele D, van Speybroeck L (2010) Ins and outs of systems biology vis-à-vis molecular biology: continuation or clear-cut? Acta Biotheor 58:15–49
De Klerk GJ (1979) Mechanism and vitalism. A history of the controversy. Acta Biotheor 28:1–10
Deamer D (2010) Special collection of essays: what is life? Introduction. Astrobiology 10(10):1001–1002
Denbigh K (1951) The thermodynamics of the steady state. Methuen, London
Descartes R (2000) Philosophical essays and correspondence. Hackett, Indianapolis, IN
Döring M (2005) A sequence of 'factishes': the media metaphorical knowledge dynamics underlying the German press coverage of the Human Genome, June 2000 to June 2001. New Genet Soc 24(3):317–336
Döring M (2013) Leben systembiologisch. TA und metaphor assessment der Systembiologie. Technikfolgenabschätzung – Theorie und Praxis 21:36–42
Döring M (2014) Metaphorische Moral in aktuellen biotechnologischen Diskursen. Ein Beitrag zur Analyse normativer Annahmen in der deutschen Presseberichterstattung zur Synthetischen

Biologie. In: Junge M (ed) Methoden der Metaphernforschung und -analyse. VS Verlag, Wiesbaden, pp 215–229
Drack M (2009) Ludwig von Bertalanffy's early system approach. Syst Res Behav Sci 26: 563–572
Drack M (2013) Towards the system heuristics of Paul Weiss. Annal Hist Philos Biol 16:69–80
Drack M, Apfalter W (2007) Is Paul A. Weiss and Ludwig von Bertalanffy's system thinking still valid today? Syst Res Behav Sci 24:537–546
Drack M, Wolkenhauer O (2011) System approaches of Weiss and Bertalanffy and their relevance for systems biology today. Semin Cancer Biol 21:150–155
Driesch H (1914) The history and theory of vitalism. Macmillian and Company, London
Dupré J, O'Malley M (2007) Metagenomics and biological ontology. Stud Hist Philos Biol Biomed Sci 38(4):834–846
Elsasser W (1958) The physical foundation of biology. Pergamon, Oxford
Elsasser W (1961) Quanta and the concept organismic law. J Theor Biol 1:27–58
Elsasser W (1998) Reflections on a theory of organisms. Holism in Biology. Johns Hopkins University Press, Baltimore, MD
Falk R (1986) What is a gene? Stud Hist Philos Sci 17(2):133–173
Falk R (2010) What is a gene? - revisited. Stud Hist Philos Biol Biomed Sci 41(4):396–406
Fang FC, Casadevall A (2011) Reductionistic and holistic science. Editorial. Infect Immun 79(4):1401–1404
Federoff H, Gostin L (2009) Evolving from reductionism to holism: is there a future for systems medicine? JAMA 302:994–996
Feigl H (1981a) The origin and spirit of logical positivism. In: Cohen R (ed) Inquiries and provocations. Springer, Dordrecht, pp 21–37
Feigl H (1981b) The power of positivistic thinking. In: Cohen R (ed) Inquiries and provocations. Springer, Dordrecht, pp 38–56
Fine A (1993) Fictionalism. Midw Stud Philos 18:1–18
Fisher R (1930) The genetical theory of natural selection. Clarendon, Oxford
Fleck L (2011) Denkstile und Tatsachen: Gesammelte Schriften und Zeugnisse. Suhrkamp, Frankfurt am Main
Forgó N, Kollek R, Arning M, Kruegel T, Petersen I (2010) Ethical and legal requirements for transnational genetic research. Beck, München
FOX Keller E (1992) Secrets of life, secrets of death: essays on language, gender and science. Routledge, and reposition it in the bibliography (Line 3404–3405)
Fox-Keller E (1995) Refiguring life: metaphors of twentieth-century biology. The Wellek library lecture series at the University of California, Irvine. Columbia University Press, New York, NY
Franceschelli S, Imbert C (2009) Computer simulations as experiments. Synthese 169(3): 557–574
Frigg R (2006) Scientific representation and the semantic view of theories. Theoria 55:37–53
Frigg R, Reiss J (2009) The philosophy of simulation: hot new issues or same old stew? Synthese 169(3):593–613
Fujimura J (2005) Postgenomic futures: translations across the machine-nature border in systems biology. New Genet Soc 24(2):195–225
Ganti T (2003) The principles of life. Oxford University Press, Oxford
Gentner D, Jeziorski M (1993) The shift from metaphor to analogy in Western science. In: Ortony A (ed) Metaphor and thought, 2nd edn. Cambridge University Press, Cambridge, pp 447–480
Gibbs R (1994) The poetics of mind. Figurative thought, language, and understanding. Cambridge University Press, Cambridge
Gibson D et al (2010) Creation of a bacterial cell controlled by chemical synthesized genome. Science 329(5987):52–56
Giere R (1988) Explaining science: a cognitive approach. University of Chicago Press, Chicago, IL
Giere RN (2004) How models are used to represent reality. Philos Sci 71:742–752
Goodman N (1968) Languages of art: an approach to a theory of symbols. Bobbs-Merrill, Indianapolis, IN
Greek R, Rice M (2012) Animal models and conserved processes. Theor Biol Med Model 40:1–33

Green S, Wolkenhauer O (2012) Integration in action. EMBO Rep 13:769–771
Griesemer J, Star SL (1989) Institutional ecology, 'translations' and boundary objects: amateurs and professionals in Berkeley's Museum of Vertebrate Zoology, 1907–1939. Soc Stud Sci 19:387–420
Haber F (1975) The cathedral clock and the cosmological clock metaphor. In: Fraser J, Lawrence N (eds) The Study of time II. Proceedings of the Second Conference of the International Society for the Study of Time Lake Yamanaka – Japan. Springer, Heidelberg, pp 399–416
Haraway D (2004) Crystal, fabrics, and fields: metaphors that shape embryos. North Atlantic Books, Berkeley, CA
Harrington A (1995) Metaphoric connections: holistic science in the shadow of the Third Reich. Soc Res 62(2):357–385
Harris T (2003) Data models and the acquisition and manipulation of data. Philos Sci 70:1508–1517
Hartwell L, Hopfield J, Leibler S, Murray A (1999) From molecular to modular cell biology. Nature 402:C47–C52
Hastrup K, Skrydstrup M (eds) (2012) The social life of climate models: anticipating nature. Routledge, London
Hein H (1972) The endurance of the mechanism-vitalism controversy. J Hist Biol 5(1):159–188
Hesse M (1966) Models and analogies in science. Notre Dame University Press, Notre Dame, IN
Hesse M (1970) Theories and the transitivity of confirmation. Philos Sci 37:50–63
Hesse M (2005) Forces and fields: the concept of action at a distance in the history of physics. Dover Publications, New York, NY
Hobsbawm E, Ranger T (eds) (1992) The invention of tradition. Cambridge University Press, Cambridge
Hood L (2000) What is systems biology? Institute for Systems Biology, Seattle, WA, http://www.systemsbiology.org. Accessed 18 Nov 2014
Hood L, Rowen L, Galas D, Aichinson J (2008) Systems biology at the Institute for Systems Biology. Brief Funct Genomics 7(4):239–248
Hughes R (1997) Models and representation. Philos Sci 64:325–336
Hull D, Ruse M (eds) (2007) The Cambridge companion to the philosophy of biology. Cambridge University Press, Cambridge
Ideker T, Galitski T, Hood L (2001) A new approach to decoding life: systems biology. Annu Rev Genomics Hum Genet 2:343–372
Ingalls B (2013) Mathematical modeling in systems biology. MIT Press, Harvard, MA
Jäkel O (1997) Metaphern in abstrakten Diskurs-Domänen. Eine kognitiv-linguistische Untersuchung anhand der Bereiche Geistestätigkeit, Wirtschaft und Wissenschaft. Lang, Frankfurt am Main
Johnson M (1987) The Body in the mind. The bodily basis of meaning, imagination, and reason. Chicago University Press, Chicago, IL
Johnson M (1993) Moral imagination: implications of cognitive science for ethics. University of Chicago Press, Chicago, IL
Johnson M (2007) The meaning of the body: aesthetics of human understanding. University of Chicago Press, Chicago, IL
Kaiser MI (2011) The limits of reductionism in the life sciences. Hist Philos Life Sci 33(4):453–476
Kaneko K (2006) Life: an introduction to complex systems biology. Springer, New York, NY
Kant I (1993) Kritik der Urteilskraft. Meiner, Hamburg
Kastenhofer K (2013) Two sides of the same coin? The (techno)epistemic cultures of systems and synthetic biology. Stud Hist Philos Biol Biomed Sci 44(2):130–140
Kather R (2003) Was ist Leben? Philosophische Positionen und Perspektiven. Wissenschaftliche Buchgesellschaft, Darmstadt
Katherndahl D (2014) More than metaphor! J Eval Clin Pract 20:692–694
Kay LE (1997) Cybernetics, information, life: the emergence of scriptural representations of heredity. Configurations 5:23–91

Kay LE (2000) Who wrote the book of life? A history of the genetic code. Stanford University Press, Stanford, CA
Keller EF (2002) Making sense of life: explaining biological development with models, metaphors, and machines. Harvard University Press, Cambridge, MA
Keller EF (2010) The mirage of a space between nature and nurture. Duke University Press, Durham, NC
Kitano H (2002a) Systems biology: a brief overview. Science 295:1662–1664
Kitano H (2002b) Computational systems biology. Nature 420:206–210
Klein C (2009) Reduction without reductionism: a defense of Nagel on connectability. Philos Q 59:39–53
Klipp E, Herwig R, Kowald A, Wierling C, Lerach H (2005) Systems biology in practice. Wiley-VCH, Weinheim
Klipp E, Liebermeister W, Wieling C, Kowald A, Lehrach H, Herwig R (2009) Systems biology: a textbook. Wiley-VCH, Weinheim
Klir G (1991) Facets of systems science. Plenum Press, New York, NY
Knorr Cetina K (1981) The manufacture of knowledge – an essay on the constructivist and contextual nature of science. Pergamon Press, Oxford
Knorr Cetina K (1999) Epistemic cultures: how the sciences make knowledge. Harvard University Press, Cambridge, MA
Koch I, Reisig W, Schreiber F (eds) (2011) Modeling in systems biology: the Petri Net approach. Springer, London
Kohler R (1971) The background to Eduard Buchner's discovery of cell-free fermentation. J Hist Biol 4(1):35–61
Kohler R (1972) The reception of Eduard Buchner's discovery of cell-free fermentation. J Hist Biol 5(2):327–353
Kollek R (1990) The limits of experimental knowledge. A feminist perspective on the ecological risks of genetic engineering. Issues Reprod Genet Eng 3(2):125–135
Krohs U (2004) Eine Theorie biologischer Theorien: Status und Gehalt von Funktionsaussagen und informationstheoretischen Modellen. Springer, Berlin
Krohs U (2013) Philosophical and foundational issues in systems biology. In: Dubitzky W, Wolkenhauer O, Cho K-H, Yokota H (eds) Encyclopedia of systems biology. Springer, New York, NY, pp 1698–1702
Krohs U, Callebaut W (2007) Data without models merging with models without data. In: Boogerd F, Bruggeman F, Hofmeyr J-H, Westerhoff H (eds) Systems biology: philosophical foundations. Elsevier, Amsterdam, pp 181–213
Krohs U (2004) Eine Theorie biologischer Theorien: Status und Gehalt von Funktionsaussagen und informationstheoretischen Modellen. Springer, Berlin
Kruse J, Biesel K, Schmieder C (2012) Eine replik auf: Schmitt, Rudolf (2011). Review essay: rekonstruktive und andere metaphernanalysen. Forum Qual Soz Forsch 13(1):Art. 2. http://nbn-resolving.de/urn:nbn:de:0114-fqs120124. Accessed 18 Nov 2014
Kuhn T (1993) Metaphor in science. In: Ortony A (ed) Metaphor and thought, 2nd edn. Cambridge University Press, Cambridge, pp 533–542
Lakoff G, Johnson M (1980) Metaphors we live by. University of Chicago Press, Chicago, IL
Lakoff G (1987) Women, fire, and dangerous things. What categories reveal about the mind. Chicago University Press, Chicago, IL
Lakoff G, Johnson M (1999) Philosophy in the flesh. The embodied mind and its challenge to Western thought. Basic Books, Chicago, IL
Laubichler M, Wagner G (2001) How molecular is molecular developmental biology? A reply to Alex Rosenberg's reductionism redux: computing the embryo. Biol Philos 16(1):53–68
László E (1972) Introduction to systems philosophy: toward a new paradigm of contemporary thought. Gordon and Breach Science Publishers, London
Leonelli S, Ankeny R (2012) Re-thinking organisms: the epistemic impact of databases on model organism biology. Stud Hist Philos Biol Biomed Sci 43:29–36
Li S (2009) Network systems underlying Traditional Chinese Medicine syndrome and herb formula. Curr Bioinform 4:188–196

Liebert W-A (1995) Metaphernreflexion in der Virologie. Das theoriesprachliche Lexikon der Metaphernmodelle als Sprachreflexionsmittel im Forschungsprozess. Eine exemplarische Studie am Beispiel der Aids forschung. Universität Mannheim, Mannheim
Lloyd E (1984) A semantic approach to the structure of population genetics. Philos Sci 51:242–264
Lloyd E (1994) The structure and confirmation of evolutionary theory. Princeton University Press, Princeton, NJ
Loeb J (1964) The mechanistic conception of life. Belknap, Cambridge, MA
Lu A-P, Bian Z-X, Chen KJ (2012) Bridging the traditional Chinese medicine pattern classification and biomedical disease diagnosis with systems biology. Chin J Integr Med 18(12):883–890
Maasen S, Weingart P (2000) Metaphors and the dynamics of knowledge. Routledge, London
MacCay D (1965) Man as mechanism. The open mind and other essays. Inter-Varsity Press Tinker, Leicester
MacLeod M, Nersessian NJ (2014) Strategies for coordinating experimentation and Modeling in Integrative Systems Biology. J Exp Zool B Mol Dev Evol 322:230–239
Magnani L, Nersessian N (eds) (2002) Model-based reasoning: science, technology, values. Kluwer, Dordrecht
Mambrey P, Tepper A (2000) Technology assessment as metaphor assessment. Visions guiding the development of information and communications technologies. In: Grin J, Grunwald A (eds) Vision assessment: shaping technology in 21st century society. Springer, Berlin, pp 33–51
Marteau T, Richards M (eds) (1996) The troubled helix. Social and psychological implications of the new human genetics. Cambridge University Press, Cambridge
Massoud T, Hademenos G, Young W, Gao E, Pile-Spellman J, Uela F (1998) Principles and philosophy of modeling in biomedical research. J Feder Am Soc Exp Biol 12:275–285
Maynard-Smith J (1986) The problems of biology. Oxford University Press, Oxford
Mayr E (1997) This is biology: the science of the living world. Harvard University Press, Cambridge, MA
Mayr E, Provine W (eds) (1988) The evolutionary synthesis: perspectives on the unification of biology. Harvard University Press, Cambridge, MA
Mazzochi F (2012) Complexity and the reductionism-holism debate in systems biology. Wiley Interdiscip Rev Syst Biol Med 4(5):413–427
Mesarovic M (1968) Systems theory and biology—view of a theoretician. In: Mesarovic M (ed) System theory and biology. Springer, New York, NY, pp 59–87
Mesarovic M, Sreenath S, Keene J (2004) Search for organizing principles: understanding in systems biology. Syst Biol 1:19–27
Mesarovic M, Takahara Y (1972) Mathematical theory of general systems. Academic, Waltham, MA
Mesarovic M, Takahara Y (1975) General systems theory. Academic, New York, NY
Mesarovic M, Takahara Y (1988) Abstract systems theory. Springer, New York, NY
Monod J (1970) Le hasard et la nécessité: essai sur la philosophie naturelle de la biologie moderne. Seuil, Paris
Monod J, Bornek E (1971) Of microbes and life. Columbia University Press, New York, NY
Morange M (2000) The history of molecular biology. Harvard University Press, Cambridge, MA
Morange M (2008) The death of molecular biology? Hist Philos Life Sci 39(1):31–42
Morange M (2009) Life explained. Yale University Press, New Haven, CT
Morgan L (1923) Emergent evolution. Henry Holt and Company, New York, NY
Morrison M (2009) Fictions, representations and reality. In: Suárez M (ed) Fictions in science. Philosophical essays on modelling and idealisation. Routledge, London, pp 110–135
Murphy M, O'Neill L (eds) (1997) What is life? The next fifty years: speculations on the future of biology. Cambridge University Press, Cambridge
Nagel E (1960) The meaning of reduction in the natural sciences. In: Danto A, Morgenbesser S (eds) Philosophy of science. Meridian Books, Cleveland, OH, pp 288–312
Nagel E (1961) The structure of science: problems in the logic of scientific explanation. Harcourt, Brace and World, New York, NY

Nagel T (1998) Reductionism and antireductionism. In: Bock GR, Goode JA (eds) The limits of reductionism in biology, vol 213, Novartis foundation symposium. John Wiley and Sons, Chichester, NY, pp 3–14
Nerlich B, Clarke D (2003) Anatomy of a media event: how arguments clashed in the 2001 human cloning debate. New Genet Soc 22(1):43–59
Nerlich B, Dingwall R, Clarke D (2002) The book of life: how the competition of the Human Genome Project was revealed to the public. Health 6(4):445–469
Nerlich B, Elliott R, Larson B (eds) (2009) Communicating biological sciences: ethical and metaphorical dimensions. Ashgate, Aldershot
Nicholson D (2012) The concept of mechanism in biology. Stud Hist Philos Biol Sci 42:152–163
Nicholson D (2013) Organisms≠machines. Stud Hist Philos Biol Sci 44:669–678
Noble D (2008a) The music of life: biology beyond genes. Oxford University Press, Oxford
Noble D (2008b) Claude Bernard, the first systems biologist, and the future of physiology. Exp Physiol 93(1):16–26
Olby R (1971) Schrödinger's problem: what is life. J Hist Biol 4(1):119–148
O'Malley M, Dupré J (2005) Fundamental issues in systems biology. Bioessays 27:1270–1276
O'Malley M, Calvert J, Dupré J (2007) The study of socioethical issues in systems biology. Am J Bioeth 7(4):67–78
O'Malley M, Soyer O (2012) The roles of integration in molecular systems biology. Stud Hist Philos Biol Biomed Sci 43(1):58–68
Ofran Y (2008) The two cultures and systems biology: how philosophy starts where science ends. Eur Leg Towar New Paradig 13(5):589–604
Oparin A (1924) The origin of life. Moscow Worker Publisher, Moscow
Oppenheim P, Putnam H (1958) The unity of science as a working hypothesis. In: Feigl H et al (eds) Minnesota studies in the philosophy of science, vol 2. Minnesota University Press, Minneapolis, MN, pp 3–36
Ouzounis C, Mazière P (2006) Maps, books and other metaphors for systems biology. Biosystems 85(1):6–10
Palsson B (2011) Systems biology: simulation of dynamic network states. Cambridge University Press, Cambridge
Parry S, Dupré J (eds) (2010) Nature after the genome. Wiley-Blackwell, Hoboken
Patel M, Nagl S (2010) The role of model integration in complex systems modeling. Springer, New York, NY
Paton R (1996) Metaphors, models and bioinformation. Biosystems 38:155–162
Peacocke A (1976) Reductionism: a review of the epistemological issues and their relevance to biology and the problem of consciousness. Zygon 11:307–334
Peacocke A (1985) Reductionism in academic disciplines. Society for research into higher. Education & NFER-Nelson, Guildford
Piaget J (1954) The construction of reality in the child. Basic Books, New York, NY
Pickering A (1995) The mangle of practice time, agency, and science. University of Chicago Press, Chicago
Pigliucci M (2014) Between holism and reductionism: a philosophical primer on emergence. Biol J Linn Soc 112:261–267
Polanyi M (1958) Personal knowledge: towards a post-critical philosophy. University of Chicago Press, Chicago, IL
Polanyi M (1966) The tacit dimension. University of Chicago Press, Chicago, IL
Polanyi M (1967) The growth of science in society. Minerva 5(4):533–545
Polanyi M (1968) Life's irreducible structure. Science 160(3834):1308–1312
Polanyi M (1977) Meaning. Chicago University Press, Chicago, IL
Porsch H (1986) Michael polanyi: a critical exposition. State University of New York Press, Albany, NY
Putnam H (1981) Reason, truth and history. Cambridge University Press, Cambridge
Putnam H (1988) Representation and reality. MIT Press, Cambridge, MA
Putnam H (1991) Repräsentation und realität. Suhrkamp, Frankfurt am Main

Putnam H (1993) Von einem realistischen standpunkt. Schriften zu sprache und wirklichkeit. Rowohlt, Reinbek
Rheinberger H-J (1997) Toward a history of epistemic things: synthesizing proteins in the test tube. Stanford University Press, Stanford, CA
Rheinberger H-J (2009) Vererbung. Fischer, Frankfurt am Main
Rheinberger H-J (2010) An epistemology of the concrete: twentieth-century histories of life. Duke University Press, Durham, NC
Richardson R, Stephan M (2007) What physicalists should provide us with … In: Penco C, Beaney M, Vignolo M (eds.), Explaining the mental: naturalist and non-naturalist approaches to mental acts and processes. Newcastle: Cambridge Scholars Publishers, pp 207–221
Roll-Hansen N (1984) E.S. Russell and J.H. Woodger: the failure of two twentieth-century opponents of mechanistic biology. J Hist Biol 17(3):399–428
Rosch E (1973) Natural categories. Cogn Psychol 4:328–350
Rosch E (1978) Principles of categorization. In: Rosch E, Lloyd BB (eds) Cognition and categorization. Lawrence Erlbaum Associates, Hillsdale, NJ, pp 27–48
Rosch E, Mervis C, Wayne D, Johnson D, Boyes-Bream P (1976) Basic objects in natural categories. Cogn Psychol 8:382–439
Rosen R (1970a) Dynamical systems theory in biology. Wiley Interscience, New York, NY
Rosen R (1970b) Optimality principles. Rosen Enterprises, New York, NY
Rosen R (1978) Fundamentals of measurement and representation of natural systems. Elsevier Science, Amsterdam
Rosen R (1985) Anticipatory systems: philosophical, mathematical and methodological foundations. Pergamon Press, Oxford
Rosen R (2000) Essays on life itself. Columbia University Press, New York, NY
Rosenberg A (1997) Reductionism redux: computing the embryo. Biol Philos 12(4):445–470
Ruse M (1988) The philosophy of biology today. State University of New York Press, Albany, NY
Sattler R (1986) Biophilosophy. Analytic and holistic perspectives. Springer, New York, NY
Schaffner KF (1969) The Watson-Crick model and reductionism. Br J Philos Sci 20(4):325–348
Schaffner KF (1993a) Discovery and explanation in biology and medicine. University of Chicago Press, Chicago, IL
Schaffner KF (1993b) Theory structure, reduction, and disciplinary integration in biology. Biol Philos 8:319–347
Schaffner K (2002) Reductionism, complexity and molecular medicine: genetic chips and the globalization of the genome. In: van Regenmortel M, Hull D (eds) Promises and limits of reductionism in the biomedical sciences. John Wiley and Sons, Chichester, NY, pp 323–351
Schaffner K (2007) Theories, models, and equations in systems biology. In: Boogerd C, Bruggeman F, Hofmeyr J-H, Westerhoff H (eds) Systems biology: philosophical foundations. Elsevier, Amsterdam, pp 145–162
Schatzki T (1996) Social practices. A Wittgensteinian approach to human activity and the social. Cambridge University Press, Cambridge
Schatzki T (2001) Practice in theory. In: Schatzki T, Knorr Cetina K, Savigny E (eds) The practice turn in contemporary theory. Routledge, London, pp 1–14
Schatzki T, Knorr Cetina K, Savigny E (eds) (2001) The practice turn in contemporary theory. Routledge, London
Schmitt R (2000) Skizzen zur Metaphernanalyse. Forum qualitative Sozialforschung 1(1). http://www.qualitative-research.net/fqs-texte/1-00/1-00schmitt-d.htm. Accessed 18 Nov 2014
Schmitt R (2005) Systematic metaphor analysis as a method of qualitative research. Qual Rep 10(2):358–394
Schmitt R (2010) Metaphernanalyse. In: Mey G, Mruck K (eds) Handbuch Qualitative Forschung in der Psychologie. VS Verlag, Wiesbaden, pp 676–691
Schmitt R (2011) Systematische Metaphernanalyse als qualitative sozialwissenschaftliche Forschungsmethode. metaphorik. de 21:47–82

Schmitt R (2014) Eine Übersicht über Methoden sozialwissenschaftlicher Metaphernanalysen. In: Junge M (ed) Methoden der Metaphernforschung und -analyse. Springer VS, Wiesbaden, pp 13–30
Schrödinger E (2012) What is life? With mind and matter and autobiographical sketches. Cambridge University Press, Cambridge
Semino E (2008) Metaphor in discourse. Cambridge University Press, Cambridge
Seth BP, Thaker VS (2014) Plant systems biology: insights, advances and challenges. Planta 240:33–54
Shore B (1996) Culture in mind. Cognition, culture, and the problem of meaning. Oxford University Press, Oxford
Smuts J (1926) Holism and evolution. Viking, New York, NY
Snobelen S (2012) The myth of the clockwork universe: Newton, newtonianism, and the enlightenment. In: Jacobs N, Firestone C (eds) The persistence of the sacred in modern though. University of Notre Dame Press, Notre Dame, IN, pp 149–184
Sober E (1993) Philosophy of biology. Oxford University Press, Oxford
Stange K (2005) The end of "naïve reductionism": rise of systems biology or renaissance of physiology? Am J Physiol Cell Physiol 288(5):C968–C974
Steen G, Biernacka E, Dorst A, Kaal A, López-Rodríguez C, Pasma T (2010) Pragglejaz in practice. Finding metaphorically used words in natural discourse. In: Low G, Todd Z, Deignan A, Cameron L (eds) Researching and applying metaphor in the real world. John Benjamins, Amsterdam, pp 165–184
Sterelny K, Griffiths P (1999) Sex and death: an introduction to philosophy of biology. University of Chicago Press, Chicago, IL
Stöckler M (1991) A short history of emergence and reductionism. In: Agazzi E (ed) The problem of reductionism in science. Kluwer, Alphen aan den Rijn, pp 71–90
Suárez M (2009) Scientific fictions as rules of inference. In: Suárez M (ed) Fictions in science. Philosophical essays on modelling and idealisation. Routledge, London, pp 158–178
Suárez M, Solé A (2006) On the analogy between cognitive representation and truth. Theoria 55:27–36
Suppes P (1960) A comparison of the meaning and uses of models in mathematics and the empirical sciences. Synthese 12:287–301
Suppes P (1962) Models of data. In: Nagel E, Suppes P, Tarski A (eds) Logic, methodology and philosophy of science: proceedings of the 1960 International Congress. Stanford University Press, Stanford, CA, pp 252–261
Sweetser E (1990) From etymology to pragmatics: metaphorical and cultural aspects of semantic structure. Cambridge University Press, Cambridge
Tabery J (2008) R.A. Fisher, Lancelot Hogben, and the origin(s) of genotype-environment interaction. J Hist Biol 41:717–761
Tin C, Poon C-S (2014) Integrative and reductionist approaches of modeling of breathing. In: Batzel J, Bachar M, Kappel F (eds) Mathematical modeling and validation in physiology. Applications to cardiovascular and respiratory systems. Springer, Heidelberg, pp 89–103
Toepfer G (2005) Der begriff des lebens. In: Toepfer G, Krohs U (eds) Philosophie der biologie. Eine einführung. Suhrkamp, Frankfurt am Main, pp 157–174
Tutton R, Corrigan O (eds) (2004) Genetic databases: socio-ethical issues in the collection and use of DNA. Routledge, London
Ukrow R (2004) Nobelpreisträger Eduard Buchner (1860-1917). Ein leben für die chemie der gärung und - fast vergessen – für die organische chemie. Technische Universität Berlin, Berlin
Ulett M (2014) Making the case for orthogenesis: the popularization of definitely directed evolution (1890–1926). Stud Hist Philos Biol Biomed Sci 45:124–132
Ullah M, Wolkenhauer O (2007) Family tree of Markov models in systems biology. IET Syst Biol 1:247–254
van Regenmortel M (2004) Biological complexity emerges from the ashes of genetic reductionism. J Mol Recognit 17:145–148

Verpoorte R, Choi Y, Kim H (2005) Ethnopharmacology and systems biology: a perfect holistic match. J Ethnopharmacol 100(1–2):53–56
Vico GB (1990) Prinzipien einer neuen Wissenschaft über die gemeinsame Natur der Völker. Teilband 1 und 2. Meiner, Hamburg
Vigotsky L (2012) Thought and language. MIT Press, Cambridge, MA
von Bertalanffy L (1932) Theoretische Biologie. Gebrüder Bornträger, Berlin
von Bertalanffy L (1949) Das biologische Weltbild. Francke, Bern
von Bertalanffy L (1950a) The theory of open systems in physics and biology. Science 111:23–29
von Bertalanffy L (1950b) An outline of general systems theory. Br J Philos Sci 1:139–164
von Bertalanffy L (1968) General system theory. Brazillier, New York, NY
Vrba E, Gould S (1986) The hierarchical expansion of sorting and selection: sorting and selection cannot be equated. Paleobiology 10:146–171
Waddington C (1968) Towards a theoretical biology, vol 4. Edinburgh University Press, Edinburgh
Waddington C (1975) The evolution of an evolutionist. Edinburgh University Press, Edinburgh
Wang M, Lamers R, Korthout H, van Nesselroij J, Witkamp R, van der Heijden R, Voshol P, Havekes L, Verpoort R, van der Greef J (2005) Metabolomics in the context of systems biology: bridging traditional Chinese medicine and molecular pharmacology. Phytother Res 19:173–182
Weiss P (1925) Animal behaviour as system reaction: the orientation towards light and gravity in the resting postures of butterflies. Biol General 1:167–248
Weiss P (1940) The problem of cell individuality. Am Nat 74:34–46
Weiss P (1973) The science of life. Futura Publishing, New York, NY
Westerhoff H, Palsson B (2004) The evolution of molecular biology into systems biology. Nat Biotechnol 22(10):1249–1252
Westerhoff H, Winder C, Messiha H, Simeonidis E, Adamczyk M, Verma M, Bruggeman F, Dunn W (2009) Systems biology: the elements and principles of life. FEBS Lett 583:3882–3890
Wiener N (1948) Cybernetics or control and communication in the animal and machine. MIT Press, Boston, MA
Williams E (2003) A cultural history of medical vitalism in enlightenment. Ashgate, Farnham
Williams R (1956) Biochemical individuality. The key for the genetotrophic concept. John Wiley and Sons, New York, NY
Wolkenhauer O (2001) Systems biology: the reincarnation of systems theory applied in biology? Brief Bioinform Henri Stewart Publ 3:258–270
Wolkenhauer O (2007a) Interpreting Rosen. Artif Life 13:1–2
Wolkenhauer O (2007b) Defining systems biology: an engineering perspective. J Biotechnol 2:329–351
Wolkenhauer O (2014) Why model? Front Physiol 5(Art 21):1–5
Wolkenhauer O, Green S (2013) The search for organizing principles as a cure against reductionism in systems medicine. FEBS J 280(23):5938–5948
Wolkenhauer O, Mesarovic M (2005) Feedback dynamics and cell function: why systems biology is called systems biology. Mol Biosyst 1(1):14–16
Wolkenhauer O, Green S (2013) The search for organizing principles as a cure against reductionism in systems medicine, in: The FEBS Journal 280/23, 5938–5948
Woodger J (2001) Biological principles: a critical study. Routledge, London
Ziman J (2000) Real science: what is and what it means. Cambridge University Press, Cambridge
Zucker A (1981) Holism and reductionism in molecular biology. A view from genetics. J Med Phils 6(2):145–164

Chapter 3
Systems-Oriented Approaches in Biology: System Biologist's Narratives of Present, Past, and Future

Martin Döring, Anne Brüninghaus, Regine Kollek, and Imme Petersen

Abstract Systems biology has been framed as a newly emerging paradigm in biology conceived to overcome the theoretical and methodological shortcomings of previous approaches such as molecular biology. Framed as an approach, its history has to date rarely been addressed which means the historical analysis of its theoretical roots and ancestors still remain in the dark. This chapter aims at partly filling this gap by analyzing the imagined presents, pasts, and futures of systems biology as seen through the systems biologist's eyes. For this to be done, a narrative analysis is applied to written sources and expert interviews conducted with system biologists in Germany. The analysis reveals considerably different pictures of imagined present, pasts and futures between the written and interview data. It becomes apparent that despite current attempts to establish a common definition of systems biology considerable differences of what it represents exist. More important, however, is the fact that an ahistoric perspective prevails among many system biologists interviewed. Albeit historical references to so-called predecessors appear now and then, we discuss the danger of a prevailing ahistoric narrative in systems biology. A solution to this problem is a still missing conceptual historiography of systems biology that holds the potential to provide clarification of definitional fuzziness and the relevance of a historically grounded understanding of its conceptual importance in current biology. Only the knowledge about imagined presents, pasts and futures can help us better understand the present condition of systems biology and contribute to substantiating its conceptual deficits.

Keywords History of systems biology • Narratives • Oral history • Sociology of the future

---

M. Döring (✉) • A. Brüninghaus • R. Kollek • I. Petersen
Research Centre for Biotechnology, Society and Environment (FSP BIOGUM), University of Hamburg, Hamburg, Germany
e-mail: doering@metaphorik.de

M. Döring et al., *Contextualizing Systems Biology*,
DOI 10.1007/978-3-319-17106-7_3

Systems biology is a new scientific approach that has its historical roots in a variety of disciplines and approaches such as genetics, biology, theoretical biology, mathematical biology, general systems theory, and the computer sciences. Mainly conceived as a functional successor of so-called Omics approaches, systems biology emerged after the successful structural analysis of the human and other organisms' genome at the end of the twentieth and the start of the twenty-first century. The plethora of data produced on genomes, transcriptomes, proteomes, and other subsystems of biological entities instigated a process of reflection on how to handle methodologically and analytically use these masses of data that were increasingly difficult to manage and to understand. New methodical and theoretical approaches were needed to systematize, integrate, and functionally interpret these data. The question, however, remains as to how systems biology could be endowed with a history or connected to different developments in the history of biology. Bearing in mind that the word "history" stems etymologically from "story," we consequently investigate in this chapter the stories published and told by systems biologists. The objects of analysis are narrative structures and their elements (Propp 1928) used to depict the present, past, and future of systems biology because their use conveys an "ordering of events, actions and elements of experience in a communicative structure" (deSilvey 2012, 33). In brief, narratives construct certain versions or representations of an event or a development (Cobley 2001, 237). They should not be understood as "objectively 'true' or historically accurate" (deSilvey 2012, 34) entities but as means of perceiving and ascribing meaning "through the promotion of selected story-lines" (deSilvey 2012, 34): they are "perceived sequences of non-randomly connected events" (Toolan 2001, 6). With regard to the data analyzed in this chapter one might say that they are individual representations of probably the same events and developments. Although diverging elements appear, converging structures exist too: the metaphorical use of the word *revolution* in the quotes taken from Westerhoff and Palsson (2004, 1249) and Aggarwal and Lee (2003, 175) depicts a change or turning point in the biosciences which is used to explain causally the advent of systems biology. Such narrative elements have also been described as important constituting components of an object and as processes of meaning making that couple past, present and future developments. Thus, narratives to be understood as imaginative and meaning-making structures, generate a reality (Bruner 1991) that reflects certain perspectives on developments ranging from individuals (McAdams 1993) to the constitution of scientific disciplines (Midgley 2003). Importantly though, narratives are also dynamic entities that are open to deconstruction as can be demonstrated by the following more critical narrative on the advent of systems biology:

> Recent advances in '-omics' technologies have allowed biologists to take snapshots of cellular components in a variety of situations, which should help us to understand how biological systems operate. It has been claimed that the interpretation of these snapshots is best provided by a shift to 'systems' thinking and the discipline of 'systems biology'. Lauded as 'the 21st century science' (http://www.systemsbiology.org/), 'a revolution' (Anderem 2005) and 'a paradigm shift in modern life science research' (Aggarwal & Lee, 2003), systems biology 'promises to revolutionize our understanding of complex biological regulatory systems' (Kitano, 2002). But revolutions in science are rare and revolutions in biology rarer still. (Bothwell 2006, 6)

Bothwell assesses a representative set of important linguistic elements—such as "revolution, twenty-first century science", "paradigm shift", or to "revolutionize"—used by different authors to develop a narrative about the advent and possible impact of systems biology on biological research in general. The language used suggests a superlative performance and impact of systems biology which is in the last sentence and the remainder of the text, however, rescinded.

In sum, one could say that the present, past, and future of systems biology is based on narratives as meaning-making elements that situate the discipline in the context of biology. Consequently, the aim of this chapter consists in taking a closer look at the stories written in articles and told in interviews by system biologists about the present, past and future of systems biology. This is interesting as the analysis reveals a representative set of individualized narratives with which scientists integrate their work into the more or less official storylines of systems biology and on the other hand narratives of the past and present meaning-making structures that bear an impact on future conceptions of it. Hence, our approach is based on a narrative analysis, as previously outlined, and conceptually combined with insights from oral history (Tomes 1991; Charlton et al. 2007; Leavy 2011), the sociology of the future (Bell and Mau 1971; Selin 2008), and the sociology of expectations (Brown 2003; Brown and Michael 2003; Borup et al. 2006). All these fields of research converge in the fact that they share an interest in how pasts, presents, and futures are constructed, imagined, and combined via narrative representations and practices. Mainly based on combinations of qualitative methods, their aim is to trace converging, mutually enhancing or conflicting patterns of interpretation by which individuals of a social group such as a scientific discipline create legitimate pasts, presents, and futures.

With regard to systems biology, we used a combined method of document analysis and expert interviews to merge approved historical visions as to be found in research papers and more colloquial constructions of systems biology's past, present, and future. For this to be done, we developed a representative corpus of scientific reviews, scientific articles, edited volumes and monographs of leading scientists with the help of the *pubmedpubmedreminer*.[1] After having cross-checked all publications on the ISI-citation-index to quantitatively secure their intradisciplinary dispersion, all entries of this database were first closely read to get an insight into the prevailing discourses and their structure in systems biology. During a second round, emphasis was put on revealing any historical and future- related references such as names of scientists to be thought of as predecessors, scientific institutes, disciplines, approaches, turning points in science or outlooks of how systems biology should progress and what kind of developments are necessary to bring systems biology forward. In doing so, certain rhetorical strategies could be tackled that mainly used prominent names such as Ludwig von Bertalanffy, Paul A. Weiss, or Robert Rosen

[1] The *pubmedpubmedreminer* is a search tool with which a thematically centered search on the PubMed database is possible. The tool not only provides information about the temporal dispersion of publications but also lists important authors and indicates the themes addressed in scientific articles. See http://hgserver2.amc.nl/cgi-bin/miner/miner2.cgi. Accessed January 3, 2015.

as narrative anchoring points together with depicting high-throughput technologies as the innovation that reinstigated interest in systems approaches in biology. This kind of approach, furthermore, disclosed that scientific articles reflecting or even analyzing the conceptual history or the historical roots of system biology are scarce. Only Drack, Apfalter, Wolkenhauer and Poivreau more deeply engaged with the conceptual history of systems theory in biology and its innovative impact on current systems biology (Drack 2009, 2013; Drack and Wolkenhauer 2011; Drack and Apfalter 2007; Drack et al. 2007) whereas a small amount of articles written by systems biologists such as, for example, Bothwell (2006) or Marcum (2008) deal with the conceptual history of systems biology. The relative scarcity of historical accounts as well as a seeming lack of awareness about the (long) history of systems biology among the ones engaged in it motivated us to structure our interview guide around three areas to be addressed: the definition of systems biology of the respective interview partner, questions about historical predecessors and their impact on the scientist's work currently undertaken, and a vision exercise about systems biology's future at the end of the interview. This approach was chosen for the interviews to develop a present perception and definition of systems biology in the interviewee against which diachronic, for example, past and future-related elements relevant for him or her were discussed. The underlying rationale of this method was informed by the rationale developed in oral history which aims at collecting and studying historical information (Yow 1994; Thompson 2000) as generated by the interviewee. All interviews were tape-recorded, transcribed, analyzed, and categorized as outlined in grounded theory[2] (Corbin and Strauss 2008; Charmaz 2006).

The aforementioned approach is interesting inasmuch as current systems approaches in biology hold a lot of promissory narratives that depict systems biology as going beyond an additive and deterministic understanding of biological functions and systems. Due to this narrative and its structure, scientific attention consequently moved away from a reductionist bottom-up approach and redirected attention to multilevel complexity and multidirectional dynamics of biological functions and processes. Such an approach was not entirely new as its conceptual foundations were already laid between the 1920s and 1950s by proponents of theoretical biology and general systems theory such as von Jacob von Uexküll (1920), Paul A. Weiss (1973) or Ludwig von Bertalanffy (1968) (see Sect. 3.3 for an overview of different system conceptions). The core of a systems conception in biology is that biological entities are characterized inter alia by organized complexity which is based on non-linear coupling and multiple interactions between individual parts and processes. Furthermore, such entities are in close and dynamic exchange with their environment(s). Basic for such a systems conception was a methodical holism implying that wholes—no matter whether they are single cells or organisms—are not only defined by their material parts and the properties thereof, but also by the

[2] Grounded theory is an important approach used in the social sciences that aims at systematically analyzing qualitative data by combining different methods to guarantee an acute in-depth analysis that explains social patterns of interpretation and behavior.

structural relations between them. Later on, comparable approaches were formulated in theoretical biology during the 1960s for example by Robert Rosen (1970) or Conrad Hal Waddington (1968). They developed, to some extent, the theoretical and mathematical prerequisites for the new systemic approach in biology, but lacked the instrumental and technical possibilities available today. Especially the improvements in the fast generation of experimental data coming along with new high-throughput technologies were not accompanied by theoretical and methodical progress which could have led to a better understanding of biological functions or systems.

This is exactly the narrative turning point where new systems biology came into play since modern computer technology opened up new ways for modeling complex systems. Generally speaking, the systems biology of the twenty-first century aims at providing a more comprehensive or even holistic understanding of biological processes or entities. Such an understanding is conceived to hold the potential to overcome the conceptual and operational shortcomings inherent in molecular biology (see Sect. 2.5 for an overview of concepts of holism). Although such assertions have to be treated with care, historical narratives can be found in scientific articles that genealogically conceptualize systems biology as the offspring of molecular biology. Westerhoff and Palsson (2004, 1249), for example, depict the development of systems biology in their introduction to a scientific review as follows:

> More recently, the genomics revolution has catapulted molecular biology into the realm of systems biology. In unicellular organisms and well-defined cell lines of higher organisms, systems approaches are making definitive strides toward scientific understanding and biotechnological applications.

Here, a causal or even linear image of the advent of systems biology is developed: The impact of this process is emphasized by the metaphor[3] "catapulted" which evokes images of extreme acceleration, speed and impact. The argumentation depicts the "genomics revolution" as the active part or catapult, molecular biology as the object to be fired and the "realm of systems biology" as the entity that engulfs the smaller projectile. The imagery used clearly highlights the considerable impact of this development in the field of biology. Aggarwal and Lee (2003, 175) apply a more sober and neutral language to depict the advent of systems biology:

> Life science-based research has undergone a revolutionary change in the past few years, with a shift in the focus of cellular studies with a reductionist approach towards an integrative approach. This shift has been driven by technology. The new integrative approach investigates 'complex' systems, which cannot be completely understood by investigation of individual components in isolation. Systems biology is a new field of science that develops a system-level understanding by describing quantitatively the interaction among all the individual components of the cell. The ultimate aim of such an approach is to develop computational models of these complex systems so that the response of the biological system to

[3] Metaphors are figures of speech that help to conceptualize abstract entities. They can be found in everyday as well as in scientific language and are conceived as meaning generating devices in language *and* thought. In narratives, they often develop a semantic framework elaborated upon by the surrounding language (see the interpretation of the example).

any kind of perturbation, for example, environmental disturbance, genetic mutation etc, can be predicted.

A first glance at Aggarwal's and Lee's text section taken from their introduction exhibits a more neutral tone because they simply use fewer metaphors and apply a factual and technical language to depict what they call recent changes in the life sciences. However, their narrative too starts with the revolution metaphor which is combined with the noun "change" (Kuhn 1962). Thus, the original meaning of revolution—to turn—is semantically conceptualized as a turning alteration or modification, and the object that is turned is the life sciences. The metaphor used clearly lacks the dramatic moment as encountered in the preceding excerpt but its narrative starts with a comparable metaphorical framework that emphasizes change and then depicts the content of change.

In summary, the aim of this chapter consists in depicting a multifaceted history of systems biology by analyzing representations of its current pasts, presents and estimated futures. The combination of written evidence and interview data offers an enlivened insight into the procedures and ways of creating a historical anchoring. In the next section (Sect. 3.1) we will now see how systems biology is synchronically defined. In view of this contemporary background, we will then turn to the imagined pasts (Sect. 3.2) before envisioned futures (Koselleck 2004) are studied (Sect. 3.3). The final section (Sect. 3.4) provides a summary of the findings, draws attention to past legacies, and interpretatively analyses linkages between past, present, and future that contribute to the current condition of systems biology.

3.1 What Is Systems Biology? Narratives of the Present

As already seen, systems biology emerged at the end of the last century and took up speed at the start of the twenty-first century. Often depicted as a new way of doing biology and genealogically framed as the offspring of former biological approaches, systems biology has also been described as a "whole-istic approach" (Chong and Ray 2002, 1161) or as way of doing "physical biology" (Zewail 2008) superseding the conceptual shortcomings of molecular biology. Its main aim is thus to develop a "biology at the system level" (Kitano 2002, 1662) by focusing on the organismal and dynamics of cellular functions. To achieve this goal, computational and mathematical modeling approaches represent an essential ingredient in systems biology and triggered an influx of other disciplines such as engineering, physics, computer science and mathematics into biology. Hence, the combination of physical, mathematical, engineering, and biological approaches created an interdisciplinary field whose proponents theoretically subscribe to a whole-istic systems approach that investigates the interaction of (bio-)chemical parts and processes on and between different scales, following the principles of methodological holism. In a nutshell, systems biology aims at a functional clarification of underlying processes of emerging properties of biological systems such as cells, tissues or whole organisms.

In addition to such more or less comprehensive explanations, systems biology semantically remains a many-sided concept as it is quite often specified differently. It has been described as a "comprehensive quantitative analysis of the manner in which all the components of a biological system interact functionally over time" (Anderem 2005, 511), as a "quantitative understanding of biological systems to an extent that one is able to predict systemic features" (Bork and Serrano 2005, 507) or as the "study of the behavior of complex biological organization and processes in terms of molecular constituents" (Kirschner 2005, 503). Bearing in mind that all these quotes appeared in the same issue of the scientific journal *Cell*, the variability of what systems biology represents is quite astonishing if not opaque. So why do explanations differ so much? Why is it so difficult to define systems biology? It seems that there is an apparent mismatch between different conceptual levels inherent in systems biology that touch upon aspects of the disciplines involved ("biology to computer science"), the scale of analysis ("genome-wide to small scale networks"), the methods applied ("computational to experimental") and the kind of analysis ("deterministic to probabilistic") undertaken (Gomes 2009, 4). The theoretical, methodological and practical integration of these different levels is a considerable interdisciplinary task, especially with regard to underlying concepts such as engineering principles, network theory, abstract mathematics, graph theory, nonlinear thermodynamics, physics and biology which require conceptual integration (Gomes 2009, 6). Consequently, theoretical and methodological tensions between these endpoints are unavoidable. This has also been depicted as the yin and yang of systems biology (Nature Blogs 2007) which comprises "the integration of experimental and computational approaches [...]; the balance between genome-wide systematical approaches [...] and smaller-scale quantitative studies; top-down versus bottom-up strategies to solve systems architecture and functional properties" (Nature Blogs 2007). This methodological diversity is merged with theoretical reference to a multidimensional system-level approach and the modeling of the processes taking place between them. Hence, definitions of what systems biology is oscillate in this complex matrix. Several attempts have been undertaken to reconcile them. The initiative *Systems Biology in the European Research Area* (ERASYS BIO) for instance, metaphorically depicts systems biology as the Big Bang in Biology (ERASYS BIO 2007, 6) and defines it as follows.

> Systems biology aims at understanding the dynamic interactions between components of a living system, between living systems and their interaction with the environment. Systems biology is an approach by which biological questions are addressed through integrating experiments in iterative cycles with computational modelling, simulation and theory. Modelling is not the final goal, but is a tool to increase understanding of the system, to develop more directed experiments and finally allow predictions. (ERASYS 2007, 6)

Almost all previously outlined aspects can also be found in this definition and it clearly frames systems biology as a complex and ambitious research endeavor. Although the ERASYS definition is an outcome of a professional discussion and a consulting process, other quotes of individual scientists or research groups taken from the scientific literature display a comparable underlying narrative framework.

This can, for example, be seen in Morel et al.'s (2004, 651) definition of systems biology taken from their introduction:

> Applied systems biology is the integrated analysis of genetic, genomic, protein, metabolite, cellular, and pathway events that are in flux and interdependent. It necessitates the use of a variety of analytic platforms as well as biostatistics, bioinformatics, data integration, computational biology, modeling, and knowledge assembly protocols.

In this definition, the scale of analysis is first indicated ("genetic," "genome," "protein," etc.) and then combined with an outline of scientific disciplines ("biostatistics and bioinformatics") and methods to be applied ("computational biology" and "knowledge assembly protocols"). The explanation is thus based on an underlying narrative sequence that combines *scale-scientific discipline-methods*. It is interesting though that the category of scale is represented quite precisely on the word level where as the scientific disciplines involved and the methods to be applied are depicted in generic terms. This aspect is also corroborated by the marker "a variety of analytic platforms" that introduces this sequence. Other combinations are also possible such as the following.

> Systems biology studies biological systems by systematically perturbing them (biologically, genetically, or chemically); monitoring the gene, protein, and informational pathway responses; integrating these data; and ultimately, formulating mathematical models that describe the structure of the system and its response to individual perturbations. (Ideker et al. 2001, 343)

Here, we find a comparable introductive narrative sequence that starts with a short depiction of what scientists do in systems biology and then develops a conceptual structure that combines scale, methods and—although implicitly by referring to mathematical models—the scientific discipline involved. Again, the level of scale is explicitly outlined with lexical items such as ("gene," "protein," etc.) and generic terms ("mathematical modeling" and "structure of the system") remain quite unspecific. It thus appears typical for this kind of narrative to depict precisely the scale to define the unit of analysis whereas the methods and disciplines to provide analysis remain, at least in the introduction of Morel et al.(2004, 651), relatively general with regard to their semantic content.

Other conceptual narratives put emphasis on different elements. Zhang et al. (2010, 386), for example, start with a methodological argument and then give details about scale:

> Systems biology, which is characterized by mathematical models, can be applied as a scaffold to integrate omics datasets and this approach is therefore well suited for study of large and complex regulatory networks. (Zhang et al. 2010, 386).

This means that the argumentation is conceptually based on an *analytical* ("mathematical models")—*methodological* ("mathematical models")—*scale* ("regulatory networks") narrative that semantically frames systems biology as an approach. Such an underlying storyline can also be deducted from the next introductory quote in which systems biology is contrasted with synthetic biology:

> While systems biology emphasizes application of computational techniques for obtaining insights into the mechanism of various biological processes, synthetic biology endeavours to develop de novo biological circuits to engineer the behaviour of living systems. (Matsuoka et al. 2009, S394)

Here, the phrase "computational techniques" clearly emphasizes methodological aspects and then elaborates with "the mechanism of various biological processes" on the scale to be analyzed. Albeit the section is used to demarcate systems from synthetic biology, the language used to depict the main scope of systems biology semantically refer to a generic level and the narrative appears to be conceptually based on a *method-scale* combination.

A comparable narrative is also inherent in the following quote which displays a *method* ("quantitative analysis")—*disciplinary* ("mathematical models")—*scale* ("biological systems") structure. The language used relies again on terms that give a generic description of systems biology bearing in mind that the subsequent sentence specifies to some extent the content of the previous sentence.

> Systems biology is the quantitative analysis, often through the use of predictive mathematical models, of biological systems. Often it involves collection, analysis, and integration of whole genome scale data sets with the objective to gain a quantitative phenotypic description of the biological system. (Otero and Nielsen 2009, 440)

Reflexive perspectives on systems biology exist as well. Accordingly, the manifold definitions of systems biology are critically mentioned from time to time as in the following excerpt. These are, at least to some extent, coupled with a self-analysis that aims at finding intersections between the different ways of doing systems biology. The quote is based on a storyline that uses methods ("mathematical modeling," "mapping intersections," and "quantification of dynamic responses") and scale ("cellular components," "living cells") as the narrative foundations to outline what systems biology is or how it should be seen.

> There are many definitions of systems biology, but most of these contain elements such as mathematical modelling, global analysis (or ome analysis), mapping of interactions between cellular components, and quantification of dynamic responses in living cells. In most cases the objective of systems biology is to obtain a quantitative description of the biological system under study, and this quantitative description may be in the form of a mathematical model. (Rokem et al 2007, 1283)

In sum, one could say that the varying narrative structures or storylines based on methodological aspects and elements of scale contribute to depicting systems biology as a new approach. This aspect surfaces from 2005 onwards and becomes an important aspect especially addressed in many scientific articles. Here systems biology is explicitly portrayed as an approach as the following quote shows.

> Systems Biology, therefore, can be seen to stand for an *approach* to bioresearch, rather than a field or a destination. This approach consciously combines reduction and integration from the outset of research and development activities, and it necessarily involves going across spatial scales of structural and functional integration (i.e., between the parts and the entity). There is no inherent restriction on the level at which 'the system' may be defined. In fact, there is no such thing as *the* system because structures that are parts of one system (say, a mitochondrion in a cell) may form systems in their own right at a different level of integration (for example, in the contexts of electron transport chains and ATP synthesis). The focus of Systems Biology can be, but is not required to be, at the single-cell level (a predominant target so far). As an approach, Systems Biology is equally applicable to small or large biological entities. (Kohl et al 2010, 25)

Kohl et al. (2010, 25) appear to depict a clearer insight into what systems biology represents by providing a comparison in the first sentence. This rhetoric strategy is based on an opposition which is elaborated upon afterwards. The following sentences exhibit a narrative structure that provides theoretical aspects ("reduction and integration") with a timeline ("right from the onset"), dimensions of scale ("mitochondria in a cell," "single cell," "part") and the coverall term system is explained by illustrating generic terms with concrete examples ("structures that are parts of one system say, a mitochondrion in a cell may form systems"). Based on this underlying narrative structure, this exemplary quote applies a rhetoric strategy on the word level that combines generic terms, and immediately frames their semantic scope by using specific words. This strategy provides a clearer and fuller picture of what systems biology could or is supposed to represent than a pure recital of scale, disciplines and methods. At the same time, it should be noted that any new scientific discipline or approach needs a certain time to define itself and to establish a more or less coherent narrative identity. Thus, it appears quite normal that earlier quotes are semantically fuzzy because the disciplinary identity of systems biology was still in the making. Once a stable but flexible narrative framework has been established semantic and theoretical differentiation take place.

Comparable narrative structures, though slightly differing, were also found in the interviews with scientists working in systems biology. Here, the descriptions of what systems biology is, were, compared to printed reflections, generated more spontaneously and through interaction. In some cases, they depend on the degree of professional experience and status. In almost all cases, the introductory question of what system biology is, instigated a thought process like the following.

> Well, I am not a friend of definitions because I have not to pass exams any longer – thank god. Systems biology is two things for me: On hand one aims at addressing a problem from a more complex perspective and to avoid to look at it from a restricted small perspective. For example, we look at the regulation of transcription and we investigate many levels which regulate transcription. We do not have one method and we ask what could contribute to a better understanding? Well we try to apply global methods to tackle different levels to get a more complex picture and not a small part. (Scientist B)
>
> German original: Ich bin kein Freund von Definitionen weil ich, Gott sei Dank nicht mehr Prüfungen ablegen muss. Systembiologie ist für mich zweierlei. Einmal, dass man sich bemüht, komplexer ein thematisches Problem zu erfassen und zu sehen, und nicht nur sehr eingeschränkt auf kleiner Ebene draufzuschauen, sondern sich zum Beispiel mit der transkriptionellen Regulation zu beschäftigen. Wir sehen uns eben sehr verschiedene Level an, durch die die Transkription reguliert wird. Wir haben nicht eine Methode, mit der wir arbeiten, sondern wir fragen: Was kann zu einem besseren Verständnis beitragen? Also wir bemühen uns, mit verschiedenen globalen Methoden verschiedene Ebenen zu erfassen, damit wir ein komplexeres Bild tatsächlich bekommen und nicht nur einen kleinen Ausschnitt.

What we can find in this quote is a discursive strategy in which systems biology is divided into two parts: One is the already encountered systems view that aims at generating a comprehensive understanding. Albeit the second aspect is not outlined, the interviewee puts considerable effort into explaining what the application of systems perspective requires by using the concrete examples of the regulation of gene transcription. In contrast to the examples taken from scientific articles, a rather theoretical aspect is addressed ("addressing a problem from a more complex perspective and to avoid to look at it from a restricted small perspective") and then combined with the scale of analysis ("the regulation of transcription and we investigate many levels which regulate transcription") and complemented with a methodical allusion ("global methods"). We here see a theory-scale-method narrative.

A shorter definition is provided by another interviewee who initially ties the notion of systems biology to his area of research:

> Well, I understand systems biology as the analysis of biological systems, especially biochemical reaction networks arising through mathematical modeling in close cooperation with quantitative cell biology. I mean that quantitative cell biology really becomes a part of mathematical models which then really enables a systematic analysis of biochemical networks. (Scientists E)
>
> German original: Also unter Systembiologie würde ich verstehen, also äh die Analyse biologischer Systeme, speziell biochemischer Reaktionsnetzwerke, bedingt durch mathematische Modellierung in enger Zusammenarbeit mit quantitativer Zellbiologie. Dass wirklich auch so quantitative Zellbiologie in die mathematischen Modelle eingeht, wodurch dann wirklich eine systematische Analyse biochemischer Netzwerke möglich ist.

After a short introduction a storyline comes into view that starts with specifically portraying a scale of analysis ("biochemical reaction networks"), methodical aspects ("mathematical modeling") and an additional discipline to be involved ("quantitative cell biology"). The narrative structure scale-method-discipline was not yet encountered whereas systems biology is framed as a discipline that aims for interdisciplinary cooperation.

In addition to such specific depictions of systems biology, more general explanations appear as the following.

> Well, for me two aspects of systems biology are important: That is the analysis of interaction of system components. You know, biology and biotechnology have in earlier periods studied single biological components in detail. And all the new technologies, tools and methods enable the real interaction of components today. This means the dynamics of interaction among components, which means from the gene to proteins and metabolites. (Scientist G)
>
> German original: Also für mich sind zwei Punkte wichtig, äh was das Fach Systembiologie anbetrifft. Das ist zum einen die Untersuchung von Interaktionen von Systemkomponenten. Sie wissen ja, in der Biologie oder Biotechnologie hat man meistens früher hauptsächlich die einzelnen Komponenten im Detail untersucht und durch die die ganzen neuen Technologien, Werkzeuge und Methoden ist man heute in der Lage, die wirkliche Interaktion der Komponenten zu untersuchen kann. Das bezieht sich insbesondere auf die Dynamik der Interaktionen zwischen Komponenten, das heißt vom Gen über Proteine bis zu Metaboliten.

This quote outlines the interaction of system components as the important characteristic of systems biology. Subliminal reference to molecular biology ("studied single biological components in detail") in the development of systems biology is made and then combined with different methodical aspects and analytical scales. Hence, the storyline starts with an implicit theoretical reference to demarcate systems biology from preceding developments in biology to then using methodical and analytical scales to characterize in more detail what systems biology represents. The language used, however, remains quite general and only becomes less abstract on the level of analytical scale. In doing so, the quote holds many characteristics revealed in the analysis of scientific articles.

A more personal relation to systems biology becomes visible in the following quote. Here, theoretical and biographical elements appear and exhibit an emotional involvement of what systems biology means for the interviewee ("which I find very exciting and my biggest approach").

> I mean the word systems biology contains biology. This is my broader approach which is fascinating to me because it deals with biology. For me, it is mainly a summary of all kinds of sciences dealing with biology. I mean mathematics, chemistry and physics have always been part of biology. This approach has now been intensified and for me it rather is a description of biology with the help of these more systematized sciences. It is a great challenge for me because biology otherwise remains something which is difficult to grasp. (Scientist A)
>
> German original: Also ich meine, ja, das Wort Systembiologie beinhaltet natürlich Biologie, also das ist halt auch mein größter Zugang, das ich da am spannendsten finde, dass es sich halt mit der Biologie beschäftigt. Aber für mich ist es halt hauptsächlich eine Zusammenfassung von allen möglichen Naturwissenschaften, die sich mit Biologie beschäftigen. Ich meine Mathematik, Chemie, Physik stecken ja schon immer irgendwo in der Biologie drin. Aber jetzt wird es sozusagen noch, würde ich sagen, intensiviert, und für mich ist es eher 'ne Beschreibung der Biologie mit Hilfe dieser, wie soll ich sagen, länger festgelegten und mehr systematisierten Wissenschaften. [...] Und das ist für mich so die größte Herausforderung. Weil sonst ist Biologie immer irgendwas relativ schwer Greifbares.

In addition to the more personalized narrative elements, systems biology is rendered as an all-embracing discipline ("a summary of all kinds of sciences dealing with biology") that structurally benefits from more systematic approaches inherent in neighboring disciplines. As a result, a picture of systems biology as an interdisciplinary merger is given that restructures biology. This aspect is narratively expressed by a discipline-oriented storyline. In doing so, the quote represents an exception because its storyline is based on conceptualizing the theoretical interplay of systems biology and other disciplines.

Reflections on how to demarcate systems biology from biology emerge in the course of some interviews. The preceding excerpt is taken from an interview with a young scientist, however, the following quote from an interview with a senior scientist gives a sophisticated account on the different scales systems biology relies on and how it relates to biology in general.

> Well, there are different definitions and I think quite different conceptions of what systems biology should achieve. For me, it is a science that tries to reduce complex life processes as ageing or growth for example to the smallest units available. The magic word would in fact be: understand the molecular processes underlying complex physiological processes. This means that the smallest functional unit in the system would be the molecule. Other sciences analyze how molecules are constructed – Chemistry – and Physics goes further in searching for subatomic components. Biology ends for me with regard to systems biology on the level of molecules. And this is the challenge. Well, this distinguishes systems biology, I would say, from traditional biology. (Scientist C)
>
> German original: Also da gibt es verschiedene Definitionen und ich denke sehr verschiedene Auffassungen was Systembiologie eigentlich leisten sollte. Für mich ist das eine Wissenschaft, die versucht, komplexe Lebensvorgänge, wie zum Beispiel Altern oder Wachstum auf ihre, wenn man so will, letzten Endes kleinsten Einheiten zu reduzieren. Also auf molekulare Prozesse. Das Zauberwort wäre tatsächlich: Löse komplexe physiologische Zusammenhänge auf in die zugrunde liegenden molekularen Prozesse. Das heißt,

> die kleinste funktionelle Einheit in dem System wäre das Molekül. Andere Wissenschaften gehen noch weiter. Die gucken sich an, wie ein Molekül zusammengebaut wird – Chemie –, und die Physik geht noch weiter und sucht nach subatomaren Bausteinen. Die Biologie endet für mich in der Systembiologie beim Molekül. Und das ist die Herausforderung. Also das unterscheidet sie, würde ich sagen, von der traditionellen Biologie.

This account of what system biology is ostensibly exhibits a concept of molecular biology and gives a differentiated account about the different conceptual and theoretical levels. Narratively based on the concept of scale ("molecular processes underlying complex physiological processes and the functional unit in the system would be the molecule") and the disciplines informing systems biology ("chemistry and physics"), a molecular picture of systems biology is given that demarcates it from traditional biology but not from molecular biology.

Standard descriptions exist as well that might function as an orientation to outline what systems biology is. Thus, the following excerpt from an interview explicitly makes reference to the ERASYS BIO[4] definition.

> Well, in a nutshell, systems biology is the science that aims to understand the functions of biological systems as emerging characteristics resulting from the interaction of components of biological systems. And a definition that I endorse and like is the one provided by ERASYSBIO which depicts systems biology as an approach and not as a discipline. It is an approach with which one tries to describe the spatial and temporal interaction between components of the cell, between cells and cell populations and their interaction with the environment. (Scientist I)
>
> German original: Also man kann in Kurzform die Systembiologie als Wissenschaft beschreiben, die versucht, die Funktion biologischer Systeme als emergente Eigenschaften aus der Interaktion der Komponenten lebender Systeme heraus zu verstehen. Und eine Definition, die ich unterstütze, die ist die von ERASYS BIO. Die gefällt mir sehr gut. Sie beschreibt die Systembiologie als ein Ansatz, also nicht als eine Disziplin, sondern ein Ansatz, mit dem man versucht die räumliche und zeitliche Interaktion zwischen Komponenten in der Zelle, zwischen Zellen, Zellpopulationen und Zellen und ihrer Interaktion in der Umgebung zu beschreiben.

What we can see here is an interpretation of systems biology as an approach. In outlining this as the main characteristic, the interviewee alludes to methodical aspects and then addresses the level of scale by depicting the units of analysis ("components of the cell, between cells and cell populations, their interaction with the environment"). Albeit a storyline based on method and scale is used here to illustrate what systems biology is, one has to bear in mind that the quote explicitly relates to the influential ERASYS BIO definition, which expresses something like an institutionally negotiated consensus of what systems biology is.

In summary, the analysis carried out in this section revealed the sometimes varying narrative structure that defines what systems biology currently represents. Bearing in mind that statements from different points in time and via different methods were

---

[4] ERASYS BIO is a funding initiative consisting of a consortium of European funding bodies, ministries and project management agencies. Their mission is to carry out strategic collaboration in the funding of systems approaches to biological research. See ERASYSBIO 2013, http://www.erasysbio.net/. Accessed January 3, 2015.

gathered and analyzed for our analysis, four structural narrative elements were found that considerably contribute to establishing a semantic framework for defining systems biology. These sometimes opposing subconcepts, such as disciplines involved, scale of analysis, methods to be applied, and the kind of analysis, were sometimes differently combined and applied a more or less restricted set of lexical items. They developed what one might call a conceptual order or a communicative scaffold displaying slightly varying constructions of what systems biology is supposed to be. What became, furthermore, apparent in the storylines detected in both datasets is that the prevailing image of systems biology as an approach is mainly based on the two subconcepts of scale of analysis and methods to be applied. The scale of analysis especially was in many cases clearly depicted via lexical items—such as gene, protein, cell, cell populations—whereas reference to methods was made with the help of linguistic constructions such as mathematical modeling, computational approaches, and the like. Aspects dealing with the disciplines involved in the systems biology project and the kind of analysis appear to a minor extent and consequently do not exert a considerable impact on the conceptual framing of systems biology. However, all kinds of structures permeate the datasets and exhibit, especially in the interview data, a skillful way of balancing and reconciling the tensions between the sometimes opposing narratively structured conceptual cornerstones. Their reconciliation is still open and the conceptual void between them gives a picture about the different theoretical and methodological aspects of crucial importance and to be addressed by systems biology. In conclusion, the narratives in this section produced by system biologists gave insight into the conceptual structure underlying their notion of systems biology and provided a reflexive picture of it. Such synchronic aspects that relate to current definitions and conceptualizations of systems biology should, however, be complemented by imagined pasts of systems biology to characterize narrative anchoring points that define a start. This is done in the following section.

3.2 Imagined Pasts: Narratives of Bygone Times

The previous chapter provided an impressionistic outline of how current systems biology is narratively constructed. Such a synchronic perspective should be complemented by studying how the history of systems biology is constructed by its members to anchor its existence in the past. The aim of this section consequently does not consist in providing a full historical outline of systems biology mentioning its predecessors and outlining its conceptual history. This has partly been done by others[5], although a comprehensive and systematic historical overview is still lacking.

[5] See Brauckmann (2000), Drack (2009, 2013), Drack and Wolkenhauer (2011); Drack and Apfalter (2007), Drack et al. (2007), Pouvreau (2007), Pouvreau and Drack (2007) and Green and Wolkenhauer (2012).

There are, however, short pieces devoted to the history of systems biology that analyze its relation to developments and predecessors in the history of biology (see, e.g., Bothwell 2006) and trace systems approaches in biology back to the start of the last century or even beyond (see also Chap. 3). The remaining gap cannot be filled here as such an enterprise would go far beyond the scope of this section which is primarily devoted to the narrative construction of system biology's past as seen through scientist's eyes. Such an investigation has not been carried-out to date and appears to be a rewarding endeavor as it might to some extent disclose the strategies of a historical self-assurance because any process of building a disciplinary identity requires a diachronic anchoring from where its narrative can start. That history matters as a narrative starting point and chronological contextualization can occasionally be seen in scientific reviews and, though to a considerably lesser extent, in scientific articles on systems biology. Thus, strategies of diachronically contextualizing systems biology, as in the following two examples, appear now and then.

> This paper elucidates the scope of and issues in systems biology [...], an emerging discipline that attempts to understand organisms at the system level. Systems biology is both an old and new field in biology. It is an old field because system-level understanding has been proposed and tried in the past. It was perhaps originated by Norbert Wiener who proposed the concept of cybernetics and devised mathematical formulae for physiological systems nearly 40 years ago (Wiener 1965). A precursor to Wiener was the concept of homeostasis by Cannon (1933). The philosopher von Bertalanffy attempted to establish a general systems theory (von Bertallanfy 1968), but it was too abstract to be a serious scientific discipline. Concepts such as robustness and feedback control were already discussed at that time and extensively investigated. (Kitano 2002, 1)
>
> In order to explore the essence of Systems Biology – a notion that, in spite of its broad appeal, is still lacking a definition – it may be helpful to start by considering the meaning of each of the two words. "Biology" is easy to define: it is the science [...] that is concerned with living matter [...]. Although perhaps less well appreciated in the biological field, the term "system" is equally well defined, as "an entity that maintains its existence through the mutual interaction of its parts." Systems research, therefore, necessarily involves the combined application of "reductionist" and "integrationist" research techniques, to allow identification and detailed characterization of the parts, investigation of their interaction with one another and with their wider environment, and elucidation of how parts and interactions give rise to maintenance of the entity. (Kohl et al. 2010, 25)

Both quotes, although different in structure and content, situate systems biology at the end of historical development. Kitano (2002) aims at providing a full sketch of preceding conceptual developments by personalizing cybernetics, homeostasis, and general systems theory and personalizes with prominent scientists. Kohl et al (2010) start with a linguistic approach. Their etymological slant uses Greek letters rhetorically and aims at suggesting historical depth and narratively connects systems biology to it. In doing so, systems biology is anchored and positioned in intellectual history. The strategies informing both quotes, however, differ considerably as Kitano provides a chronological historical account with important names and metonyms whereas Kohl et al (2010) aim rather at giving a conceptual account. As already said, both quotes represent paradigmatic examples for supplying historical accounts that emerge in the scientific literature from time to time. Their function as points of orientation are, however, limited as many scientific articles do not reflect the history of systems biology. A good example for a historically limited perspective is the paper by Chuang et al. (2010, 721) who depict a decade of systems biology between 2000 and 2010 without any reference to its conceptual history or possible predecessors. These results provoke two questions. First, is there no interest in the discipline's history and in historical self-reflection among system biologists? Second, how do systems biologists anchor and relate their work historically? To analyze these questions, we asked systems biologist how they appraise the history of the discipline and what kind of conceptual predecessors or developments are important for them and their work. As a result, a more or less deeper historical consciousness emerged from the interviews which, at first glance, could be correlated with the experience of and position held by the interviewee. It was, hence, expected that older and more experienced scientists know more about the historical depth of the discipline. This hypothesis was mainly corroborated although a smaller group of younger scientists from different disciplines offered astonishing accounts of how their work and conceptual mindset was influenced by predecessors. Such an account could, for example, be seen in the following quote taken from an interview with a group leader.

> Ok, I would say that in the discipline there is not much historical awareness about its predecessors. Colleagues are more focused on their daily work and not so much on where the concepts or ideas they apply stem from. For me personally, the work of Monod has been influential and provides something like a red threat in what we do in our group. And institutionally, I would say that the funding from the German Ministry of Science was essential – but I think you should ask older colleagues involved in the process of developing this funding initiative. (Scientist K)
>
> German original: Ok, ich würde sagen, dass es in der Disziplin selber nicht wirklich ein historisches Bewusstsein über Vorläufer gibt. Die Kollegen sind mehr auf ihre tägliche Arbeit fokussiert und nicht wirklich auf die Frage woher die Konzepte kommen. Für mich selber waren die Arbeiten von Monod sehr einflussreich und die durchziehen wie ein roter Faden auch die Arbeit in unserer Gruppe. Und institutionell war, so würde ich sagen, war die Förderung durch das BMBF grundlegend – aber ich glaube, da sollten sie lieber einen Kollegen fragen, der in die Entwicklung dieser Förderlinie involviert war.

What is interesting about the quote is that the work of Jacques Monod, who discovered the genetic control of enzyme and virus synthesis, is mentioned here as influential. In fact, the name Monod stands as *pars pro toto* not only for the scientist's own work but also for the concepts and work carried out in the group leader's working environment. Although the conceptual connections between Monod and the interviewee are not depicted in detail, a diachronic as well as a synchronic connection is developed both bearing an influence today. This aspect appears in the use of the image "red threat" which literally connects past and present. Another important point appears in the sentence in which the immersion in daily work is addressed. Here, one reason is given why a lack of historical awareness still prevails in systems biology. In sum, the interview is based on a narrative that paradigmatically connects one influential scientist with the daily scientific work undertaken: Past and present are conceptually merged.

The quite differentiated theoretical flashback encountered in the previous quote could be contrasted with the following.

Well, this surely comes for example from oscillations which are a well-known phenomenon in physics. And the question arises whether mechanical oscillation should be compared with biochemical oscillation because oscillations play an important role in biology. Yes, and I got well into touch with this question and then I came into contact with statistical physics. And there was the development of non-linear dynamics which took up speed with the development of computing technologies. You cannot solve systems analytically and this was really the boom in this area which started with the development of high-powered computers. And I have experienced this during my studies. (Scientist E)

German original: Also, weil das kommt sicher daher, weil Oszillationen ein bekanntes Phänomen in der Physik sind. Und natürlich kommt dann die Frage, mechanische Oszillation mit biochemischer Oszillation zu vergleichen, weil Oszillationen eine wichtige Rolle in der Biologie spielen. Ja, und da bin ich mit in Berührung gekommen, und darüber hinaus kam ich auch in die statistische Physik. Da gab es ja dann die Entwicklung der nichtlineare Dynamik und die hat sich dann auch nachher rasant entwickelt mit der Entwicklung von Rechentechnik. Weil man die Systeme ja nicht analytisch lösen kann und das war der Boom, der in dem Gebiet mit der Entwicklung leistungsfähiger Computer einsetzte. Und das hab ich grad so in meinem Studium miterlebt.

What we witness here is a rather personal account of a variety of experiences as indicated in the final phrase ("And I have experienced this during my studies"). The logic of this narrative is thus not purely based on the cognitive reflection of predecessors and concepts, but rather on the personal immersion and the developments within a scientific field. Thus, the disciplinary developments ("oscillation," "statistical physics,","non-linear dynamics"), technological innovations ("computing technologies"), and personal experience ("And I have experienced this during my studies") are blended to build up a rather individualized historical narrative on systems biology.

Comparable aspects are also raised in the following quote in which—and here it differs from the previous one—a rather strategic procedure towards systems biology appears slightly peppered with historical references.

I decided at the end of the 1990s to concentrate in my studies on the systems theory – electrical engineering, control engineering, well systems theory – of biological systems. And I took a look around and saw that many books on biology dealt with the identification and characterization of molecules, genes and proteins, but not with the functioning of the cell. I thought that that these processes are dynamic processes that happen there, and exactly these dynamic processes are central elements in systems theory when I talk about systems theory. I always mean dynamic systems. I have also read some Bertalanffy and Rosen, and this was a motivation for me. I became aware that these functions of the cell and the behavior of cells are dynamic processes. (Scientist J)

German original: Ich habe Ende der 1990er beschlossen, mich mit meinem Studium der Systemtheorie – Elektrotechnik, Regelungstechnik, also Systems Theory – auf biologische Systeme zu konzentrieren. Und habe mich umgeschaut und gesehen, dass in Biologiebüchern sich überwiegend mit der Identifizierung und Charakterisierung von Molekülen, Genen und Proteinen beschäftigt hat, aber nicht mit der Funktion der Zelle. Und dann habe ich gedacht, das sind dynamische Prozesse, die dort stattfinden, und genau diese dynamischen Aspekte sind ja das zentrale Element der Systemtheorie, also wenn ich von Systemtheorie spreche, dann meine ich immer die Theorie dynamischer Systeme. Ich habe auch einiges von Bertalanffy und Rosen und so gelesen und das war so meine Motivation, dass ich gesehen habe, dass die [...] diese Funktionen der Zelle und das Verhalten der Zellen dynamische Prozesse sind.

The quote exhibits a complex narrative that provides a personal and strategic account about the motivation to work in the area of systems biology. The merging of the competences acquired and the conceptual transfer of this kind of knowledge about biology in the context of a biographical narrative represents an epistemic strategy. It is informed by concepts encountered in the works of predecessors that present a historic depth: the interplay of biographical narratives with historical predecessors ("Ludwig von Bertalanffy" and "Robert Rosen") can lead to shedding new light on old problems ("I became aware that these functions of the cell and the behavior of cells are dynamic processes"). The narrative thus combines biographical aspects ("I decided at the end of the 1990s to concentrate in my studies on systems theory—electrical engineering, control engineering, well systems theory—of biological systems") with an exploration of the field of research ("many books on biology dealt with the identification and characterization of molecules, genes and proteins but not with the functioning of the cell") and historical references ("I have also read some Bertalanffy and Rosen").

In addition to the more biographical narratives encountered in the data, also ahistorical or (positively expressed) pragmatic storylines exist such as the following:

> I don't mind because the word systems biology is not important. Yes, we get money for it and this is important. But I am more of an experimenter and systems biology came into view due to certain questions we asked and which we could not solve alone. We first needed cooperation partners who were able to model and then wanted cooperation partners who helped us to apply unbiased global methods. (Scientist L)
>
> German original: Das ist mir ziemlich egal, weil für mich das Wort Systembiologie nicht das Entscheidende. Natürlich gibt es dafür Geld und das ist schon entscheidend. Aber ich bin Experimentator und wir sind eigentlich zur Systembiologie dadurch gekommen, dass wir ganz bestimmte Fragestellungen hatten, die wir nicht alleine lösen konnten. Und da haben wir erstens Kooperationspartner gebraucht, die modellieren könne und zum andern haben wir Kooperationspartner gebraucht, die uns geholfen haben, globale ungebiaste Methoden anzuwenden.

Here any diachronic reference is intentionally avoided and pushed aside with reference to approach problems pragmatically and solve them. The narrative displays in detail the strategies and reasons why and for what purpose certain collaborators were needed. The important message is contained in the phrase "systems biology is just a word for me" which neglects any diachronic content tied to it and at the same time pretends to dissociate itself from it. The narrative thus categorically negates the relevance of any historical predecessors, underlines the relevance of a pragmatic approach with a self-reference ("I am more of an experimenter") and legitimates it with reference to emerging research questions ("systems biology came into views due to certain questions we asked").

Albeit a certain degree of ahistoricity could be tackled in the scientific literature on systems biology and in the expert interviews, some of them exhibit positions in which their own research is already historically framed. Such an account can be seen in the following interview excerpt.

> Well, I claim that my group does systems biology since 25 years, well for a long time before the notion had been established. We are doing the same things as we did 25 years ago, but we are doing them more efficiently because we have better computers today and know more.

> This means, however, that we have chosen from systems biology a certain part or approach which I would call mechanistic systems biology. We really try to understand systems on the previously outlined concept based on the modeling of molecular processes. Well, we start bottom-up and aim at predicting the complex behavior of the model. (Scientist O)
>
> German original: Also ich behaupte, meine Gruppe macht seit 25 Jahren Systembiologie, also lange Zeit, bevor der Begriff etabliert worden ist. Wir machen dieselben Dinge wie vor 25 Jahren, allerdings machen wir sie heute effektiver, weil wir bessere Computer haben, und natürlich auch mehr wissen inzwischen. Allerdings bedeutet das, dass wir von der Systembiologie uns einen ganz bestimmten Teil, also so ein bestimmtes Gebiet der Systembiologie gewählt haben, was ich nennen würde „mechanistische Systembiologie". Das heißt, wir versuchen wirklich, die Systeme in dem Sinne, wie ich es zuvor gesagt habe, zu verstehen äh auf der Basis der Modellierung von molekularen Prozessen. Also wir kommen von unten und wollen als Ergebnis des Modells das komplexe Verhalten vorhersagen.

The interesting aspect revealed in this interview excerpt consists in the claim that systems biology already existed in practice. This challenges the general idea of a linear model of history and refers to an unsynchronized and multifaceted historical development path that is narratively structured in the form of temporal back reference ("Well, I claim that my group does systems biology since 25 years, well for a long time before the notion had been established") synchronic technological improvement ("because we have better computers today") and growth of knowledge ("and know more"). This underlying narrative pattern informs to some extent a historically situated self-assurance and at the same time characterizes the interviewee and the group as predecessors.

To summarize, we have encountered a variety of paradigmatic quotes that display to some extent the oral history of systems biology as seen through systems biologists' eyes in Germany. The narrative analysis of the data disclosed biographical, experience-based, ahistorical and self-historicizing narrative structures to depict an individually experienced history of systems biology. This is most likely due to the fact that to date a canonical historical account of systems biology is still lacking. This fact might also explain why many of the historical reconstructions in the interviews start with biographical, experience-based or self-historicizing narratives with which the scientist's own disciplinary identity is situated and a starting point for the interview question is generated. Only the ahistorical narrative avoids any kind of historical contextualization and legitimates this with reference to practical needs and scientific problems that await solution. Although now and then reference is made to important predecessors such as Ludwig von Bertalanffy, Robert Rosen, or Jacques Monod, only seldom is their possible theoretical impact on the discipline elaborated upon and assessed. What gathers more attention though is the fact that technical innovation opened up new possibilities and paths of research. Thus, one could say that the linkages between past and present as encountered in the interviews conducted rather represent a rooting of systems biology in biographical and complementing narrative structures and clearly exhibit the need for a conceptual historiography of systems biology that anchors it in time. Such work would help to avoid its definitional fuzziness and develop a historically grounded understanding of its conceptual importance in current biology. As long as systems biology remains in what one might call an out-of-history spectrum as encountered in the ahistorical narrative, it will theoretically rest on wonky grounds. In sum, the narratives analyzed in this section could form a starting point for a more elaborate study about systems biology's past because there is no present and future without awareness about the past. Hence, only a secure foundation of the past can contribute to developing a consistent present and future because imagined pasts relate to envisioned futures. These are analyzed in the next section.

3.3 Imagined Futures: Narratives of a Systems Biological Future

The previous section dealt with the past as seen through system biologist's eyes. This diachronic perspective is now complemented by an analysis of the narratives systems biologists used to anticipate and portray the future of systems biology to provide a full picture of its imagined timeline. Such an investigation has to date seldom been undertaken and the only futures depicted can occasionally be found in the final sections of review papers and scientific articles where the authors often talk about the challenges still to be met by system biology. These future tasks comprise, generally speaking, a sheer unbelievable amount of problems to be addressed ranging from the maintenance of databases to methodological problems of systems research, or from the integration of data into certain models to the potential impact of systems biology on health care (Weston and Hood 2004). In view of this variety of important future topics it might turn out to be useless to investigate how system biology may develop in the future. We, however, would like to underline that the future does not represent a hidden or completely unforeseeable dimension or process. On the contrary, the future—or more precise different futures—are mainly imagined, developed and forged in the present, and to reveal their complexities might help us get a clearer picture of which way into the future should be taken. They could be conceived as narrative constructs that rhetorically develop anticipated futures and promises which are often socially shared and help to allocate symbolic, political and economic resources to realize these tasks (Brown and Michael 2003, Smart and Martin 2006; Hedgecoe and Martin 2003). Although written forms of futures in scientific papers, reviews or reports of all sorts often outline general aspects in a more strategic format, the narrated futures imagined and depicted by scientists on the ground of their daily work represent intermingling entities that merge strategic with everyday and conceptual aspects. Hence, the narratives about the future of systems biology which came up in the course of the interviews developed a sophisticated and imagined future as they were orally generated. They do not appear as strategically determined as visions encountered in scientific articles and reviews. Good examples for the latter are the following two quotes taken from journal articles:

> A systems biology approach to apply new high throughput technologies will be required to efficiently fulfill the promise of personalized molecular medicine. New clinical trial designs

> are needed to rapidly evaluate the hundreds of targeted therapeutics and potential biomarkers that are under preclinical evaluation. At our institution, we are implementing an effort designated Project T9 (10,000 therapies, 10,000 tests, and 10,000 treatments). We will characterize aberrations in 10,000 patients for all of the genes shown to be mutationally activated in 5 % of any major cancers. Information will be used to determine the frequency of mutation and commutation events, to correlate mutations with patient outcomes, and to direct patients to targeted therapy trials aiming the aberrations present in their tumors. This is designed to develop paradigms and approaches to bypass the challenges associated with wide spread implementation of biomarker-driven personalized molecular medicine. (Gonzalez-Angulo et al. 2010, 2782–2783)

Entitled "The Future of Personalized Medicine in Oncology: A Systems Approach," Gonzalez-Angulo et al. (2010) develop a narrative in their concluding remarks which is mainly based on a claim in the first sentence ("fulfill the promise of personalized molecular medicine") and future developments ("paradigms and approaches to bypass the challenges"). In between, requirements are outlined that pave the way towards the promise. The narrative thus applies a promise-requirements-future developments structure at the end of the article which uses strong lexical items such as paradigm and a metaphorical *source*-path-obstacle-goal schema ("paradigms and approaches to bypass the challenges associated") to depict the way into the future.

A comparable structure can be seen in the following quote taken from Hood et al. (2012, 10).

> There are general challenges to bringing P4 medicine to patients. First, we must invent the systems strategies, technologies, and analytical tools necessary to implement the P4 medicine vision in practice. Second, P4 medicine poses a host of challenges to society – ethics, privacy, confidentiality, legal, economic, regulatory, national policy, etc. These social challenges represent the greatest barrier to implementation of P4 medicine.

This excerpt taken from an article entitled "Revolutionizing Medicine in the Twenty-First Century Through Systems Approaches" is opening the concluding section by narratively setting the focus for the whole text that follows and on which it elaborates. Hence, what we encounter here is a structure that starts with an aim ("bringing P4 medicine to patients,, "implementation") and then elaborates again on an underlying source-path-*goal* schema[6] whose aim is outlined at the end. Hence, we encounter a step-by-step procedure that is narratively constructed into an aim: first requirements, second requirements, and future developments structure. The final sentence concludes with the source-path-obstacle-goal pattern by metaphorically reframing social challenges as obstacles ("social challenges represent the greatest barrier to implementation"). What we can take from these two excerpts is the fact that future-related narratives are based on a specific storyline that uses a varying promise/aim-requirements-future developments structure that relies on a metaphorically induced source-path-goal schema. In brief, future developments are framed with the experiential help of walking along a path.

[6] The source-path-goal schema represents a typical structure that almost always is inherent in narratives. Metaphorically used, it applies a start, a middle ground and an aim to any kind of development and helps to understand them in terms of a way.

Interestingly though, comparable narrative structures do not appear in the conducted interviews that could also be understood as oral narratives. Here, aspects are raised that are mainly concerned with the role systems biology might play in the future and a differentiated outline of a path into the future seldom comes into view. This aspect becomes visible in the following quote.

> Everyone will do it [systems biology] and no one will use the word. [...] I think every period in history has its typical methods. And we are simply lucky that these global and unbiased methods emerged that fast. [...] You can look at the cell through a variety of parameters and experiments and we get a more comprehensive picture for which colleagues in the past had to work hard and with extreme care and did not get to it. (Scientist B)
>
> German original: Da wird 's jeder machen und keiner mehr das Wort in den Mund nehmen. [...] Also ich denke, jede Zeit hat so ihre Methoden, wie sie rangeht. Und wir haben einfach das Glück, dass jetzt gerade rasant solche globalen ungebiasten Methoden entstanden. Man kann eben durch sich zig Parameter und Experimente in die Zelle schauen und ein viel umfassenderes Bild bekommen, das Kollegen früher selbst mit ganz viel Fleiß und Akribie nicht bekommen haben.

The excerpt offers a historical interpretation which refers to some sort of *Zeitgeist* by indicating that "every period in history has its typical methods." This aspect is also alluded to in the concluding sentence where a relation is developed between approaches now and then: Mainly based on imagery that relies on visibility ("look at the cell through a variety of parameters," "comprehensive picture") aspects of scientific insights due to different methods are depicted. On a narrative level, the quote is based on stating that systems biology becomes the standard ("everyone will do it [systems biology] and no one will use the word"). This statement is legitimated with a look into the past ("every period in history has its typical methods") and a comparison between now and then. On this foundation, systems biology is rhetorically depicted as the future in biology or biology as the future of systems biology.

Comparable aspects are also raised in the following quote in which the notion of systems biology is conceived of as being supplemented by life sciences. Systems biology is again portrayed here as a basic ingredient in the future research landscape of biology. The narrative is, however, differently structured as it depicts an image of biology that is rather driven by computer technology.

> Yes. Ok, if someone does not invent a new word I would not intentionally stick to the notion [systems biology]. I would imagine that we will do not call it systems biology in 15 years but perhaps biology again or life sciences. But it is clear that it is an integral part. Well, all will become immensely complex and technical, also in biology; and it is impossible to do experiments without a computer. Almost all of our facilities are connected to a computer. This simply does not enable the classical biology of 20–30 years ago. Consequently, I simply see an all – comprising systems biology. (Scientists A)
>
> German original: Ja. Gut, also wenn sich bis dahin nicht wieder jemand ein neues Wort dafür einfallen lässt, würde ich mich jetzt an dem Begriff nicht festhalten wollen. Also ich könnte mir vorstellen, dass es in 15 Jahren nicht mehr Systembiologie, sondern auch vielleicht wieder Biologie oder Lebenswissenschaften heißt. Aber es ist ganz klar, dass es nicht mehr wegzudenken ist. Also es wird alles so wahnsinnig komplex und technisiert, auch in der Biologie, man kann kein einziges Experiment mehr ohne Computer machen. Fast jedes unserer Geräte ist an einen Computer angeschlossen. Allein das ermöglicht ja

schon nicht mehr die klassische Art von Biologie, wie man sie vor 20 bis 30 Jahren sich hat vorstellen können. Von daher sehe ich einfach nur eine noch mehr vereinnahmende Systembiologie […].

A general flashback to the biology of 20–30 years ago sets a timeline and situates systems biology in the present and the vanishing of the notion in the future. The narrative structure consequently depicts it as the standard biology of the future by using a claim ("I would imagine that we will do not call it systems biology in 15 years but perhaps biology again or life sciences"), technology ("is impossible to do experiments without a computer") and historicizing ("this simply does not enable the classical biology of 20–30 years ago") sub-structure as legitimating elements.

A rather more differentiated account is given in the next quote where a methodological and conceptual perspective is outlined on the basis that systems biology has become an essential for biological research.

Well, I think, quite objectively, that it [systems biology] is an area of research which we would not like to miss any longer. It will be an established area of research regardless of the funding it will receive or not. It is the imperative to conduct research from an integrative and interdisciplinary point of view. And this is not a fashion trend; this develops from the objective state of science. I think we have reached a point where disciplines have to merge. And this is the reason why I think that systems biology will continue and grow. (Scientist P)

German original: Ja, also ich denke ganz objektiv das ist 'n Gebiet, das kann man nicht mehr missen. Also das wird sich weiter etablieren. Egal, ob es nun sehr oder nicht so stark gefördert wird. Es ist der Zwang, integrativ und interdisziplinär zu forschen. Und das ist keine Moderichtung, sondern das ist ganz objektiv aus dem Stand der Wissenschaften heraus entwickelt. Es ist ein Punkt erreicht, wo Gebiete sich vereinen müssen. Und insofern glaub ich, wird die Systembiologie weiter bestehen und wird sogar noch wachsen.

Systems biology is in the last sentence metaphorically depicted as a growing entity ("I think that systems biology will continue and grow"). This conclusion relates to the introductory sentence of the quote where the claim is made that systems approaches will become a basic ingredient of biological research. This claim is legitimated on conceptual grounds ("it is the imperative to conduct research from an integrative and interdisciplinary point of view") and an assessment ("I think we have reached a point where disciplines have to merge"). This conviction drives the interviewee to conclude that systems biology will become an established discipline. What is striking about the future depicted here is that it, as all examples, starts with a claim but then illustrates the need for and the reasons why systems biology will be standard on conceptual grounds. The narrative revolves around this anchoring point and exhibits a structure based on *claim-conceptual renovation* aspects without historicizing them. This structure is rhetorically underlined by using the metaphor of fashion trend to stabilize its relevance.

Comparable aspects are also raised in the following excerpt. They form the core of the quote in which the relevance of interdisciplinary approaches is highlighted.

Well, first of all, I think that it [systems biology] is a necessary and compelling approach because a kind of rethinking has started through and with it. It's a fact that the questions we ask with this approach make an interdisciplinary approach simple necessary. And this will be the case in 10–15 years. And yes, the practical problems I described will surely be

> hurdles for the one or the other to work in this area. [...] And as we said: from genes, proteins to cells, to networks, and then from networks to cell connections in the direction of tissues and higher forms of organization, that simply is natural progression. Technology keeps improving. (Scientist F)
>
> German original: Ja, also zunächst mal glaube ich ja, dass es äh ein zwingend notwendiger Ansatz ist, und ein notwendiges Umdenken stattgefunden hat. [...] Aber Tatsache ist, dass die Fragen, die wir uns stellen, diesen interdisziplinären Ansatz einfach zwingend notwendig machen. Und das wird auch noch in 10 bis 15 Jahren der Fall sein. [...] Und klar, die praktischen Probleme, die ich beschrieben habe, die werden sicherlich eine Hürde für den einen oder anderen sein, in dieses Gebiet zu gehen. [...] Und ansonsten alles andere, so wie wir auch gesagt haben, von Genen, Proteinen her auf Zellen, auf Netzwerke, dann von Netzwerken auf Zellverbindungen, jetzt geht man in Richtung Gewebe, höhere Organisation, das ist einfach ein natürlicher Fortschritt. Die Technologien werden immer besser. (Scientist F)

This concluding quote binds important aspects of systems biology together. Based on conceptual grounds, systems biology is portrayed as the outcome of a process of self-reflection ("a kind of rethinking has started through and with it"). Although it is conceived of as playing an important role in the future, this is legitimated on theoretical grounds ("and this will be the case in 10–15 years") and with a differentiated outline of the different scales it addresses and will address ("from genes, proteins to cells, to networks, and then from networks to cell connections in the direction of tissues and higher forms of organization"). The development is, however, depicted in terms of a natural progression but then technologically reframed ("Technology keeps improving"). We thus encounter a narrative that is structured as follows: *conceptual improvement-temporal justification-analytical scales*.

In sum, we have encountered in this section two different kinds of depicting system biology's future: written evidence that deploys narratives based on structures such as promise-requirements-future developments whereas the interview data on the other hand display structures that present systems biology as established on more or less differentiated conceptual narratives to legitimate its future existence. Although the first two paradigmatic examples alluded to methodological and conceptual innovation based on historical references, the last two quotes rather depicted a theoretically driven image of systems biology reflecting on the generation of new questions in the context of new technologies and the resulting need for conceptual innovation via interdisciplinarity. This aspect has either been represented metaphorically as a natural progression or as a point of development requiring the merging of disciplines to answer new and more complex questions in biology. Thus, one could say that the futures of systems biology are depicted on different narrative grounds and that newly emerging theoretical and methodological challenges in an interdisciplinary framework are used to legitimate system biology's existence. At the same time they also appear to anticipate the need for further theoretical and methodological integration and amendment among the disciplines currently involved. From this we conclude that—in order to avoid an nonreflective path into the future as expressed, for example, in the first quote—a deeper and perhaps reflexive immersion into the construction of current futures in systems biology might help us better understand not only its foundations, but also its conceptual consequences for future development. To summarize, the narrative foundations of imagined futures could contribute to developing a reflexive knowledge that avoids the traps of business as usual. It holds the potential to contribute to conceptually identifying anticipated and socially shared areas of possible innovation and facilitates discussion whether these make sense before symbolic, political and economic resources are allocated.

3.4 Conclusion: Between Present, Past, and Future

The previous sections aimed at depicting how systems biology is framed among systems biologists, what kind of past systems biologists ascribe to their discipline and what kind of future they estimate for it. To partly unravel and depict the conceptual grounding of these areas, a narrative approach was applied to written and oral (interview) data to disclose the linguistic structures informing these different areas. Bearing in mind that the two datasets were differently generated and exhibited differing and overlapping narrative structures, we found that the representation of systems biology is based on a varying narrative structure that develops a conceptual framework which frames systems biology mainly as an approach, and not as a discipline.[7] This image was conveyed with the help of narrative subconcepts of scale and methods to be applied although other narrative structures appeared and permeated the data. It, furthermore, became apparent that there is a conceptual tension among the conceptual cornerstones of scale, discipline, method, and analysis. Definitions of systems biology seem to oscillate between these cornerstones and their reconciliation requires further steps towards theoretical and methodological integration and amendment.

Contemporary definitions of systems biology remained to some extent fuzzy; the analysis of narratives depicting its history suggests that the same is true for the perceptions of the past. Especially the lack of a canonical historical account was acknowledged that could function as a reference point for situating and estimating innovations brought about by systems biology. Even though historical references to so-called predecessors appear now and then, we discussed the danger of a prevailing ahistoric narrative; its solution requires the development of a conceptual historiography, at least to some extent, clarification of definitional fuzziness, and provision of a historically grounded understanding of its conceptual importance in current biology. Only the knowledge about the past can help us better understand the present condition of systems biology and possibly contribute to substantiating its conceptual deficits.

The imagined futures encountered meandered between promise-requirements-future developments narrative structures in the written form, and a rather theoretically driven narrative of systems biology in the interviews. It became apparent that

[7] It is important to refer here to the differences of the two datasets. The written sources underlie a rigorous review process whereas the data taken from interviews rather represent ad hoc constructions between the interviewee and the interviewer. They display spontaneous and freer associations that provide a nonrestrictive insight into conceptualizations of system biology's present, past, and future.

the past and future potential of systems biology is seen in the questions it addresses, the interdisciplinary cooperation and integration it requires, and the development of conceptual and methodological innovation on which it relies. This narrative arrangement formed what we call a subliminal anticipative narrative which holds the potential to avoid a rather business as usual way of doing science. A reflexive immersion into the narrative construction of futures could help us better understand the conceptual shortcomings and unintended consequences of current developments in and of systems biology, and contribute to a negotiable but also shared vision of what systems biology will turn out to be before symbolic, political and economic resources are allocated. This might help to disclose implications underlying future visions and provide food for thought for a continuing discussion of systems biology's future directions.

In sum, we aimed in this chapter at analyzing how systems biology's present, past, and future as imagined by systems biologists. It became clear that the discipline might profit from such a reflexive endeavor to develop a disciplinary identity and to explore theoretical and methodological issues currently at stake. The narrative structures that developed pasts, presents and futures laid open the conceptual foundations they rely on and from which further investigations into present pasts or futures can start.

References

Aggarwal K, Lee K (2003) Functional genomics and proteomics as a foundation for systems biology. Brief Funct Genomic Proteomic 2(3):175–185
Anderem A (2005) Systems biology: its practice and challenges. Cell 121(4):511–513
Bell W, Mau J (1971) The sociology of the future: theory cases and annotated biography. Tussell Sage Foundation, New York, NY
Bork P, Serrano L (2005) Towards cellular systems in 4D. Cell 121(4):507–509
Borup M, Brown N, Konrad K, van Lente H (2006) The sociology of expectations in science and technology. Technol Anal Strateg Manag 18(3–4):285–298
Bothwell J (2006) The long past of systems biology. New Phytol 170(1):6–10
Brauckmann S (2000) Eine Theorie für Lebendes? Die synthetische Antwort Ludwig von Bertalanffys. PhD thesis, University of Münster
Brown N (2003) Hope against hype: accountability in biopasts, presents and futures. Sci Stud 16(2):3–21
Brown N, Michael M (2003) A sociology of expectations: retrospecting prospects and prospecting retrospects. Technol Anal Strateg Manag 18(3–4):3–18
Bruner J (1991) The narrative construction of reality. Crit Inq 18(1):1–21
Cannon, W. (1933): The wisdom of the body. New York: W. W. Norton
Charlton T, Myers L, Sharpless R (2007) History of oral history: foundations and methodology. AltaMira Press, Lanham, MD
Charmaz K (2006) Constructing grounded theory: a practical guide through qualitative analysis. Sage, London
Chong L, Ray B (2002) Introduction to the special issue: whole-istic biology. Science 295(5560):1661
Chuang H-CH, Hofree M, Ideker T (2010) A decade of systems biology. Annu Rev Cell Dev Biol 10(26):721–744

Cobley P (2001) Narrative. Routledge, London
Cannon W (1933) The wisdom of the body. New York: W. W. Norton
Corbin J, Strauss A (2008) Basics of qualitative research: grounded theory procedures and techniques. Sage, London
deSilvey C (2012) Making sense of transience: an anticipatory history. Cult Geogr 19(1):31–54
Drack M (2009) Ludwig von Bertalanffy's early system approach. Syst Res Behav Sci 26:563–572
Drack M (2013) Towards the system heuristics of Paul Weiss. Ann Hist Philos Biol 16:69–80
Drack M, Apfalter W (2007) Is Paul A. Weiss and Ludwig von Bertalanffy's system thinking still valid today? Syst Res Behav Sci 24:537–546
Drack M, Wolkenhauer O (2011) System approaches of Weiss and Bertalanffy and their relevance for systems biology today. Semin Cancer Biol 21:150–155
Drack M, Apfalter W, Pouvreau D (2007) On the making of a system theory of life: Paul A Weiss and Ludwig von Bertalanffy's conceptual connection. Q Rev Biol 82(4):349–373
ERASYS (2007) Systems biology in the European research area. Forschungszentreum Jülich
Gomes J (2009) All about systems biology. http://web.iitd.ac.in/~jgomes/Systems%20Biology%20Lecture%20-%20SCFBIO%20Sep%2018,%2009.pdf. Accessed 12 Mar 2014
Gonzalez-Angulo A, Hennessy B, Mills G (2010) The future of personalized medicine in oncology: a system biology approach. J Clin Oncol 28(16):2777–2783
Green S, Wolkenhauer O (2012) Integration in action. Embo Reports 13:769–771
Hedgecoe A, Martin P (2003) The drugs don't work: expectations and the shaping of pharmacogenetics. Soc Stud Sci 33(3):327–364
Hood L, Balling R, Auffray CH (2012) Revolutionizing medicine in the 21st century through systems approaches. Biotechnol J 7(8):1–14
Ideker T, Galitski T, Hood L (2001) A new approach to decoding life: systems biology. Annu Rev Genomics Hum Genet 2:343–372
Kirschner M (2005) The meaning of systems biology. Cell 121(4):503–504
Kitano H (2002) Systems biology: a brief overview. Science 295:1662–1664
Kohl P, Crampin E, Quinn T, Noble D (2010) Systems biology: an approach. Clin Pharmacol Ther 88:25–33
Koselleck R (2004) Future's past. On the semantics of historical time. New York: Columbia Univeristy Press
Leavy P (2011) Oral history: understanding qualitative research. Oxford University Press, Oxford
Marcum J (2008) Does systems biology represent a Kuhnian paradigm shift? New Phytol 179:587–589
Matsuoka Y, Gosh S, Kitano H (2009) Consistent design schematics for biological systems: standardization of representation in biological engineering. J R Soc Interface 6:393–404
McAdams D (1993) The stories we live by: personal myths and the making of the self. The Guildford Press, New York, NY
Midgley M (2003) Myths we live by. Routledge, London
Morel N, Holland J, van der Greef J, Marple E, Clish C, Loscalzo J, Naylor S (2004) Primer on medical genomics. Part XIV: introduction to systems biology—a new approach to understanding disease and treatment. Mayo Clin Proc 79(5):651–658
Nature Blogs (2007) The seven stones: what is systems biology. http://blogs.nature.com/sevenstones/2007/07/what_is_systems_biology_3.html. Accessed 14 Jan 2014
Otero J, Nielsen J (2009) Industrial systems biology. Biotechnol Bioeng 105(3):439–460
Pouvreau D (2007) La tragédie dialectique du concept totalité: Une biographie non officielle der Ludwig von Bertalanffy (1901–1972) d'après ses texts, sa correspondence et ses archives. Bertalanffy Centre for the Study of System Science, Vienna
Pouvreau D, Drack M (2007) On the history of Ludiwg Bertalanffy's "General Systems Theory" and on its relationships to cybernetics – Part 1: Elements on the origins and genesis of Ludwig Bertalanffy's "General Systemology. Int J Gen Syst" 36(3):281–337
Propp V (1928) Morphology of the folktale, vol 2. University of Texas Press, Austin

Rokem J, Lantz A, Nielsen J (2007) Systems biology of antibiotic production by microorganisms. Nat Prod Rep 24(6):1262–1287
Rosen R (1970) Dynamical systems theory in biology. Wiley Interscience, New York, NY
Selin C (2008) The sociology of the future: tracing stories of technology and time. Sociology Compass 2(6):1878–1895
Smart A, Martin P (2006) The promise of pharmacogenetics: assessing the prospects for disease and patient stratification. Stud Hist Philos Biol Biomed Sci 37:583–601
Thompson P (2000) The voice of the past: oral history. Oxford University Press, Oxford
Tomes N (1991) Oral history in the history of medicine. J Am Hist 78(2):6007–6617
Toolan M (2001) Narrative: a critical linguistic introduction. Taylor and Francis, London
von Bertallanfy L (1968) General system theory. Brazillier, New York, NY
von Uexküll J (1920) Theoretische Biologie. Verlag von Gebrüder Paetel, Berlin
Waddington C (1968) Towards a theoretical biology, vol 4. Edinburgh University Press, Edinburgh
Weiss P (1973) The science of life. Futura Publishing, New York, NY
Westerhoff H, Palsson B (2004) The evolution of molecular biology into systems biology. Nat Biotechnol 22(10):1249–1252
Weston A, Hood L (2004) Systems biology, proteomics, and the future of health care: toward predictive, preventative and personalized medicine. J Proteome Res 3(2):179–196
Wiener N (1965) Cybernetics or control and communication in the animal and the machine. Cambridge MA: MIT Press
Yow V (1994) Recording oral history: a practical guide for social scientists. Sage, London
Zewail A (2008) Physical biology: from atoms to medicine. Imperial College Press, London
Zhang J, Vemuri G, Nielsen J (2010) Systems biology of energy homeostasis in yeast. Curr Opin Microbiol 13(3):382–388

Chapter 4
Systems Biology, Information Technology, and Cancer Research

Imme Petersen, Regine Kollek, Anne Brüninghaus, and Martin Döring

Abstract The plethora and heterogeneity of data on biological processes have caused a change in approaches to data handling and processing by using high-performance computing and informatics. Infrastructures based on information and communication technology (ICT) have been developed to facilitate data management, access, and sharing of data on biological structures and processes on which systems biology is based. Although such infrastructures are essential for research and collaboration, they are often not regarded as being part of knowledge production. In contrast to this, we hypothesize that ICT infrastructures are not mere service facilities to support research activities, but enable, and restrict doing systems research at the same time. Based on a case study in systems cancer research, we argue that the understanding and modeling of biological systems is profoundly shaped by ICT and their underlying conceptualizations. In addition, individual scientists and research institutions cede the responsibilities of the activities associated with standardization, integration, and management of data. From the perspective of the sociological Actor-Network-Theory, our analysis also showed that such ICT infrastructures will become new powerful actors for knowledge production and within the knowledge-producing community of systems biology. Individual scientists and research institutions often neglect the challenges related to standardization, integration, and management of data that complicates and sometimes impedes innovation and translation of new developments into practice. This implies that standardization and integration in systems biology are as important as data generation.

Keywords Scientific practice • ICT infrastructure • Data management • Integration • Standardization • Case study

The development of systems biology has only been possible through the application of information and communication technology (ICT) to handle the large volume and variety of data about molecular processes in cells and organisms. Databases and

I. Petersen (✉) • R. Kollek • A. Brüninghaus • M. Döring
Research Centre for Biotechnology, Society and the Environment (FSP BIOGUM), University of Hamburg, Hamburg, Germany
e-mail: imme.petersen@uni-hamburg.de

M. Döring et al., *Contextualizing Systems Biology*,
DOI 10.1007/978-3-319-17106-7_4

infrastructures based on ICT were established for systems biological research to support systematization and integration of data on genomes, transcriptomes, and proteomes. To understand better the practices of data handling and data use in systems biology or, in a broader sense, to get an idea of systems biology in the making, we investigate how systems-oriented research is organized and performed in an ICT-based research environment. Even though ICT infrastructures are often considered as service facilities to ease research, we hypothesize that the understanding and modeling of biological systems is deeply shaped by ICT and their underlying design and conceptualization. Our second hypothesis is that the application of ICT enables and restricts doing systems research at the same time.

Therefore, we were looking for influences, dependencies, translations, and potentials that have occurred when systems biology has met ICT for doing research. As far as we are aware, little is known about the complexity and dynamics of the relationship between systems research and information technology. We therefore decided to use an exploratory research strategy known as case study. By carefully describing an individual case, the study aims at giving a deep insight into the subject of the chosen case and at drawing indications from it for further hypothesis creation on the subject. We empirically explored the challenges of organizing and doing systems-oriented research in an ICT environment in the applied field of systems medicine. In general, systems medicine is the making of systems biology in the making, as it implements systems biological approaches in medical concepts, research, and practice. The case under study was an international research project in which an integrated European ICT infrastructure was designed and developed in support of the systems-oriented research community in oncology.

After a short introduction into the case under study and the methods used in the empirical analysis, we present the empirical results in regard to the needs and demands of coordinating data collections in a computational environment (Sect. 4.1). In the second part, we trace the development of in silico oncology to understand better the underlying ideas regarding data analysis in systems-oriented research in cancer (Sect. 4.2). In focus is a knowledge discovery tool called the oncosimulator that was built in the course of the project. Based on the analysis of the results and outcomes of the case study, we retrace the current status of ICT in systems-oriented research and assess the potential of such an approach (Sect. 4.3). In the last section, we discuss what kind of function and role ICT infrastructures may in fact play in systems-oriented research in oncology in the future (Sect. 4.4).

4.1 Computers, Cancers, and Clinics: Coordinating Systems-Oriented Research in Oncology

High-throughput production of genomic data has confirmed that cancer can be regarded as a system: moving from single gene-based molecular investigation to molecular network research is seen as the most promising track to discover the mechanistic underpinning of cancer. It is assumed that cancer generally arises from disease-perturbed networks and that different network perturbations lead to different cancers (Lin et al. 2005). Additionally, it seems likely that network perturbations change with cancer progression. Such cancer-related, perturbed networks are currently under study to understand better how the cancer genome functions as a complex biological system in individual patients. The shift of interest from the identification of individual cancer components to the ways that these components interact has led to an explosion in the number of different types of data generated from the patients. The following data types are acquired.

- Molecular data types often referred to as Omics data (e.g., DNA variations, RNA, proteins, metabolites)
- Epigenetic data (e.g., DNA methylation patterns)
- Clinical data collected on clinical case report forms (e.g., symptoms, histology, administered treatment, treatment response)
- Imaging data (e.g., MRI, CT, ultrasound)
- Pathology data and other laboratory data

In order to understand better or even intervene more effectively in cancer and its development, these different data types are assembled. How these components involved in the processes under investigation might relate to and react with each other is systematically explored and formalized in mathematical models (e.g., Wolkenhauer and Green 2013). In this sense, data integration describes a dynamic process in which different data types and methods as well as disciplinary explanations and approaches are combined (O'Malley and Soyer 2012, 59).

The data types that are collected to build models of formalized relations and interactions are usually managed and stored in separate databanks in different geographical sites. Integration of data coming from such databanks raises, however, questions concerning data protection and audited data access.[1] Further problems have occurred: first, data coming from different sources vary significantly in terms of the contexts and circumstances they were gathered and stored (e.g., location, national law, history of data collection, context of application); second, efficient integration of different data sets is often also hampered by conflicting terminology and classification (e.g., Meier and Gehring 2008).

In regard to Omics data, many external technological platforms aim at solving this problem. They offer quality assessment of single Omics data types, which is needed to control the variations in the large number of biological and experimental parameters involved in data production. For example, to profile DNA methylation at least five different techniques exist that capture slightly different aspects of this process referring to the epigenetic modification or reprogramming of DNA (Rakyan et al. 2011). Accordingly, it is still a bioinformatic challenge to analyze a single type of Omics data because there are different approaches to raise data and different platforms to safeguard them.

The hurdles or even inability to share data and technologies is considered to be a bottleneck of the research process as it hampers efficient research collaborations

[1] For further information on data protection see, for example, Forgó et al. (2010).

(Swertz and Jansen 2007). Researchers even have problems integrating data from different technologies within a single laboratory. As a result, "clinicians or molecular biologists often find it hard to exploit each other's expertise due to the absence of a cooperative environment which enables the sharing of data, resources or tools for comparing results and experiments, and a uniform platform supporting the seamless integration and analysis of disease-related data at all levels" (Tsiknakis et al. 2006, 248).

One of the major challenges of systems medical research is, however, to translate laboratory findings into clinical treatments. It aims at finding ways to tailor therapy to the molecular characteristics of individual patients that can be used for precise diagnosis and as targets for novel treatments. Hence, the patient's biological profile arising from different molecular techniques has to be combined with clinical data relevant to the development, treatment, and prognosis of cancer (Abu-Asab et al. 2013). As of today, there is still no common methodology for integrating data types such as genomic and clinical data or proteomic and imaging data (Green and Wolkenhauer 2012). However, in translational research a few success stories already exist. For instance, gene expression profiling was used to classify tumors into subgroups representing distinct disease states that respond differently to currently used therapies. Finally, these experiments were successful in predicting the likelihood of chemotherapy benefit for patients with low-grade breast cancer and to quantify the likelihood of recurrence (Symmans et al. 2010; Lee et al. 2010; Desmedt et al. 2011).

4.1.1 Description of the Case Study and the Empirical Approach

After introducing systems-oriented research in cancer as the field of study, we now give a short description of the case under study and the empirical approach to study the scientific practice of systems biology.

4.1.1.1 Case Study

To elaborate on the relationship between systems-oriented research and information technology we chose an exploratory research strategy known as case study. The case—which can be persons, events, decisions, periods, projects, or institutions—is empirically inquired within its real-life context by using multiple sources and one or more methods (Thomas 2011). By intensely looking at an individual case, conclusions can only be drawn about that case in its specific context. From this it follows that emphasis is placed on exploration and description and not on testing generalizable hypotheses. However, case studies aim at giving deep insights into the subject of the chosen case and drawing indications from it to allow further elaboration and hypothesis creation on the subject (Yin 2009).

This is why we chose to analyze empirically the conception and realization of an ICT infrastructure in the domain of cancer research (see Box 4.1). The case under study is the research project "ACGT—Advancing Clinico-Genomic Trials on Cancer: Open Grid Services for Improving Medical Knowledge Discovery" funded by the 6th Framework Program of the European Commission under the Action Line "Integrated Biomedical Information for Better Health" (FP6/2004/IST-026996). From February 2006 until July 2010, 26 research groups from 12 European countries and Japan designed and developed an integrated technological platform in support of postgenomic, multicentric clinical trials targeting two major cancer diseases, namely, breast cancer and pediatric nephroblastoma, a childhood cancer of the kidneys (see Box 4.2).

Box 4.1: **Distinction Between and Meanings of ICT Infrastructure, Platform, Architecture, and Environment**

In this chapter, we use the terms *infrastructure, platform, architecture,* and *environment* to emphasize different meanings of information and communication technology (ICT) regarding the case study. In order to be as clear as possible in our terminology, we use the following definitions.

ICT **infrastructure:** The term describes the technology as a new phenomenon in science. The ACGT infrastructure is an example for the phenomenon having individual characteristics.

ICT **platform**: The term emphasizes the utilization of an ICT infrastructure by the users.

ICT **architecture:** The term refers to the technological components of a distinct infrastructure and how these components are interconnected in technological terms.

ICT **environment:** The term emphasizes the broader context of how ICT is integrated into science not focusing on an individual case.

Box 4.2: **Definitions of Genomics, Postgenomics, Molecular Technologies, and Clinical Trial**

Again, in order to be as clear as possible in our terminology, we use the following definitions.

Genomics/postgenomics: *Genomics* is part of genetics that applies recombinant DNA, DNA sequencing methods, and bioinformatics to sequence, assemble, and analyze the structure and the function of genes and genomes and studies their expression and regulation. *Postgenomics* refers to any fields of study that is only possible after the genome of an organism is published.

(continued)

Postgenomic research investigates which genes are active at particular times and under different environmental conditions (gene expression), for example, how genes are transcribed into messenger RNA, the chemical that carries the instructions for forming proteins (transcriptomics), how genes are expressed as proteins (proteomics), and how they influence the chemicals that control our cellular biochemistry and metabolism (metabolomics).

Molecular technologies: Molecular technologies are used to characterize, isolate, and manipulate the molecular components of cells and organisms. Thus, molecular technologies are the basic tools to study genetic information. Polymerase chain reaction (PCR) is the most basic molecular technology. It is used to produce multiple identical copies of DNA fragments. Other key technologies include DNA sequencing methods used to determine the order of the four bases (Adenine, Guanine, Cytosine, Thymine) in a strand of DNA and DNA microarrays that visualize the gene expression of an organism at a particular stage (expression profiling).

Clinical trial/study: Clinical trials are prospective biomedical or behavioral research studies on patients or volunteers that are designed to answer specific questions about biomedical or behavioral interventions, such as drugs, vaccines, biological products, surgery procedures, radiology procedures, devices, behavioral treatments, process-of-care changes, or preventive care. *Multicentric trials* are conducted in several locations (e.g., clinical centers). *Clinico-genomic trials* explicitly approach the integration of genomic data with clinical data in medical research.

The ultimate initial aims of the ACGT consortium were: (1) to design experiments for obtaining coherent and consistent medical and biological data, while avoiding various types of biases and errors; (2) to develop methods for integrating heterogeneous (e.g., genomic, medical) data sources, including the use of ontologies that facilitate mapping and information retrieval; (3) to develop methods for selection, checking, cleaning, and pre-processing of combined genomic-medical data; and (4) to incorporate collaborative approaches to data analysis, inasmuch as biomedical statisticians and data miners in genomics and medicine have been following different methodologies and dedicated, often proprietary, tools (ACGT 2005, 9).[2]

To address these different goals, different tools and services were developed and implemented in the ACGT platform. The main technical components of the ACGT infrastructure are the following (ACGT 2005, 11f).

[2] As two of the authors were part of the ACGT consortium, we had access to unpublished working papers, for example, ACGT (2005) Annex 1—Description of Work, Proposal.

- *Biomedical technology Grid layer*: The Grid technology comprises the basic technology for scheduling and brokering of resources. This layer-based architecture offers seamless mediation services for sharing data and data processing methods and tools, and advanced security tools according to European legal and ethical regulations.
- *Distributed data access and applications*. A set of compatible software services based on Web services provide uniform data access to distributed and heterogeneous data sources, that is, clinical data, eHealth records, microarrays, SNP data, and the like.
- *Ontologies and semantic mediation tools*. Formalized knowledge representations (ontologies) are required to facilitate semantic data integration as well as annotation and data analysis of large-scale biomedical data. The ACGT infrastructure offers a reference ontology for the field targeted by the ACGT project.
- *Clinical trial management system*. The clinical trial builder based on an ontology-driven software aims at helping to set up new clinico-genomic trials easily, to collect different types of data, and to put researchers in the position to perform cross-trial analysis.
- *Technologies and tools for in silico oncology*. The oncosimulator models tumor growth and therapy response in silico. The aim here is to create patient-specific computer simulation models of the biological activity of malignant tumors and normal tissues in order to optimize the therapeutic schemes and to contribute to the understanding of the disease at the molecular, cellular, and higher levels of complexity.
- *Grid-enabled application layer*. The data-mining Grid services support and improve complex knowledge discovery processes and knowledge extraction operations. Ways are sought for enabling easy integration and reuse of existing bioinformatics services into the ACGT infrastructure. Analytic services, for example, literature mining, visualization of results, and so on are also been implemented.
- *The integrated ACGT architecture*. Integration of applications requires a composite service that orchestrates other services in order to interoperate in a workflow. The ACGT workflow editor organizes and ensures that data formats are compatible and that semantic relationships between objects shared or transferred in workflows are clear. This creates any easy-to-use workflow environment so that researchers can design their discovery workflows.

The project was structured into a number of interrelated milestones representing the different components and tasks: (1) user requirement analysis and specifications; (2) technologies and services; (3) trust and security; (4) clinical trial implementation, verification, and demonstration; and (5) project management (ACGT 2005, 49f). The major milestones of the project representing the achievement of the objectives and goals were assigned to one of the 16 work packages. Work package (WP) 1 was responsible for the formation of the management structure of the project and the coordination of the project activities. WP 2 explored the user needs and requirements. WP 3 and WP 4 were assigned to build the Grid architecture; in WP 5 the data access services were constructed, and in WP 6 data mining and knowledge discovery tools were developed. The ontologies for mediation services and the clinical trials and applications were assigned to WP 7 and technologies and tools for in silico oncology were built in WP 8. The ethical and legal requirements have been addressed to WP 10 and the security requirements to WP 11, and a separate work package (WP 9) integrated the various tools to synthesize the integrated ACGT architecture. The evaluation and validation of the infrastructure was done in WP 13, the training aspects arising from clinico-genomic integration in WP 14 and the dissemination in WP 15. Last, WP 16 was responsible for the market investigation and the exploitation plan of ACGT (ACGT 2005, 121).

At the end of the ACGT research project, a first prototype of an ICT infrastructure was delivered that facilitated the integrated and secure access to heterogeneous data sources (e.g., distributed clinical trial databases). Furthermore, the ACGT prototype provided a range of reusable, open source analytical tools for the analysis of such integrated, multilevel clinico-genomic data. The data analyses were supported by discovery-driven analytical workflows. Finally, these research activities complied with existing ethical and legal regulation (ACGT 2005, 10; Bucur et al. 2011, 1120).

4.1.1.2 Empirical Approach

Our empirical approach was stimulated by and based upon the fact that two of the authors (Regine Kollek and Imme Petersen) participated in the ACGT project as consortium members from February 2006 until July 2010. Coordinating the ethical framework, we collaborated in particular with the partners being responsible for legal and security requirements in WP 10 and 11 as well as with the clinical partners in WP 12 and attended the consortium meetings taking place every 6 months. After the project ended in July 2010, we conducted guided interviews with selected project participants.

To select interview partners, we wanted first to identify the most relevant actors within the ACGT consortium keeping the consortium and the project running. We assumed that the ACGT consortium was a network of actors working together on the joint task of developing an integrative ICT infrastructure. Accordingly, the most relevant actors were the ones working most intensely in cooperation with other ACGT participants. As publications in such large research projects are usually based on joint work, we conducted a bibliometric analysis of the collaboration for internal publications (deliverables) and external publication (peer-reviewed articles, books, conference proceedings).[3]

As a first step, we identified the ACGT participants and counted their coauthorships for deliverables. Deliverables are usually created within the work packages; however, we wanted to cover all kinds of cooperation. Therefore, we counted the amount of cooperation per coauthor across work packages as well. We summed up

[3] Many of the deliverables are still available online. ACGT, http://acgt.ercim.eu/documents/public-deliverables.html. Accessed December 11, 2014.

the amount of coauthorships for individual actors and added to it the amount of cooperation across work packages. Resulting from that, we had a data cluster that comprised coauthorship and cooperation within the ACGT consortium. Additionally, we checked if all actors having designated tasks within the project (e.g., work package leadership, project management, quality control) were included in our sample. This was the case. The first most active 20 project participants were chosen; 18 scientists consented to an interview (13 computer scientists (IT), 4 biomedical researchers such as biologists, biostatisticians, and clinicians (BioMed), and 1 lawyer (LAW). They were queried using a theme-structured interview guideline. The interviews were focused on the participants' personal experiences as well as their judgments regarding the ACGT project. The interview guideline was structured into four sections addressing the following topics: (1) experiences of scientific and practical cooperation in the ACGT project (in particular, interdisciplinary negotiations); (2) experiences regarding the realization of the ACGT infrastructure (in particular, tasks and challenges); (3) judgments regarding the project outcome and science policy; and (4) judgments regarding the anticipated profit of ACGT for cancer research and systems biology.

The interviews were digitally recorded, anonymized, and literally transcribed. Interviews with German participants were conducted in German; if cited, these interview passages were translated afterwards into English. The empirical results are based on qualitative content analysis by using the software MAXQDA 11. First, the interviews were paraphrased and sequenced. Then, we created headings (categories) for individual statements and compiled topically similar statements. This resulted in main headings characterizing the topics that were jointly discussed in the interviews (Meuser and Nagel 1991). Below, the interview citations are characterized by the professional background of the interviewee. Heuristically, the citations are used to describe facts, circumstances, and situations in a narrative and comprehensive language and to prove personal statements and judgments.

In addition to the interview material, we analyzed the content of internal ACGT documents accessed via the ACGT intranet (e.g., descriptions of work, progress reports, newsletter volumes, reviews, meeting minutes, deliverables, and conference presentations) and publications that were published by the interview partners. Internal documents reveal original goals, project progression, self-representation, and evaluation by external reviewers, whereas publications offer more background information regarding the research being done in the ACGT project.

4.1.2 Needs and Demands for Coordinating Systems-Oriented Research in Oncology

In order to process and share data from heterogeneous data sources, ICT was applied to systems research and its fields of application. In the late 1990s and early 2000s, the first digital databases were established to support systematization and integration of data from research of a given domain, for example, a model organism or a disease, in a formalized manner (Leonelli and Ankeny 2012, 30). In cancer research, one of the most prominent examples of such databases is the cancer Biomedical Informatics Grid (caBIG) launched by a US government program.[4] In 2004, the National Institutes of Health started implementing an open-source ICT infrastructure for data sharing among organizations by developing software tools, data-sharing policies, and common standards and vocabularies to facilitate data sharing.[5] The systems-oriented research community in cancer and further globalized scientific communities have appreciated such solutions for capturing and sharing data and information, in particular if the established digital databases allow and promote cooperation with other databases, thus providing a platform for community building. "There are clear pragmatic advantages to this form of digital technology, which include ease of access on a global basis, the ability to maintaining and update them dynamically and at relatively low cost, the ability to simultaneously access various types of information for comparison, the open access to all interested researchers, and so on" (Leonelli and Ankeny 2012, 31). Systems-oriented research in oncology especially benefits from ICT support, as one of the interviewees of our case study pointed out:

> I think cancer research does have a higher degree of complexity. Because of the heterogeneity of cancer in general, because it is not a single disease, there are so many diseases that are completely molecularly and genetically different. And also the data that is being collected is very heterogeneous, very complex. It requires a lot of work to analyze it, to understand it, to use it. So I think that complexity was basically suggesting the need for [ICT] solutions. And also the fact that at some point people see the need to collaborate and to work together. So they do it on this multicentric trials, they want to share data, they want to analyze data together and so I think it is a very good domain, a very good model for trying to set up this multidisciplinary type of projects. (Interview [I] 2, IT)

In the context of clinical oncology, classical clinical trials are conducted in various phases, whereas each phase has a different objective involving different groups of patients, for example, medicalized group and randomized control group. Post-genomic clinical trials, however, cannot be conducted with current methodologies; they are characterized by the fact that molecular technologies are used—sometimes different kinds of molecular technologie in one clinical trial—and that large data sets are needed for statistical analysis. The need for statistically relevant data sets is challenging, because some biologically distinct patient groups may be represented only in small numbers. Hence, some predictions are limited inasmuch as robust classifiers that work well for predicting outcome in well-represented patient populations may, in fact, not work well in underrepresented groups. For example, one of the important prognostic biomarkers for breast cancer is the status of the estrogen receptor (ER). The majority of tumors examined within the good-prognosis group are ER-positive, but it is not yet clear whether the ER-negative group will not

---

[4] caBIG, http://cabig.cancer.gov. Accessed January 6, 2014.

[5] In 2011 a report on caBIG raised significant questions about effectiveness and oversight. As a consequence, its budget and scope were significantly trimmed. In May 2012, the National Cancer Informatics Program (NCIP) was created as caBIG's successor program. caBIG, http://cbiit.nci.nih.gov/ncip/about-ncip. Accessed February 21, 2014.

develop cancer later on and how well this predictor works in a larger cohort of ER-negative patients (van't Veer et al. 2002). In order to find this out, large patient cohorts representing the whole spectrum of a given cancer are needed for molecular profiling and statistical analysis. Such numbers are only accessible in multicentric international trials.

Today, the amount of data is increased by complex research designs with large patient cohorts. However, large-scale data sets are also due to the application of high-throughput technologies. For instance, one biological sample can be used to generate many kinds of data in parallel, such as genome sequence, patterns of gene expression, metabolite concentrations and fluxes, and so on. Furthermore, because of the continuous development of molecular technologies (e.g., next-generation sequencing methods), new types of data are continuously introduced into research. Generally, all the data are immediately integrated into data collections as databases and infrastructures are used to store as much data as possible (Ankeny and Leonelli 2011). In systems-oriented research in oncology, very heterogeneous data types such as Omics data, clinical data, imaging data, and pathology data are brought together and managed by ICT support.

> Well, I think that the initial idea – if you go back to the vision of the project which was influenced by similar visions and initiatives that have been already established in the United State- was the fact that due to the developments of the time, the new types and size of data generated through developments in the biological domain, molecular biology, and the new types of technology generating tons of new types of data—proteomics and other types of data—we realized that the key problem was the fact that there were a lot of inefficiencies in the pipeline of trying to bring together diverse types of data, diverse tools of technology that need to exist in analyzing those data, and support more efficient ways of distributing teams that by nature are involved in such interdisciplinary types of research, clinicians like molecular biology, computer scientists, etc. So there are a lot of inefficiencies in the process of semantics, harmonizing the data, and the representation of the data, developing shared tools. Therefore, to support this concept of open source sharing of tools, avoiding reinventing the wheel, etc., since every specific lab invests in developing their own computational solutions and platforms. And therefore, the vision and the ultimate objective was to establish an infrastructure that would attempt to move forward toward a more efficient way of managing data, sharing data, sharing tools, and enabling distributed collaborators to work as a virtual type of an organization supported by an information technology solution. (I7, IT)

The quotation indicates that the starting point to bring the ACGT project into being were not only new tasks and approaches in data management due to the high-throughput production of postgenomic data. The interviewee mentioned in the same breath that the consortium members have been aware of obstacles and inefficiencies in the analysis of such data as their variety and volume increase tremendously.

4.1.3 The Development of ICT Infrastructures

The vision and ultimate objective of the ACGT project was the development and realization of an ICT infrastructure that aims at meeting the needs and demands of postgenomic cancer research. Above all, ICT infrastructures have to facilitate integrated access to heterogeneous data sources, for example, data from international distributed clinical trials, in vitro experiments, scientific literature, or Omic platforms.

Integration as a prerequisite of sharing data is often conceived of as the major problem or at least as a major challenge in systems biology (e.g., O'Malley and Soyer 2012, 61). In the following section, we take a closer look at the activities of facilitating integration of different data types and databases to build up an effective ICT infrastructure supporting systems medical research. Five challenges have been identified that have to be addressed to create an ICT infrastructure and, hence, to make data integration and sharing possible: (1) challenges of data acquisition, (2) ethical–legal challenges, (3) challenges of interdisciplinarity, (4) technological challenges, and (5) challenges of standardization. As discussed in the conclusion of this section, the analysis of the challenges points at the track of how the integration of data and, looking at the broader context, the understanding of biological systems is deeply shaped by ICT and their underlying design and conceptualization.

4.1.3.1 Challenges of Data Acquisition

In order to investigate how disease-perturbed networks function, systems medicine needs different types of data generated from the patients (see Sect. 4.1). In addition to the different molecular data extracted from biomaterials, data are collected from clinical studies and health care. Clinical studies are usually done in clinical trials that are designed to answer specific questions about biomedical or behavioral interventions prospectively. Interventions to evaluate the effects on health outcomes, for example, include but are not restricted to drugs, cells and other biological products, surgical procedures, radiologic procedures, devices, behavioral treatments, process-of-care changes, or preventive care.[6]

Clinical data from health care are collected from patients who are diagnosed and treated in the clinic. Symptoms, diagnosis, histology, medications, treatment responses, and information related to lifestyle or environmental factors are stored on so-called patient records. According to the systems-oriented approach, it is furthermore claimed that the patient records ought to include the temporal dimension of biological parameters as well (Wolkenhauer et al. 2013, 503).

Hence, these three data types used in systems medical research are coming from different contexts and acquired for different purposes: molecular data are raised in laboratory research for investigating the molecular mechanisms of diseases; clinical study data are acquired in clinical trials for determining the safety and efficacy of interventions; and health care data are collected on a patient record for the purpose of reporting the patient's individual diagnosis and response to treatment regimes. The genesis of the data type has an impact on the validity and reliability of data as one of the laboratory researchers explained.

> If you are using data only for research, you are doing research on a large number of patients to get some general conclusions. If you want to use it as a diagnostic tool, you want a conclusion for one individual patient at a time. It has to be much more precise, much more

[6] See World Health Organization, http://www.who.int/ictrp/en/. Accessed November 30, 2014.

reproducible, much more standardized. This is most of the time done by companies, which is really done in standardized labs and not research labs. (I10, BioMed)

From the quotation it follows that treatment decisions in clinical care must rely on individualized data that have to be precise, reproducible, and standardized. Research data, on the other hand, must serve statistical calculations and don't need to have individualized validity. Patient records contain only data of an individual patient; however, they also have a problem concerning data reliability and validity as one interviewee with a clinical background highlighted in the interview.

I'd rely on the data in the clinical information systems even less than on all kinds of other data. I would only rely on data that were gathered in clinical studies that were structured prospectively. Because they were defined in these case report forms from the outset. But all the unsorted stuff that's in the clinical information systems isn't structured at all. There are pathology reports, surgery reports, but it's all simply text. And then I need good data mining tools to dig out the required information. And if it isn't entered in a medical report in a structured way, then something might be missing, well, because whoever dictates the report may dictate two different reports for two patients with the same diagnosis. (I18, BioMed)[7]

As the interviewee stressed, he only assumes from prospective clinical trials that patient data collections are structured on a reliable base. However, even these data sets do not maintain reliability as long as data are not systematically managed and curated.

[O]ur problem really is the data. If you check the volume of the tumor. Or if I have this progression [of the tumor], if I want to be able to do something meaningful with the data, then they have to be good enough that you can rely on them. The data have to, even if I have an oncology patient and I start to gather data, then I have to know two years after I've gathered them, is he alive, or has he died? And if I don't follow that up, in other words, if I don't do data curation, then after … what do I know, after a certain time, the data are useless. Well, because I'd get incorrect results again. That means, I have to involve the patient so that I get information about it, also about data curation etc., to get good results in the end. (I18, BioMed)[8]

---

[7] German original: "Auf die Daten, die in Klinik-Informationssystemen drin sind, würde ich mich noch weniger drauf verlassen als auf alle möglichen anderen Daten. Ich würde mich nur auf Daten verlassen, die in prospektiv strukturierten klinischen Studien erhoben worden sind. Weil die von vorn herein in diesen Case Report Forms definiert worden sind. Während das, was da lose in den Klinik-Informationssystemen drin steht, ist überhaupt nicht strukturiert. Da sind Pathologieberichte drin, OP-Berichte drin, aber das ist einfach Text. Und dann brauche ich gute Data Mining Tools, um daraus die nötigen Informationen rauszuholen. Und wenn ich das nicht strukturiert in einen Brief rein gebe, dann fehlt vielleicht auch etwas, ja, weil ich dann irgendwas diktiere oder derjenige als Arzt, der was diktiert, der diktiert bei dem einen Patienten, der die gleiche Diagnose hat wie der andere, zwei unterschiedliche Berichte." (I18, BioMed)

[8] German original: "[U]nser Problem sind wirklich die Daten. Wenn man guckt, wie groß ist das Tumorvolumen? Oder wenn ich diesen Verlauf [des Tumors] habe, damit ich richtig mit den Daten was anfangen kann, müssen die ja so sein, dass man sich auf die verlassen kann. Die Daten müssen, auch wenn ich einen onkologischen Patient habe und ich sammele jetzt Daten, dann muss ich aber nach zwei Jahren nach dem Sammeln immer noch wissen, lebt der oder ist der schon gestorben? Und wenn ich das nicht nachverfolge, also keine Data Curation mache, dann kann ich nach … was weiß ich, nach einer gewissen Zeit mit den Daten auch nichts mehr anfangen. Ja, weil ich wieder falsche Ergebnisse kriege. Das heißt, ich muss den Patienten mit einbinden, dass ich darüber Informationen, auch über Data Curation kriege und etc., um dann am Ende auch gute Ergebnisse zu bekommen." (I18, BioMed)

The reliability and validity of data have an interdisciplinary dimension as data originate in different contexts and are usually gathered by scientists of different disciplines. The following citation shows that epistemic and pragmatic differences impede the understanding of how reliable data can be acquired in another scientific context.

> The idea about which data clinicians can supply was completely abstruse in the beginning. […] If I have patient data on a disease, then the scientist expects, okay, I've got my case report files here, and I've defined all the important things. And then each patient has exactly the same data set. And that is precisely not the case. Take a disease, for example a brain tumor. And then the question arises … I'll give you two examples. The requirement I have to fulfill, I have to inform them now what the volume of the tumor is during the course of therapy. So I said, I can't do that. I can't tell you what the volume of the tumor is for glioblastoma. It's a highly malignant tumor. […] And then I showed [them] images of the same patient in a T1, in a T1 with a contrast agent, T2 … in four different modalities. The same plane through the skull. And all of them show the same diagnosis for the same tumor on the same day. But it looks different in each of the images. Because the different modalities show the tumor in different ways. And then there was the question: how big is the tumor in this image, how big in that image, how big in that one, then it turns out that it has four different sizes depending on which modality I use. And how big is it in reality? Nobody can answer that question. I could take the easy way out, a lot of people have done that, they deliver data, but they don't discuss the data with the people who want to use the data. So then I give them data where I say, the tumor had such and such a size at this point in time and such and such a size at that point in time. And then they do their calculations, and in the end everybody's surprised that the results are useless. Because the requirements for the data simply aren't right. And that's a very important point, that you convey the knowledge that you have in such a way that they understand: the data that we as clinicians can provide are biological data, they're completely different from mathematical data. (I18, BioMed)[9]

---

[9] German original: "Die Vorstellung, welche Daten die Kliniker liefern können, war am Anfang völlig abstrus. […] Wenn ich hingehe und Daten von Patienten zu einer Krankheit habe, dann erwartet der Scientist okay, ich habe hier meine Case Report Files und habe definiert, was es alles gibt. Und dann hat jeder Patient genau den gleichen Datensatz. Und das ist eben nicht der Fall. Ich habe eine Krankheit, wie zum Beispiel einen Hirntumor. Und dann kommt die Frage auf … Ich mache da zwei Beispiele. Die Anforderung an mich, ich habe denen jetzt mitzuteilen, wie das Tumorvolumen unter dem Verlauf der Therapie ist. Da habe ich gesagt, kann ich nicht. Ich kann euch nicht sagen, wie das Tumorvolumen beim Glioblastom ist. Also ein hoch maligner Tumor. […] Und dann habe ich von dem gleichen Patienten Bilder gezeigt in einer T1, in einer T1 mit Kontrast, T2 … in vier verschiedenen Modalitäten. Die gleiche Ebene durch den Schädel. Und alle zeigen am gleichen Tag bei Diagnose den gleichen Tumor. Aber der sieht in allen Bildern unterschiedlich aus. Weil die unterschiedlichen Modalitäten den Tumor unterschiedlich darstellen. Und dann war die Frage, wie groß ist der Tumor auf dem Bild, wie groß ist er auf dem Bild, wie groß ist er auf dem Bild, und dann kommt raus, der ist vier Mal unterschiedlich groß, je nachdem, welche Modalität ich benutze. Und wie groß ist er in der Realität? Das kann keiner beantworten. Ich könnte es mir einfach machen, das haben viele vorher gemacht, die liefern Daten, diskutieren die mit den Leuten aber nicht, die die Daten benutzen wollen. Dann liefere ich denen Daten, wo ich sage, der Tumor hat die und die Größe zu dem Zeitpunkt gehabt, hat die und die Größe zu dem Zeitpunkt gehabt. Und dann rechnen die, und am Ende wundert man sich, es kommt eigentlich nichts dabei heraus. Weil die Vorgaben der Daten einfach nicht stimmen. Und das ist ein ganz wesentlicher Punkt, dass man das Wissen, was man hat, auch so vermitteln kann, dass die verstehen, die Daten, die wir als Kliniker liefern können, sind biologische Daten, die sind völlig anders als mathematische Daten." (I18, BioMed)

To summarize, three different data types were identified in the interviews relevant to systems medical research. These three data types are acquired in different institutional and disciplinary settings (molecular research, clinical research, health care). Because of the different context of genesis, the data types used in systems medicine have different preconditions regarding data validity and fulfill different levels of reliability. Hence, according to data acquisition and management practices, the quality of data raised and stored for systems medical research can highly be variable. This is an important challenge when it comes to data integration: because of the interdisciplinary claim of systems medicine (and systems biology), a shared understanding of data reliability and validity in the respective acquisition context is obligatory before data can be integrated and related to each other.

4.1.3.2 Ethical–Legal Challenges

The meaning of data not only depends on the scientific preconditions of the specific data acquisition context but also on relevant ethical–legal requirements. From this point of view, in particular patient records are severely restricted and managed differently than research data. Normally, a patient records are only accessible to the patient's physician as well as a few other people directly involved in patient care who are obliged to maintain medical confidentiality. Clinical research conducted on patient data (and samples) is usually bound to an informed consent given for a single research purpose, that is, participating in a clinical trial, by the trial participant (Kollek 2009). Instead, molecular research data are normally authorized by a broad or blanket approval (Coebergh et al. 2006) to use the data for unlimited research purposes, such as the development or evaluation of new diagnostic tools, genetic studies, or biomarker identifications. As systems medical research aims at integrating clinical and molecular data, the different ethical–legal requirements of confidentiality and protection of data have to be served. The ethical–legal requirements for data management increase when it comes to postgenomic research in international research settings as one of the interviewees pointed out.

> For example, what we like to see is that most of the trials they are being done by different hospitals. These hospitals can be in the same country, but most of the time they are also in different countries and so you have to make sure that legally everything is okay, in the framework for the trail across the countries, because different countries can have different laws. Then you have to make sure that everything is also legally and ethically fine for the sampling and the shipping of the samples, where the analyses are being done. (I10, BioMed)

To address the ethical–legal requirements for trans-European research projects, the ACGT infrastructure provides a data protection framework, which was designed as a safety net consisting of three pillars (Forgó et al. 2010, 102). The first pillar was the development of a network of trust within the project. A legal body taking over the responsibility of the data controller (data protection authority), the involvement of an internal security authority (trusted third party), the conclusion of legally binding contracts and, finally, the ICT security tool called the Custodix Anonymisation Tool (CAT) ensured that the data would be estimated as de facto anonymous data within the research network. The second pillar ensured patient involvement. This was achieved by obligatory patient information and the requirement of informed consent as well as the establishment of a central contact point for all participating patients. The third pillar was the identification of provisions that allow the processing of personal data for research purposes. This was done on the basis of a thorough analysis of different national legislations. Taking the three pillars together, scientists can use the ACGT platform knowing that the use of data is ensured by the data protection framework.

> The goal of the data protection framework is to come to an environment where whenever you add new applications, new tools, you as an end user, as a partner, or as a patient would know that by default it would be in compliance with all laws and it would stick to the same ethics code that governs our platform. So basically you would know that it would be safe. Compliance by default that is what we want to reach. So if you plug in, and you are allowed to plug in, then everyone around you knows it must be okay. That is the achievement that we wanted to reach. (I11, IT)

In conclusion, data are scientifically as well as ethically–legally embedded into research traditions and international standards and guidelines. The ACGT consortium was very aware that different ethical–legal requirements of molecular research, clinical research, and health care have to be served to gather data from patients in the different settings. The ethical–legal framework, translated into legal contracts between the different stakeholders participating in clinico-genomic research on cancer, was fully developed after the completion of the ACGT project and is currently used in ACGT follow-up projects.

4.1.3.3 Challenges of Interdisciplinarity

Systems biology and, even more, systems medicine is a genuine interdisciplinary field targeting the modeling, understanding, and finally manipulation of living systems. From the very beginning, proponents of the new systems approach stress that systems biology must and will be able to display a holistic view on processes of life explaining how cells, tissue, and organisms interact from the workings on the molecular level (see Sect. 2.5). Hence, systems biology brings together scientists from a variety of disciplines such as mathematics, computer sciences, medicine, and biology. The interdisciplinary claim of systems-oriented approaches is, however, another prominent challenge in the development of systems biology and systems medicine as different scientific research logics, theories, methods, practices, and discourses come into play and interact. In the ACGT project, computer scientists worked together with biologists, clinicians, and experts from law and ethics. Asked for interdisciplinary problems in the interviews, the scientists often referred to misunderstandings in the communication across disciplines.

> But I think what we have to say is that, for me honestly, it took me six months to one year before all the partners could communicate a bit in their projects, because the backgrounds were so different that it takes some time to understand the vocabulary and the backgrounds of everyone before you can move together. (I10, BioMed)

> Well, it is difficult to talk with each other across different expertise fields. Because people use different technologies, people have very different objectives and so what you need in

> order to be able to communicate is a form of respect. And it is often difficult to create respect across different scientific fields or technological fields. You know, because the one feels superior to the other. That is something that you see a lot. (I11, IT)

The interviewees stressed that interdisciplinary communication is built on the acknowledgment of different backgrounds and the comprehension of disciplinary vocabulary as well as respect to listen to other disciplines. Several interviewees mentioned that it finally took up to a year to bridge the language issue and to understand and work with reference to each other.

> For me, the first year was the decisive year in which people had to learn to agree on a language so that everybody understood what was meant. And that's also something that reflects all the semantic integration in the project. If I want to use the data, then I have to generate the data in such a way that they can really be combined with one another in a simple way. And for me, in the beginning, in the first year, that was pretty—how should I say—where I thought, I don't understand a thing here. (I18, BioMed)[10]

Another interviewee explained interdisciplinary differences by referring to different disciplinary modes of thoughts. His example was the ACGT Master Ontology that was built for clinicians but did not, in his view, stand a chance in the clinic because of the differing mindsets of the ICT experts who had created the ontology and the clinicians who wanted to use it in daily work.[11]

> For a clinician, the ontology is cumbersome to use. This thing hardly has a chance of becoming widely used in clinical work, because the way of thinking of a person who develops an ontology is entirely different from how a clinician thinks. That's why in ACGT, there was already the idea to develop a tool that represents the ontology in a way that the clinician can understand it. The ontology is structured like a tree. And the clinician's thinking may also be like a tree, but he thinks in a way where the patient comes first, he wants to have the diagnosis first of all, then there are diagnostic measures, then there are therapeutic measures and so on and so forth. That means, this tree is represented in a completely different way than how the ontologist represents such a tree. And that's a clinical view of an ontology. And it's absolutely difficult to develop it. ACGT didn't make a lot of progress there. (I18, BioMed)[12]

---

[10] German original: "Das erste Jahr war für mich das entscheidende Jahr, in dem die Leute lernen mussten sich auf eine Sprache zu verständigen, so dass jeder verstand, was damit gemeint ist. Und das ist auch etwas, was ja die ganze semantische Integration in dem Projekt widerspiegelt. Wenn ich die Daten verwenden will, muss ich auch die Daten so machen, dass die sich untereinander wirklich einfach verbinden lassen. Und das war für mich so am Anfang, im ersten Jahr, ziemlich—wie soll ich sagen—wo ich dachte, ich verstehe hier überhaupt nichts." (I18, BioMed)

[11] The ontology formally represents knowledge as a set of concepts within a domain (such as a disease), and the relationships between those concepts (see the following Sect. 4.1.3.4).

[12] German original: "Die Ontologie ist für die Benutzung für einen Kliniker schwerfällig. Das ist ein Ding, das kaum eine Chance hat, in die Klinik einzuziehen, weil die Denkweise eines Menschen, der eine Ontologie entwickelt, völlig anders ist als ein Kliniker denkt. Deswegen war in ACGT schon die Vorstellung, man muss hier ein Tool entwickeln, was die Ontologie abbildet in einer Art und Weise, dass der Kliniker das versteht. Die Ontologie ist ja so baumförmig aufgebaut. Und der Kliniker denkt vielleicht auch baumförmig, aber der denkt in einer Weise, wo am Anfang der Patient kommt, der will die Diagnose erst mal haben, dann gibt es für ihn diagnostische Maßnahmen, dann gibt es therapeutische Maßnahmen und so weiter, und so fort. Das heißt, dieser Baum ist völlig anders abgebildet als der Ontologe so einen Baum abbildet. Und das ist ein Clinical View auf eine Ontologie. Und die zu entwickeln, ist absolut schwierig. Da ist ACGT nicht sehr viel weiter gekommen." (I18, BioMed)

Taking the quotations together, it becomes evident that interdisciplinary collaboration is always challenging, even if the scientists already have experience in other interdisciplinary settings. The ACGT interviewees described forthrightly that it took them the first year of the project to align terminology, mindsets, concepts, and expectations before the joint research even began. Maybe for that reason, they stress that progress and success of the ACGT project was mainly grounded in overcoming the interdisciplinary challenges and initial misunderstandings between the disciplines involved. However, the ontology was given as an example that in particular, differences affecting the mode of thought and research logic sometimes persist. This example shows that ICT as applied science in the interdisciplinary setting sometimes restricts the articulation of theories and the development of a research logic specific for systems biology.

4.1.3.4 Technological Challenges

One of the most prominent challenges frequently brought up in our interviews was technological problems regarding storing, integrating, and accessing high-throughput data. Classical approaches to solve these problems focus on syntactic interoperability which means that two or more databases have to be capable of communicating and exchanging data (Sujanski 2001). Technically, this requires a software component called a parser that analyzes input data to build the underlying data structure. The structural representation of the input, often described as an abstract syntax tree or other hierarchical structure, facilitates that different data and message formats (e.g., data-exchange protocols, programming languages) can be interconnected to an application programming interface called the data abstraction layer. Finally, the data abstraction layer is able to unify the communication between a computer application and databases by representing the data structures in a unified data and message format such as a programming language, for example, the eXtensible Markup Language (XML) or Structured Query Language (SQL).

Another critical feature to create interoperability between different data sources refers to the meaningful and accurate interpretation of the information exchanged. Here, the absence of shared terminology is one of the basic obstacles to enable communication and sharing of data (Tsiknakis et al. 2006, 248; Burgoon 2007, 404). Semantic uncertainties often refer to conflicting terminologies and classifications, or in other words, to missing agreements on terms and concepts. A basic tool to homogenize terminology and to build semantic interconnections is the ontology (Rubin et al. 2008). The ontology formalizes the meaning of terms through a set of assertions and rules that are collectively known as description logics. The ontology is concerned with what concepts are contained within the field, what information is required for each concept to have existence, and how different concepts are related to each other. Therefore, it depicts concepts within a domain (such as a disease), and the relationships between those concepts.[13] There is no need to attach any language

[13] Philosophically, ontology is a part of metaphysics dealing with questions concerning the nature of being, becoming, existence, or reality and their relations (e.g., Burkhardt and Smith 1991).

term to the classes as the ontology can be built in a language-neutral way. However, it can be done, inasmuch as naming the classes fosters the ontology's transparency to the users. For them, ontologies offer a structured knowledge repository which is used to describe the domain and can be used to reason about the entities within that domain. Bio-ontologies are already acknowledged as a relevant method for database integration in systems biology (Wierling et al. 2007). The Gene Ontology, for example, has been continuously developed since the late 1990s to classify, exchange, and compare data about gene products of a wide variety of species (Leonelli et al. 2011).

The ACGT consortium did not adapt to an already existing ontology, but decided to create an ontology that met the specific needs of the ACGT project.

> We studied and came up with solutions to […] achieving this semantic data integration based on some specific and strange assumptions, which are the following two: that data reside and are under the control and responsibility of the data producing entity, so that data belong to the lab and the group and the individuals that are responsible for producing them. And the second hypothesis we build our solution upon is that this integration is achieved through the use of a shared conceptualization of what we call the master ontology and there are appropriate techniques for utilizing this master ontology in achieving integration at the level of data. (I7, IT)

The ACGT Master Ontology *(*ACGT MO) was hand-tailored for the use in postgenomic cancer research. In particular, it structures and describes the concepts that are important in the domain of postgenomic clinical trials on nephroblastoma and breast cancer (Brochhausen et al. 2011). As the data stay in the original databases, a translation or mapping of the data is necessary to interlink the data to the ACGT platform. "We needed a mapping that says, well, within the project this is how we refer to a patient, this is how we refer to a microarray dataset, this is how we refer to … So there was an agreed list of terms and then the semantic mediator has to do the mapping; so it says, well, this entry here it actually maps to this term in a global dictionary." (I1, IT)

Within ACGT, the semantic mediator is a software tool that harmonizes data contents to make heterogeneous data acquirable to the components of the ICT system, "So that data are more than just bits and bytes, so that the other parts of the system can understand them" (I1, IT). Technically, the semantic mediator systematically coordinates data from different data sources by performing query translations from the ACGT MO to the local databases.

> Being able to computationally—because an individual will not look into a contextual description of what this algorithm does etc.—being able to computationally assess the capabilities of a specific tool, a specific service, you need to describe these capabilities: inputs, outputs, types definitions, etc. of data in a way that makes sense to a computing system. That is the essence of metadata. So in a sense rather than having the producer of this tool describing in a page or half a page verbally through text, that is algorithm. It models a specific function etc. and requires a set of input. You need to do that through elements that describe this capabilities and requirements. That is what we call metadata. (I7, IT)

Basically, the metadata help to categorize the data coming from different data sources to map or define the data for further data processing. Hence, each data type has a profile defining how the data have to be treated. Once the profile is in place, all data of the same type are processed in the same way. In this respect, the generality of the ACGT MO causes problems, because it intends to represent the whole domain of cancer, whereas the databases are normally developed with a specific goal in mind. As a consequence, when trying to explore a database by formulating queries using the ACGT MO as a guide, it is likely that the query term cannot be found in the specific database (Bucur et al. 2011, 1124).[14]

Within the ACGT MO, the user can search for terms or concepts as in a dictionary. In addition, the ontology viewer visualizes the interrelations between the concepts as a tree-like structure of the ACGT MO. In this context, it is important to notice that many existing ontologies focus on the classes or categories of the entities in a given domain. These ontologies might give a hierarchy of those entities via the basic taxonomical relation, the *is_a* relation. But only the inclusion of additional semantic relations between classes, for example, *x is part of y*, *z is adjacent to u*, *a is prior to b*, can lead to a comprehensive representation of the phenomena (Brochhausen and Blobel 2011). Taking this into consideration, users of the ACGT MO can create their own semantic tree by setting different kinds of semantic relations that are useful for their scientific observations (Brochhausen et al. 2011).

The ACGT MO finally contained more than 1,600 classes and nearly 300 properties. The ACGT consortium applied it to the Open Biomedical Ontology Foundry, which is an open source initiative to create a suite of orthogonal interoperable reference ontologies in the biomedical domain.[15] However, the ACGT MO didn't succeed in going through the quality assurance of the OBO Foundry until the end of the project, which was the last step before actually becoming an agreed ontological element of the OBO Foundry in the domain of cancer.

To conclude, the development of the ACGT MO itself was a complex and challenging task. Unsurprisingly, one of the interviewees made very clear that semantic interoperability has a much higher level of complexity than the syntactic interoperability: "The semantics is more difficult than the syntax, because understanding the syntax of something or agreeing on the syntax doesn't guarantee that you know the meaning. And mostly the meaning is harder to agree upon." (I2, IT) The ACGT consortium tried to tackle the considerable challenges such as the specification for the domain of cancer or the mapping of flexible interrelations between the different terms and concepts. However, its incomplete process of application to OBO Foundry indicates that semantic standardization is a long-lasting endeavor that needs further developmental steps with regard to approval and sustainability within research communities. Even if foundational semantic work, in particular in the domain of metadata definitions, was done in the course of ACGT, from today's perspective it seems too ambitious to set up an ontology from scratch to be used in clinical practice in time-limited research projects.

[14] ACGT MO is written in the Web ontology language OWL-DL. This ensures that all conclusions drawn from the ontology are guaranteed to be computable. For example, OWL-DL allows automatic reasoning (e.g., for consistency checking).

[15] OBO Foundry, www.obofoundry.org/. Accessed September 19, 2014.

4.1.3.5 Challenges of Standardization

The previous section showed that the ACGT consortium tackled syntactic and semantic integration of data and tools technologically. This approach usually includes the attempt to standardize such technologically driven integration processes. With regard to ICT infrastructures, standards ensure the line of communication between data and tools as well as tools and users (Hanseth et al. 1996, 410). However, the molecular technologies used to generate data for systems biological research are challenging the communication processes as these technologies are fast-evolving and, hence, may alter data that are supposed to be stored in standardized databases or infrastructures.

> And I think that most of the technologies are so immature and fast-evolving, therefore, it's practically impossible to follow this kind of development. So there are so many new data, so many new technologies everywhere. They are appearing on the landscape, there are so many of them disappearing from the labs or from the practice. So in some sense this kind of postgenomic research is hard to harmonize. (I6, BioMed)

Cutting-edge technologies keep the daily working processes in motion, in particular as reliable standards are often missing. As a result, the researchers are confronted with an almost too confusing amount of different technologies and standards referring to these technologies.

> As I already said there are lots of standards that evolve. There are standards at the lowest level of the IT, for example, Web services, the whole network exchange, how do computers exchange messages. One of the things of ACGT was the distributed system, so there were services and resources distributed all across Europe. We also had the Grid infrastructure and the Grid services set their own standards. They are still evolving and there are on top of that the genetic standards of how you express sequencing information and the clinical standards and the query language standards. If you look at all the standards that were evolved, you tend to get more and more standards that are not relevant. (I1, IT)

However, many scholars refer to the urgent need for standards to describe, format, submit, and exchange data (e.g., Green and Wolkenhauer 2012, 769). The short and insecure innovation cycles of high-throughput technologies and the interdisciplinary approach in systems medicine increase the demand for standards that are reliable and accepted across the community of users. But the technological and interdisciplinary innovations trigger multiple standard operating procedures at the same time (Auffray et al. 2009, 2). Thus, a growing number of unnecessary overlaps and duplications of standardization procedures evolves (Field et al. 2009, 234), in particular when ICT-based standards are involved as the following quotation shows.

> There is almost a standard for anything, even worse. There are more standards for the same thing in many cases. So the problem is that if you are writing programs and you want to conform to a standard, you have to understand the way that you can apply this to your own software. And so like I said, it can mean that you download a piece of software from somewhere that conforms to the standard and you interface it. But then you have to understand this API that allows you to use it. That is basically the story that I was just telling. So for many people who do not have this understanding, it is just a matter of being pragmatic. You can either spend, let's say, a week of time to try and understand how to interface something that is not yours into your own software or you can just create something for yourself in an

hour. And so everybody chooses the last option unless you are a computer scientist. If you are a computer scientist it takes you one hour to conform to the standard. (I3, IT)

Of course, anyone can claim to develop a new standard but standards necessarily need approval as one of the interviewees explained in relation to formal and de facto standard-setting procedures. Top-down standardization is initiated by standard development organizations (e.g., ISO, SEN, HL7). They are usually entitled to develop formal standards for a specific setting. The other process is the bottom-up approach, where user communities or industry trigger de facto standardization.

For me, DICOM is the indicative: the digital imaging standard, which was an effort through the American Society of Radiologists and NEMA, the National Engineering Manufacturing Association. So the industry and the users jointly developed a standard that was pushed to become a universal standard. That is very difficult to happen actually in the context of a three or even four year EU funded project. You need for exploitation, for exploiting the results of the research project, you need structures that will leave after the end of the project. (I7, IT)

As the quotation indicates, neither formal nor de facto standards were developed in the course of the ACGT project. Missing sustainable structures for exploitation of research results are one aspect of hesitant commitment to formal standard-setting procedures, in particular in time-limited research projects. Another aspect is the dynamics of the research field and the individual interests of researchers, as another interviewee pointed out.

In a fast evolving field of research, the problem is that standardization is causing delays. In order to be compatible with other, in order to maintain this kind of compatibility, you have to slow down and put some effort toward this kind of end. However, it appears that the forefront, the people who are really on the edge of developments, neglects standardization and then move on. In that sense, whenever standardization has to be done, it has to be done by, let's say, the second line of, or the second front of research that is not that ambitious, but it may be as important. So I would agree that it is important, but it appears that nobody really who has ambitious scientific questions would spend time to serve the purposes of standardization. (I6, BioMed)

However, even if scientists are usually not interested in taking responsibility for standardization processes, they need to find an agreement on the standards that ought to be used in a research project. One of the interviewees described the daily working experience of how standards are chosen in research collaborations. Commonly, joint discussions take place on how standards should be set and what should be used according to the specific research purposes.

Often there are fifty per cent split by discussions on standards on which one is the best standard. From standards in data collection to standards in normalization processing of data. I mean there is a long list and in some cases, there are different standards that do different things. And so we have to decide on the standard that is best for the specific use of the data. It is a very complex question on how to … and probably in most cases it is not good to just use one standard, because different ways of processing your data do give you different data. And they are better in some contexts rather than others. So it's a good question. And I don't have a general answer. We don't have a general process for using the standards. It's on a case basis. Of course, it is based on what other people are doing, what other census are doing. In some cases, it is not us choosing the standards, because if you are part, like often we are, of a large international study, the standards are chosen together. (I9, BioMed)

In the course of the ACGT project, the approach was to first look and review what kinds of standards already exist and to reuse as many of them as possible. Usually, those standards are preferred that are supported by large communities of users. It is assumed that those ICT tools and services that are built on broadly accepted standards will be recognized and reused in the respected research community as well. Another interviewee precisely described the problems that occurred with picking standards for a particular task in the ACGT project, namely building tools for knowledge discovery workflows.

> So, we had the question about standards regarding workflows, for example. How do I represent a workflow, how do I save it? Which data storage device do I use for the individual services, for example, how do I describe it? How do I describe whether it has a particular quality? How do I describe who built it, how can I verify if it's still okay? Well, the number of standards that you can use or that you could wish for is relatively high. The problem is just that there are simply very many of them. So, there's no single standard in the sense that really everyone uses it, but there's simply an incredible number of things that an incredible number of people have done in those areas, and where in the end, everyone picks out whatever they happen to need. That means, the only thing that really is a standard is if you decide to take a particular tool and then simply use the format that that tool uses as a standard. (I13, IT)[16]

As the interviewee precisely described it, in the process of creating new tools the ICT formats of existing tools work as standards. Therefore, newly developed tools such as the mentioned data discovery tool for workflow building can be directly linked to the chosen ICT format, which was in this case the programming language R for statistical calculations and graphs. Another good example for standardization based on ICT support is the so-called MIAME Convention describing the minimum information about a microarray experiment that must be provided to report data in microarray-based publications (Brazma et al. 2001).[17] According to the convention, the raw data have to be defined as data files produced by the microarray image analysis software. Even if the formats, annotations, and protocols are not prescribed, the convention includes a list of possible MIAME compliant software. This makes obvious that study designs to be approved have to be based on ICT. Again, ICT

---

[16] German original: "Also wir hatten die Frage nach Standards zum Beispiel bei Workflows. Wie stelle ich so einen Workflow dar, wie speichere ich den ab? Welchen Datenspeicher nehme ich bei dem einzelnen Services zum Beispiel, wie beschreibe ich den? Wie beschreibe ich, ob der eine bestimmte Qualität hat? Wie beschreibe ich, wer den gebaut hat, wie kann ich dann überprüfen, ob der auch noch in Ordnung ist? Da gibt es dann auch relativ viele Standards, die man brauchen kann oder die man sich wünschen könnte. Das Problem ist halt, dass es da einfach sehr viele gibt. Also es gibt keinen Standard in dem Sinne, dass wirklich alle ihn benutzen, sondern es gibt einfach unheimlich viele Sachen, die unheimlich viele Leute gemacht haben in den Bereichen, und ja, wo sich dann im Endeffekt jeder selber das zusammensucht, was er gerade braucht. Das heißt, das Einzige, was halt wirklich ein Standard ist, ist wenn man sich entscheidet, ein bestimmtes Tool zu nehmen und dann einfach das Format, das dieses Tool nimmt, als Standard zu nehmen." (I13, IT)

[17] See also www.fged.org/projects/miame/. Accessed June 12, 2014.

software works as a standard enabling the unambiguous interpretation of the results of the experiment. The MIAME standardization process set up by the Functional Genomics Data Society (FGDS) was very powerful in the systems biological community. One possible reason was that gene expression was successful right from the beginning in the emerging fields of systems-oriented research and many microarrays were done. Taken the aspects outlined together, standardization procedures in systems biology are triggered by the need to make data and study designs comparable to integrate and share data and finally study results. As data storing and processing are based on ICT systems, ICT also deploys the standards for data quality, annotation, and exchange.

4.1.4 Concluding Remarks

In this section, we looked at the challenges of facilitating integration of different data types and databases to build up an effective ICT infrastructure supporting systems medical research. In the interviews, we identified at least five challenges relevant to data integration realized in an ICT environment.

The challenge that prepares the ground is data acquisition which takes place in different contexts to gather the three basic data types (molecular data, clinical study data, and health care data) to be used in systems-oriented research. As shown, the preconditions of data acquisition in the laboratory and the clinic are very different. Gathering molecular research data means in the first instance to hook up with new acquisition technologies and to deal with the scale and breadth of Omics data. Gathering clinical study data means following a prospective research protocol and to set up a recruitment process including informed consent. Gathering health care data stemming from treated patients means dealing with unstructured and incomplete data sets and with high data protection standards because of medical confidentiality. The interviewees broadly discussed that the different preconditions of the data acquisition challenge data reliability and validity. Hence, it was put forward that due to the data acquisition context, reliability of the different data types is highly variable. Furthermore, the ethical–legal requirements of the data types differ according to the data acquisition context. Generally, data access is only permitted by informed consent. However, the broad consent for research on biomaterial is usually given one time for unlimited use, whereas research conducted on clinical study data is usually bound to an informed consent given for a single research purpose. Access to health care data is most restricted as these data are primarily raised for treatment decisions and not for research purposes.

Of course, restricted access to medical data for research purposes has been broadly discussed in ethical debates. However, according to the interdisciplinary approach in systems medicine, data are no longer exclusively analyzed by the scientists who had raised them. Therefore, it is necessary that the nature of the data type (e.g., reliability, ethical–legal requirements) is known beyond its specific context. The interviewees generally acknowledged that interdisciplinarity is an ambitious challenge, as misunderstandings between the participating disciplines often occur because of ignorance and differences in terminology, research practices, and logics. The interviewees stressed that progress and success of the ACGT project was mainly grounded in overcoming the interdisciplinary challenges. However, the interviews revealed at the same time that in particular, differences affecting the mode of thought and research logics sometimes persist.

The most prominent challenge for systems-oriented research is, however, the integration of the different data types coming from different data sources (e.g., O'Malley and Soyer 2012). The ACGT consortium tackled the challenge by setting up an ICT infrastructure addressing two levels of integration. In brief, syntactic integration provides the technological rack to facilitate data exchange, whereas semantic integration ensures that the data exchanged are accurately interpreted. A range of technological tools and services, such as the workflow builder or the ontology discussed earlier in Sect. 4.1.3, was developed to assist syntactic and semantic data integration. These tools and services finally prove that the challenge of data integration is exclusively approached by ICT-driven technology. This approach usually includes the challenge to standardize tools and services triggering the integration process. As the interviewees described it, new tools are directly built on ICT as the ICT formats of existing tools work as standards for the new ones.

Because of the nature and amount of data generated by different technologies, laboratories, researchers, and clinicians, data integration has triggered the implementation of multiple standard operating procedures based on ICT. In the face of the different scopes and formats, and different disciplinary origins and developments, standardization processes seem to be the central mechanism of coping with the overarching tasks of data integration for building up data collections. Not surprisingly, the data that are most successfully assembled into large data sets are genomic data which are produced through highly standardized technologies such as genome sequences or microarrays (Leonelli 2014, 5).

ICT-based standards and guidelines define what counts as reliable evidence, clear nomenclature, and commonly accepted experimental practice within the emerging field of systems medicine. This already has a sustainable impact on the handling of data: because of the computational environment, data are split into the pure data content and the data structure describing the content. Of course, the distinction of data and metadata is a phenomenon with a long tradition in biology (Edwards et al. 2011; Leonelli 2010). What is different in systems medicine (and biology) is the fact that the latter is first and foremost defined and attached by ICT. It finally forms a new body of information including formal data and message format, accurate classification, or other relevant ICT metadata. In this context, Sabina Leonelli (2014, 6) points out that the task to create the information of data classification (e.g., adding keywords, metadata, etc.) is usually one of the tasks of ICT-trained curators who have therefore gained influence on the meaning and interpretation of data in systems biological research.

To conclude, it is often stressed that the application of high-throughput technologies has made it possible to increase dramatically the amount of information that can be stored and integrated (e.g., Leonelli 2012). This assumption was verified by the experience of the interviewees. However, our case study also revealed that this quantitative shift has brought the need to standardize data for data integration and reuse. We argue that the quantitative shift has led to qualitative changes of how to handle and use large repositories of standardized data in systems medical research: the significance and meaning of data have changed by defining which part is for scientific use and which contains more or less purely technical information such as the message format. Finally, data produced by almost fully automated and highly standardized procedures are rather regarded as a computer output reaching value because of the reproducibility and reliability based on ICT (García-Sancho 2012, 26). This has resulted in the acknowledgement of data "as key scientific components, outputs in their own right" (Leonelli 2014, 9) that need to be widely disseminated. Hence, ICT environments collect and process not only data; they construct at the same time the data by assigning significance, meaning, and finally, evaluation to data and its parts. From this it follows that understanding and modeling of biological systems are deeply shaped by ICT. In the next section we look in more detail at the process of how ICT and their underlying design and conceptualization shape the modeling of cancer in silico.

4.2 Simulating Cancer In Silico: The Oncosimulator

The goal of systems biology is to model molecules, cells, tissue, organs, body systems, and whole organisms holistically. One possible access to reach this ambitious goal is to study diseases and the alterations between normal and diseased biospecimens. Models to understand and predict the genesis and development of a disease such as cancer can not only be proved in vivo and in vitro, but also be theoretically analyzed with in silico techniques. In silico as a term describes the modeling, simulation, and visualization of biological and medical processes in computers referring to any application of computer-based technologies (Michelson et al. 2006). The increasing volume of molecular data and the decreasing costs of computational power have made it possible to run more and more in silico simulations today (Deisboeck et al. 2009).

For instance, the virtual self-surviving cell modeled by Masaru Tomita is regarded as one of the first whole cell in silico models (Tomita 2001). This modeled cell consists of 127 in silico genes, 120 coming from M. genitalium and 7 coming from other microorganisms. Based on the model, it was investigated how the alterations of glucose supply and signal pathways control the knockout of distinct genes. Observing the behavior of in silico cells yields comparative insights and can lead to the discovery of causalities and interdependences by providing in silico experimental devices for hypothesis testing and predictions (Gramelsberger 2013, 157).

In the clinical context, in silico modeling might finally lead to reliable predictions as to which treatment will fail in a patient before it is applied (Graf et al. 2009, 142). Thus, in silico oncology is one of the most visionary endeavors of the ACGT project concerning the actual use of systems approaches to medical decision making. We therefore have chosen in silico technology as an example to analyze the conceptual development of tools and services for doing systems medicine. By looking at the development of an individual tool from scratch we aim in particular at investigating how ICT and their conceptualization shape systems medical research. The example under study is the oncosimulator incorporated into the ACGT infrastructure as an experimental platform. It simulates in vivo response of tumors and normal tissue to therapies based on clinical, imaging, histopathologic, and molecular data of a given cancer patient. In the long run, it aims at a better understanding of cancer at the molecular, cellular, organ, and body level and optimizing therapeutic interventions on a patient-individualized basis by performing in silico experiments of candidate therapeutic schemes.[18]

In the course of the ACGT project, Georgios Stamatakos and the In Silico Oncology Group at the Institute of Communication and Computer Systems, National Technical University of Athens[19] developed the initial version of the oncosimulator focusing on pediatric nephroblastoma, a childhood cancer of the kidneys, and in particular on a trial run by the International Society of Pediatric Oncology (SIOP) in collaboration with the Department of Pediatric Hematology and Oncology at the University Hospital of Saarland (Germany) led by Norbert Graf. For the first time, Stamatakos and his team were able to use real data before and after chemotherapeutic treatment (Stamatakos et al. 2007). This was a breakthrough to adapt the software to real clinical conditions and, at the same time, validate the software using real-world results. By using real medical data concerning nephroblastoma for a number of patients in conjunction with model parameters based on literature research, the tumor volume shrinkage has been predicted with reasonable accuracy. Up to now, the oncosimulator has been advanced and implemented in further research projects in the context of the 7th EU Framework Program such as the projects p-medicine and CHIC.[20]

4.2.1 Vision and Definition of the Oncosimulator

The genesis and progression of cancer is associated with tumor morphology, invasion, and related molecular phenomena (Sanga et al. 2007, 120). One of the grant challenges of the understanding of cancer progression is therefore to find the links between alterations and the hallmarks of cancer such as increased proliferation

[18] This work has been supported in part by the European Commission under the projects "ACGT: Advancing clinico-genomic trials on cancer" (FP6-2005-IST-026996), and "CHIC: Computational horizons in cancer: Developing meta- and hyper-multiscale models and repositories for in silico oncology" (FP7-ICT-2011-9-600841).

[19] In Silico Oncology Group, Institute of Communication and Computer Systems, National Technical University of Athens, http://in-silico-oncology.iccs.ntua.gr/english/index.php. Accessed May 3, 2014.

[20] p-medicine, www.p-medicine.eu and CHIC, http://chic-vph.eu. Both accessed May 10, 2014.

and survival, aggressive invasion and metastasis, evasion of cell death, and increased metabolism (Hanahan and Weinberg 2011). However, it has been difficult to quantify the relative effect of these links on disease progression and prognosis using conventional clinical and experimental methods and observations. For example, the primary role of angiogenesis in promoting tumor growth and invasion has been well demonstrated, whereas the results of clinical trials using drugs to suppress neovascularization have not yet yielded unambiguous results (Kuiper et al. 1998; Bernsen and van der Kogel 1999). Hence, what is needed is a method to enable prediction of tumor growth and therapy outcome through quantification of the relation between the underlying dynamics and morphological characteristics.

The fundamental assumption underlying this approach is that any biological processes are amenable to mathematic and/or algorithmic description. In this regard, the genesis and development of cancer is regarded as a disease and at the same time as a natural phenomenon. From the cancer treatment perspective, what really matters is the discrete number of the usually few tumor cells surviving treatment and their discrete mitotic status (e.g., stem cells, cells of various mitotic potential levels, differentiated cells). Therefore, such a mathematical approach must apparently take into account both the deterministic and the stochastic character of the disease (Stamatakos et al. 2007). This challenge is tackled by a multidisciplinary method integrating mathematical description and computational simulation of the multiscale biological mechanisms that constitute the phenomenon of cancer and its response to therapeutic regimes. By primarily applying discrete mathematics, the In Silico Oncology Group developed the modeling method called Discrete Event-Based Cancer Simulation Technique (DEBCaST) (Stamatakos 2011, 408).

DEBCaST is basically a top-down approach using clinical observations, including anatomic and metabolic tomographic images of the tumor, and the knowledge about the behavior of a cancer as a whole based on available physiological and biological findings. This information is required to identify subsystems of the tumor and to build a reproducible model of a specific cancer and its progression. Given that the discrete entities and quantities of a specific cancer in conjunction with their complex interdependences give rise to tumor relapse or ensure tumor control over a given time interval, the constant alignment with the clinical observations is required.

> Multi-scale cancer models, which lie at the heart of the oncosimulator, should be driven by real clinical trials. This is completely different from the standard bottom-up approach adopted by most cancer modelers. Developing a tumor model by trying to exploit what you can do and by trying to extend what you can do is of course interesting and potentially useful. But the reality itself, at least the clinical reality, expects the modeler to be adapted as much as possible to the real clinical questions, the real clinical problems as they are posed within the clinical walls, let's say, or the clinical theatres. In that sense, a top-down approach rather than a bottom-up approach seems to be better in order to address such complex problems. This is my personal approach. […] Anyway, models should be adapted and validated to real clinical trial data. The oncosimulator should undergo both retrospective and prospective clinical validation as a prerequisite to be translated into clinical practice. Models should be modular and extensible so as to be able to integrate new advances in cancer biology and clinical experience. (GS)[21]

[21] In this section (4.2), citations from Georgios Stamatakos are designated to him (GS) by mutual agreement as he was apparently interviewed regarding the oncosimulator. In all other sections, his

As described in the citation, in silico cancer modeling following the top-down approach is an iterative process: the more clinical data are supplied to the model, the more accurate it becomes in reflecting reality. However, the top-down approach is very challenging with regard to developing the cancer models. Right from the beginning, the whole range of complexity of cancer has to be accounted for. In order to include multi-scale dynamics in cancer modeling, strategies of how to pass information from a lower-scale level to a higher-scale level and vice versa are required. To solve this problem, each level is characterized by summarizing principles that can be passed to another level of complexity: "That means strategies to summarize what is happening, for example, on the molecular level, and to summarize it in one or two of the very small number of parameters, which can be understood by higher complexity levels. [...] This kind of problems had to be solved in a practical way. And we had to work very hard for this." (GS)

Instead, a bottom-up approach assembles all parts of a system starting with genes and proteins and brings them into a formal model (Michelson et al. 2006). Therefore, the discovery of each new component needs a reconfiguration of the whole model. Stamatakos (2011, 407) criticizes that this approach focuses rather on microscopic tumor dynamics mechanisms than on multilevel interdependencies and interactions. However, a careful combination of the top-down with the bottom-up approach in the clinical context has its own merits. This has led to the integration of the latter into the latest versions of the oncosimulator.

> The oncosimulator envisions to encompass all levels of bio-complexity including the molecular level, the cell level and the supercellular levels. I would say that systems biology in the traditional sense is a very important component of the all-scale approach of in silico oncology, or in silico medicine in the broader sense. We do need to simulate what is happening on the molecular level, but this is not enough and sometimes the molecular level complexities are so high that you might not even end up with something robust and reproducible. In the real world molecular pathways are very sensitive to crosstalking with other pathways. Therefore, I firmly believe that the molecular level traditionally addressed by systems biology is one of the very important levels to be taken into account in detail. But it is not enough. All other levels should also be taken into account. So you could call such an approach an all-level approach, or a multi-scale approach, or an extended systems biology approach. It is a matter of definition. (GS)

A bottom-up approach focusing exclusively on the molecular level may therefore not deal adequately with concrete and pragmatic questions of importance in the clinical setting. On the other hand, the top-down approach encompassing all bio-complexity levels of the body may be adapted right from the beginning to clinical questions. Of importance are particular questions such as the following. Can the response of the local tumor and the metastases to a given treatment be predicted in size and shape over time? What is the best treatment schedule for a patient regarding drugs, surgery, irradiation and their combination, dosage, time schedule, and duration (Graf and Hoppe 2006)?

The ACGT oncosimulator paves this way by focusing on clinical utility. It primarily aims at supporting the clinician in the process of optimizing patient-specific

citations are pseudonomized as are all others.

cancer treatment through conducting experiments in silico. Through performing in silico experiments, the likely outcomes of several candidate therapeutic schemes are evaluated based on the particular clinical (e.g., symptoms, progress of disease), imaging (e.g., MRI, PET, CT), histopathological (e.g., type of tumor), and molecular (e.g., DNA microarray) data of the individual patient (Stamatakos et al. 2006a).

> So let's start from the beginning. The clinical data, the previous treatment history, the imaging data, the body fluid samples, and the biopsy material taken from the patient when available are collected. The extracted multi-scale and inhomogeneous data are pre-processed – some of them through molecular networks, some others by exploiting disciplines such as radiobiology or pharmacology – in order to create the kind of data that the simulation module can understand. At this point, the user, who in the future is expected to be primarily the clinical doctor, describes several candidate schemes, or treatment schedules, or treatment scenarios. They introduce those scenarios into the simulation model. Following the execution of scenarios, the oncosimulator predicts the expected outcome. The outcome is evaluated by the clinician in order to eliminate any eventually not justified extremes or extremely unlikely responses. The user then selects the optimal scheme to be applied to the patient. (GS)

As Stamatakos pointed out in the last quotation, the vision of the oncosimulator is primarily based on its clinical application. Therefore, the cancer models are supposed to be adapted as much as possible to the clinical questions of importance, or in other words, to clinical reality. The predictions aim at supporting clinicians with information on the most effective treatment out of several alternatives, as well as detailed parameters on the optimal composition of a treatment scheme, including the total treatment period, the type of drugs, dose, and interval between treatments (Graf et al. 2009, 147; see Box 4.3).

In consequence, cancer modeling was set up as a top-down approach using all kinds of available clinical data and observations to simulate cancer genesis as a biological phenomenon and its progression under the influence of therapeutic regimes. In this regard, the oncosimulator is not only a clinical tool, but at the same

Box 4.3: Seven Steps How the Oncosimulator Is To Be Used in Patient-Specific Cancer Treatment (Stamatakos 2011, 411f)

Step 1: Obtain patient's individual multi-scale and inhomogeneous data. Data sets to be collected for each patient include: clinical data (age, sex, weight, etc.), possible previous antitumor treatment history, imaging data (e.g., MRI, CT, PET, etc.), histopathological data (detailed identification of the tumor type, grade and stage, histopathology slide images whenever biopsy is allowed and feasible, etc.), and molecular data (DNA array data, selected molecular marker values or statuses, serum markers, etc.).

Step 2: Preprocess patient's data. The data collected are pre-processed in order to take an adequate form allowing its introduction into the tumor-and-normal-tissue-response-simulation-module of the oncosimulator. For example, the imaging data are segmented, interpolated, and eventually fused; subsequently, the anatomic entities of interest are three-dimensionally reconstructed. This reconstruction will form the framework for the integration of the rest of the data and the execution of the simulation. In parallel the molecular data are processed via molecular interaction networks so as to perturb and individualize the average pharmacodynamic or radiobiological cell survival parameters.

Step 3: Describe one or more candidate therapeutic scheme(s) and/or schedule(s). The clinician describes a number of candidate therapeutic schemes and/or schedules or no treatment (obviously leading to free, i.e., non-inhibited tumor growth), to be simulated in silico.

Step 4: Run the simulation. The computer code of tumor growth and treatment response is massively executed on distributed Grid or Cluster computing resources so that several candidate treatment schemes and/or schedules are simulated for numerous combinations of possible tumor parameter values in parallel. Predictions concerning the toxicological compatibility of each candidate treatment scheme are also produced.

Step 5: Visualize the predictions. The expected reaction of the tumor as well as toxicologically relevant side-effect estimates for all scenarios simulated are visualized using several techniques ranging from simple graph plotting to four-dimensional virtual reality rendering.

Step 6: Evaluate the predictions and decide on the optimal scheme or schedule to be administered to the patient. The oncosimulator's predictions are carefully evaluated by the clinician by making use of their logic, medical education, and even qualitative experience. If no serious discrepancies are detected, the predictions support the clinicians in taking their final and expectedly optimal decision regarding the actual treatment to be administered to the patient.

Step 7: Apply the theoretically optimal therapeutic scheme or schedule and further optimize the oncosimulator. The expectedly optimal therapeutic scheme or schedule is administered to the patient. Subsequently, the predictions regarding the finally adopted and applied scheme or schedule are compared with the actual tumor course and a negative feedback signal is generated and used in order to optimize the oncosimulator.

time a concept of multilevel integrative cancer biology, a complex algorithmic construct, and a biomedical engineering system (Stamatakos 2011, 411). In the next section we look deeper into the model basics to see how the oncosimulator is able to serve as tool, concept, construct, and system at the same time.

4.2.1.1 The Model Basics

At the core of the simulation approach is the idea to explore the natural phenomenon of cancer. In order to describe the biological activity of a discrete tumor spatially, the oncosimulator correlates the response of normal tissue with the response of tumor tissue.

> The heart of the system is the tumor and normal tissue response simulation model. This is actually the computer code, a pretty complex simulation code, which gets as input the processed patient data and produces as output the predictions concerning the response of the tumor to concrete candidate therapeutic schemes or schedules (GS).

Based on imaging data, the tumor is simulated as a multidimensional virtual reconstruction including the eventual necrotic region and the surrounding anatomical features before, during, and after treatment (e.g., chemotherapy, radiation). The imaging data provide information on the boundaries of the gross volume of the tumor, the volume itself, and the spatial distribution of the metabolic activity of the tumor (e.g., regions where there is significant provision of oxygen and nutrients through the neovasculature and necrotic regions where there is lack of adequate vascularization and subsequently lack of adequate oxygenation and provision of nutrients).

To simulate how a discrete tumor will spatially spread, the tumor is discretized using a cubic mesh. Each elementary cube of the mesh is called a geometrical cell and is used for the description of the tumor in a statistical way (Stamatakos et al. 2002, 2006a; Dionysiou et al 2004). The geometric mesh covering the tumor region is scanned in certain time intervals (e.g., every 1 h). In each time step, the updated state of a given geometric cell is determined on the basis of a number of algorithms describing the behavior of the cells constituting the tumor. More precisely, each geometrical cell of the mesh belonging to the tumor contains a number of biological cells characterized by the cell phase in which they are found (e.g., stem cells, limited mitotic potential or progenitor cells, differentiated cells, necrotic cells). According to the adapted cytokinetic model (Stamatakos et al. 2006a, 1468), tumor cells usually pass through the following cell phases: G1 (gap 1), S (DNA synthesis), G2 (gap2), and M (mitosis). After mitosis is completed, each of the resulting cells re-enters G1 if the oxygen and nutrient supply is adequate. Otherwise, it enters the necrotic phase which finally leads to cell death. The number of biological cells constituting each phase class is initially determined according to the spatial position of the geometrical cell within the tumor and the metabolic activity in the local area.

It is generally assumed that each geometrical cell of the mesh contains a constant number of biological cells. However, in the case that the actual number of tumor cells contained within a given geometrical cell drops below a given threshold, during the simulation process a procedure starts that attempts to unload the remaining biological cells in the neighboring geometric cells. If the given geometric cell becomes empty it is assumed that the geometric cell is removed from the tumor. Therefore, an appropriate shift of a chain of geometric cells intended to fill in the vacuum leads to tumor shrinkage. This can, for example, happen after irradiation of a radiation-responsive tumor. On the other hand, if the number of tumor cells in a given geometrical cell exceeds a limit, then additional geometrical cells emerge. By an appropriate shift of a chain of geometric cells towards the boundaries of the tumor, the tumor expands (Kyriazis et al. 2008).

To simulate tumor expansion or shrinkage of a discrete tumor, a number of algorithms (operators) are periodically and sequentially adapted to the anatomic region of interest (Stamatakos et al. 2002, 1771). These algorithms are based on selected parameters[22] influencing tumor growth and response to treatment. They finally steer the simulation as one interviewee explained:

> The parameters produce different kinds of data. Therefore, different settings produce different kinds of data. It is like, you know, if you have an oven at home and you have lots of bread that is not baked. You put your oven at twenty degrees, you put in the bread and you see what happens. And then you change the oven to fifty degrees and you put in another bread and you see what happens. And then you put all these breads together again and you see the one that was at twenty degrees is still dough. It's not baked, it hasn't done anything. The one that you put at five hundred degrees is burned. And somewhere in between there is an optimum. This is only one parameter. This is temperature. But there may be other types of parameters that are also going to influence your bread. The humidity or the type of dough that you put in it. How much water did you put in the dough? How much yeast did you put in the dough? Those are different parameters so you can imagine that there are many different kinds of bread. And many different kinds of baking that you can do. And your challenge as a baker would be to find this optimum. Now in the case of oncology, it is not only five parameters, but regarding the in silico oncology simulator of Georgios Stamatakos, it was something like forty parameters. So there is a huge space that is spent and that you need to search for the optimum solution. (I3, IT)

Some of the parameter values used for modeling, such as cell-cycle duration or the necrosis rate of differentiated cells, have been based on literature reviews for particular tumor types. Others have been defined based on exploitable medical data and logic. The latter concerns those parameters for which only qualitative data are available. In cases where those different values are available for a given parameter, all values are considered in different instances of the model. Generally only those parameters relevant to the particular type of tumor are used. For instance, in regard to the scenario of pre-operative chemotherapy in nephroblastoma, 30 clinical parameter values are listed covering cell-cycle dynamics to treatment modalities (Graf et al. 2009, 144).

The selected parameters have to be interrelated and ranked based on their effect on the treatment outcome. Some restrictions regarding the interrelation of several parameters (e.g., higher resistance of stem cells to treatment in relation to progenitor cells) are known from the literature. They are further exploited by the simulation process itself.

---

[22] The term parameter is used in mathematics for a quantity or arbitrary constant whose value varies with the circumstances of its application. In a more common sense, parameters are any factor forming one of a set that defines a system and determines (or limits) its performance (Yourdictionary, www.yourdictionary.com/parameter. Accessed July 7, 2014).

> The basic idea is that a patient comes in, has a tumor, you use the oncosimulator to predict what the tumor will do when you give treatment, for example, if you radiate it, or if you give it chemotherapy. That has many very different parameters. You can radiate longer or with a higher power. You can radiate daily or semi-daily or maybe weekly. You can give chemo or not. And if you give chemo, there are two different kinds of chemo that you can give. Do you give them both? Do you give them one after another? Do you combine radiation with chemo? Or do it after each other? Or do you first do chemo and then radiation, etc.? So that is a huge parameter space that you build up. Now what we were doing is create visualizations of all these different settings. So you have a number of settings that you keep static. There is one setting that you change a little. And then you can see in the visualization what the shape of the tumor would do. So it could grow. It could stay the same. Or it could shrink. And based on that you could give an indication. Okay, I understand now, if I change this knob in this direction, the tumor will shrink. So okay, this is what I want. This is more optimal. So this is a good change for me. So now I am going to change a different parameter. If you change that parameter, your visualization will change again, but now in two dimensions. So you had this change of one parameter. Then you change a different parameter, so there are two parameters that you change. This is two-dimensional. And then you get different solutions. So you can see what the effect of the second parameter is in combination with the first parameter. If you see this, then you hope to find the pattern. Again, it can basically shrink in a second direction, or it can stay the same in the second direction, or it can increase in the second direction. If you know this you can see what the correlation is of these two parameters based on this visualization and then you include a third parameter, etc. So you combine all these parameters in a visual interface that will show you how the simulation is influenced by changing the setting of one parameter, or two parameters, or maybe three parameters. (I3, IT)

Therefore, further validation, adaptation, and optimization take place following each simulation. The real response of the patient to the treatment is compared with the predicted response and this result is utilized as feedback in order to improve the simulation model. "That means that the more patients that have been addressed by the oncosimulator, the better its predictive potential is expected to become" (GS).

Summing up, the ACGT oncosimulator interprets tumor information according to mathematical measures and models to predict the composition and the shape of the patient-specific tumor and the response to therapeutic regimes over the course of time. Viewing biological processes through the lenses of mathematics means in the first place to quantify living matter that might be affected and determine it in time and space. This has two crucial consequences: this approach relies on as much data as being available, and it relies on parameters that are mathematically applicable.

4.2.2 Genesis and Development of the Oncosimulator

The genesis and development of the oncosimulator are deeply embedded in the academic career of Georgios Stamatakos, the teamwork in the In Silico Oncology Group at the Institute of Communication and Computer Systems (ICCS), National Technical University of Athens (NTUA), and interdisciplinary and international collaborations. In the very beginning, after Georgios Stamatakos had passed his master's degree in bioengineering at the University of Strathclyde (Glasgow, UK) and his PhD in biophysics at the NTUA, the work began with the desire to take off into a new research field and the advice of a supervisor.

> Actually, after my PhD thesis, which was on bio-electromagnetics, Professor Nikolaos Uzunoglu suggested to me that I should extend my research interests by doing something regarding radiation therapy. In that period, there was a good collaboration at the Athens Technical University with the Klinikum Offenbach in Germany […] concerning, for example, the use of electromagnetic fields in order to enhance radiation therapy (hyperthermia). In that way, I started working on radiobiology and radiobiological modeling. But this kind of interaction with the Klinikum Offenbach (Prof. Nikolaos Zamboglou) helped me to get more concrete. And more clinically oriented, let's say. And of course, by trying to utilize a previous expertise in particular I loved the mathematics somehow, I proceeded to the formulation of this concept of the oncosimulator. And then of course, there were a number of students I directed in their diploma thesis or PhD thesis, who all helped to contribute to the implementation of this concept. (GS)

At National Technical Uninversity of Athens, the In Silico Oncology Group was set up in 1997 initially working on a number of simulation models regarding tumor response to treatment both in vitro and in vivo. Again, international interaction and cooperation encouraged Stamatakos and his team to move on with the idea of developing the oncosimulator.

> At this point, I would like to particularly mention the very important help we got from Werner Düchting from the University of Siegen in Germany. Actually, Werner worked before us concerning the simulation of tumor growth in vitro. His modeling work referred to small tumors. That's before the creation of new blood vessels. And that was pretty inspiring for us to move to the in vivo simulation using imaging and multi-scale data as we did. So even before ACGT, there have been quite extensive multinational interactions. I would also call them intercontinental interactions concerning this approach. The contribution of Norbert Graf, professor of pediatric oncology and hematology at the University Clinic of the Saarland, has been of paramount importance. (GS)

In the beginning, the research focused on one specific cancer type. That was glioblastoma multiforme which served as the first paradigm. Afterwards, the In Silico Oncology Group tried to reuse and exploit parts of the algorithms, codes, and the major philosophy of the approach in other cancer types. In addition to glioblastoma, in silico simulation has up to now been applied to breast cancer, lung cancer, leukemia, nephroblastoma, and cervix cancer.

From the very beginning, two preconditions have been defined before envisaging the introduction of in silico methods into the clinic: first, every prediction of an in silico simulation has to be compared with the reality; and second, every in silico simulation has to be part of a clinical trial in which the clinical, imaging, biochemical, and genomic data are systematically acquired (Graf et al. 2009, 142). The first simulation using real trial data was based on a clinical trial outcome of the Radiation Therapy Oncology Group, a clinical cooperative group funded by the National Cancer Institute (Philadelphia, USA). Using the clinical study on hyperfractionated radiation therapy and bis-chlorethyl nitrosourea in the treatment of malignant glioma (Werner-Wasik et al. 1996), the In Silico Oncology Group at NTUA simulated the hyperfractionation of two different radiation doses.[23] In regard to shrinkage and regrowth of the tumor, the simulation predicted the real clinical trial outcome in advance (Stamatakos et al. 2006b). The study has revealed that trial participants who received the higher doses had survival superior to the patients receiving the lower

[23] One time it was the total dose of 48 gy/day and another time the total dose of 81.6 gy/day.

doses. This was the first breakthrough in the development of the oncosimulator with regard to clinical validation. However, more breakthroughs had to follow. The next step was the development of the first integrated version of the oncosimulator in the course of the ACGT project.

> Regarding the oncosimulator, an initial version of the entire integrated system was produced. We started the clinical adaptation and validation process, but this of course will take some years to be completed. Nevertheless, I do believe that we ended up with something pretty concrete. Of course, it is a first version. But still, it is an integrated and complete first version, at least as far as the scientific and technological components are concerned. The clinical aspects, as I mentioned, are much, much more time consuming due to the requirements of the clinical trials, both retrospective and prospective. But as I mentioned, the oncosimulator is being improved and extended within the context of the new projects. (GS)

The initial idea of working on radiobiology has finally resulted in a concrete endeavor of developing a clinically adaptable tool to simulate tumor response to treatment. However, the onward development of the oncosimulator was only possible because Georgios Stamatakos has continuously pursued his vision. Additionally, he has at all times found colleagues who supported his ideas and worked with him together on realizing his vision. In this regard, as we show in the next section, the ACGT project was a very important working environment implementing the oncosimulator into a broader scientific community.

4.2.2.1 Interdisciplinary Challenges

Described as a "really multidisciplinary construct" (GS), the development of the oncosimulator has required expertise from many different domains. First of all, mathematics is cited as being at the core of the whole endeavor. In particular, methods and strategies from discrete mathematics are used to simulate natural phenomena that have a discrete character, for example, the discrete number of tumor cells or the discrete phases of the cell cycle. These discrete entities and quantities in conjunction with their complex interdependences may give rise to predictions of tumor relapse or tumor control over a given time interval (Stamatakos 2011, 409). In addition, strategies of continuous mathematics (e.g., continuous functions, differential equations) are used in order to tackle specific aspects of the models such as pharmacokinetics and cell survival probabilities. Currently more continuous mathematics-based oncosimulators have also been developed.

> Since it is a multidisciplinary, scientific, and technological system, it implies that you need, first of all, mathematics. Of course, you need biology. You need expertise in various domains. But mathematics will always light up the heart of these types of systems. And mathematics can be found in any conceivable technological and scientific domain. My view is to try to somehow reuse, extend, and enhance, if possible, already known mathematical methods and, of course, to suggest new ones as well. But at least I personally believe that the effort to somehow extend mathematical methods and tools used in other scientific domains can much accelerate the whole process. (GS)

From a technological point of view, integrating the dynamic and multidimensional visualization of both the medical input and the simulation predictions is particularly challenging. In addition to virtual reality visualization techniques, further technological components were needed to build a first integrative version of the oncosimulator.

> Just to mention that we need components dealing with image processing, internal code parallelization, code acceleration, the execution of the models on several computer architectures including cluster, execution, and nowadays cloud execution, and so on. Grid execution was the one mostly adopted by ACGT. [...] This is a need for the simulator, but there is a need for computer resources, for example, Grid resources. There must be a data management system and, of course, quite a complex interaction of those modules. (GS)

Regarding these scientific and technological tasks and challenges, an interdisciplinary approach was mandatory. The In Silico Oncology Group is therefore composed of scientists with backgrounds in mathematics, informatics, and electrical and computer engineering. In addition, they all need to have an interdisciplinary and visionary mindset.

> Of course, there is need for anybody involved in this effort to broaden their horizons. Everybody has to read a lot about scientific fields unfamiliar to them. I would like also to mention the very important contribution of my PhD student, Dimitra Dionysiou, currently a senior researcher in the In Silico Oncology Group, who was actually the first student working on the pre-oncosimulator stage that I directed many years ago. She was very passionate with that idea and, of course, she had to read a lot and do rather unconventional work. (GS)

In addition to the continuous work of the In Silico Oncology Group, interdisciplinary cooperation is often stressed in the interviews. In particular, collaborations within the ACGT project were highlighted in regard to technological solutions of the integrated oncosimulator. One of the interviewees explained how the oncosimulator profited from the architecture of the ACGT infrastructure

> So then we moved to more work together with Georgios Stamatakos. He had a problem. His problem was that he had a simulator that would allow you to simulate the effects of a tumor treatment within patients. But the problem was that for every simulation there are a lot of different settings that you can identify that would all make sense, but you would never know which one was the optimal solution. And we had the technological idea that we could use his simulation and sort of try out many different combinations of settings. And use all the computational resources in this Grid infrastructure to do calculations on this simulation in parallel. And then take out the best solution and provide this to an oncologist as a possible treatment for specific patients. That was the idea. So in principle, Georgios Stamatakos, if he had the same question, he would need years of real time to compute an optimal solution. And our idea was to provide an infrastructure that would allow him to do this in, let's say, a couple of hours or maybe even in a couple of minutes by doing all these simulations in parallel and then taking the best solution and provide this, as I said, as a candidate for treatment. (I3, IT)

Even if this cited example illustrates that the oncosimulator has benefited tremendously from the interdisciplinary approach of ACGT, one of the major challenges was in fact to overcome interdisciplinary misunderstanding in the beginning of the project. In the very beginning, for many ACGT partners it was even difficult to understand the idea of modeling cancer.

> I still remember the first presentations of Georgios Stamatakos talking about modeling cancer and developing models, etc. Looking back 9, 10 years ago, it wasn't easy for clinicians to understand what he was talking about. You know, can we develop models? Can we model that? I still remember debates trying to use parallels from, let's say, twenty years ago. We couldn't really predict the weather, but we gradually developed models. So we opted for a more systemic approach in trying to develop models that can predict. And today through trial and error we see that our predictions are more accurate or totally accurate. And we had to make each other understand on what we imply on terminologies such as predictive models, develop models of cancer evolution, etc. (I7, IT)

> Georgios Stamatakos, kick-off meeting in Nice, he introduces an oncosimulator, he simulates the disease in a computer. Afterward, I went up to him and said, 'what nonsense, what he's doing is nothing but utopia'. That may exist in 100 years, etc. But on the other hand, I found it incredibly interesting, so for me it was something where I thought, okay, let's wait and see. And actually, a pretty good relationship emerged from that, so that we are actually developing this oncosimulator further from the clinical side, and it has become a really close collaboration. (I18, BioMed)[24]

Despite the skepticism in the beginning, the collaboration within ACGT in regard to the development of the oncosimulator was fruitful and lasting. Finally, "I think that practically all members of ACGT were optimistic and we did our best in order to contribute to the shaping and the construction, let's say, of an initial version of this basic science and technology integrative systems biology system" (GS). In particular, Georgios Stamatakos appreciated the optimistic attitude of his ACGT partners, as one of the major problems regarding the development of the oncosimulator was in the beginning of his project the skepticism of other scientific disciplines.

> One of the major problems, maybe historically the most important problem, was the reluctance of biologists and clinicians to accept the possibility that such a tool would ever be of clinical use and would ever be translated to clinical practice. And that was not entirely inexplicable in the sense that both biology and medicine at that time were mainly based on empirical knowledge. Of course, there had been a number of biomedical engineering devices, but the idea of bringing together so diverse disciplines and knowledge coming from areas spanning from image processing, let's say, to molecular dynamics or in the spatial scale, let's say, from nanometres to metres. That's in time from nanoseconds to years. That sounded at least in the beginning too futuristic. More a dream than something of any realistic content. Nevertheless, I was not taken back by such a very critical, let's say, approach. (GS)

In this regard, the ACGT project was one of the first working environments where Georgios Stamatakos received approval from the systems-oriented community in oncology. Close and still ongoing interactions and collaboration within the frame of ACGT began during this time. It can be concluded that the first integrative version of the oncosimulator was only possible because of the ACGT environment. In particu-

[24] German original: "Georgios Stamatakos, Kick-Off-Meeting in Nice, er stellt seinen Oncosimulator vor, er simuliert die Krankheit im Computer. Da bin ich nachher zu ihm gegangen und habe gesagt, so ein Schwachsinn, was er da macht ist nichts anderes als Utopie. Das gibt es mal in 100 Jahren vielleicht, etc. Aber auf der anderen Seite fand ich das unheimlich interessant, so dass es für mich etwas war, wo ich gedacht habe okay, jetzt gucken wir mal. Und daraus ist eigentlich eine ganz gute Beziehung entstanden, so dass wir tatsächlich diesen Oncosimulator von der klinischen Seite weiterentwickeln, und da ist eine richtig enge Zusammenarbeit draus entstanden." (I18, BioMed)

lar, the collaboration with ACGT partners from the biomedical domain opened the path to work with real clinical data and to start clinical validation on a systematic basis. In retrospect, the Community Research and Development Information Service (CORDIS) of the European Commission highlighted the oncosimulator as one of the EU-funded project success stories.[25]

4.2.2.2 Continuing Research After ACGT

Even if the first integrative version of the oncosimulator was built during the ACGT project and clinical validation has started by using real clinical trial data the oncosimulator was still located in the stage of research after the ACGT project ended in 2010. However, the In Silico Oncology Group was able to continue its work on the oncosimulator in several research projects funded by the 7th EU Framework Program.

In the research project p-medicine, the oncosimulator advanced and expanded in regard to the cancer types (acute lymphoblastic leukemia in addition to nephroblastoma and breast cancer) and the treatment protocols (chemotherapy, targeted therapy, radiotherapy, and combinations). As in the ACGT project, clinical trial data are used in order to optimize and validate the simulation models.[26]

In the research project MyHealthAvatar, the target is not the modeling of cancer or distinct cancer types, respectively, but of personal health in a broader sense. The in silico models of the In Silico Oncology Group are integrated into an ICT infrastructure that aims at collecting, sharing, and offering access to long-term and consistent personal health status data through an integrated in silico environment.[27]

In the research project DR THERAPAT, the digital radiation therapy patient platform is built up to integrate available knowledge on tumor imaging, image analysis and interpretation, radiobiological models, and radiation therapy. The goal is a coherent, reusable, multi-scale digital representation. Radiation therapy was chosen as the application to prove the integration of those concepts because inherently imaging plays a major role in radiation therapy planning and delivery, so the imaging information is available as input for various models, and the delivery process is relatively well understood, making model validation easier compared to, for example, chemotherapy.[28]

In the research project TUMOR that aims at implementing a cancer model repository, the In Silico Oncology Group focuses on multilevel cancer models which address more aspects of the natural phenomenon of cancer.[29]

Finally, the In Silico Oncology Group is the coordinator of the research consortium of CHIC. This research project proposes the development of clinical trial-

---

[25] CORDIS, http://cordis.europa.eu/result/brief/rcn/6061_en.html. Accessed June 1, 2014.

[26] p-medicine, http://www.p-medicine.eu/. Accessed June 1, 2014.

[27] MyHealthAvatar, http://www.myhealthavatar.eu/. Accessed June 1, 2014.

[28] DR THERAPAT, http://drtherapat.eu/. Accessed June 2, 2014.

[29] TUMOR, http://tumor-project.eu/. Accessed June 2, 2014.

driven tools, services, and infrastructures that will support the creation of multi-scale, integrative cancer models. One important focus is on the standardization of model description and model fusion. The creation of such elaborate and integrated models is expected to sharply accelerate the clinical translation of multi-scale cancer models and oncosimulators following their prospective clinical validation.[30]

Because of the collaboration on a range of research projects funded by the 7th EU Framework Program, the In Silico Oncology Group was able to include further cancer types (e.g., leukemia, lung cancer, prostate cancer) into the oncosimulation as well as further treatment protocols. In addition, clinical validation of the oncosimulators advanced according to the increasing access to clinical trials and real patient data. From a theoretical point of view, multiscale cancer modeling progressed towards more and more integrative models. In particular the last listed research project CHIC indicates this next step in the development of in silico oncology. The so-called hyper-models are defined as choreographies of component models, each one describing a biological process at a characteristic spatiotemporal scale. The component models are related to hyper-models defining the relations across scales and integrative models can become component models for other integrative models (Stamatakos et al. 2013).

In CHIC, the next steps of in silico oncology are already targeted: the development of an infrastructure that will support accessibility and reusability of mathematical and computational hyper-models. The standardization of cancer model and data annotation allowing multiscale hyper-modeling is one of the preconditions to be fostered in the future. In addition, the secure access to already existing data, models, and analysis tools is estimated as a necessary requirement in the development of in silico oncology. Accordingly, the set-up of extensive, in silico oriented repositories (e.g., hyper-models, hyper-model driven clinical data, distributed metadata, in silico trials) are demanded to keep track of the development of simulating cancer in silico.

4.2.3 Concluding Remarks

The description of the ACGT oncosimulator has shown that the development of an innovative technology is a story deeply connected with very different incidences. The most important step of the oncosimulator's storyline is, of course, its beginning: the initial idea to support clinicians with predictions on the most effective treatment out of several alternatives. This ultimate research objective has come up very early in the emerging field of in silico oncology. Just as important as the vision itself is the initiator, Georgios Stamatakos, who has consequently developed the first idea and expanded it to a vision of a biomedical technology that entails a comprehensive concept of multilevel integrative cancer biology, a complex algorithmic construct, and a biomedical engineering system. Furthermore, Stamatakos has at all stages of

[30] CHIC, http://www.chic-vph.eu/. Accessed June 2, 2014.

research found colleagues who supported his ideas and accompanied him on the long path realizing his vision.

Fundamental to the ongoing story is also that Stamatakos and his team pursued the initial version of the oncosimulator and, from the beginning, have subordinated any decision made in the course of research to this original vision. Even though this basic philosophy has sometimes led to take very difficult tracks, Stamatakos and his team pursued their ideas. For instance, the vision of the oncosimulator is primarily based on its clinical application. Therefore, the cancer models are supposed to be adapted as much as possible to clinical reality. As a consequence of this, cancer modeling was set up as a primarily top-down approach using all kinds of available clinical data and observations to simulate cancer genesis as a biological phenomenon and its progression under the influence of therapeutic regimes. At this stage of research, Stamatakos and his team have already reached at one of the core challenges of systems biology: the multilevel integration of biological processes. Again, Stamatakos chose a pragmatic approach to overcome this problem. To move on in the development of the oncosimulator, each level was characterized by summarizing principles as a set of parameters that can be passed back and forth between different levels of complexity.

Another far-reaching decision made at the start was to give priority to strategies and methods of particular mathematics. By viewing biological processes through the lenses of particular mathematics, all natural phenomena that might be affected were quantified and as much data being available are collected and used in constructing the models. The formalization of those mathematic models is realized by ICT. Even though real clinical data play such a prominent role in oncosimulation, the mathematization and formalization of biological processes at the same time limits their digital reconstruction. In the models, those phenomena are primarily being considered which have a discrete character, for example, the discrete number of tumor cells. In addition, the mathematization and formalization limit the application of parameters that are basically steering simulation and prediction. In the simulations, only those parameters are being considered that are mathematically and digitally applicable. To sum, the analysis of the oncosimulator's underlying conceptualization has shown that the consistent application of mathematic modeling formalized by ICT has far-reaching consequences on doing research: those concepts (e.g., discrete character, applicable parameter) shape the research process from scratch and restrict it at the same time.

The use of mathematical methods and tools in other scientific domains and its ascribed supremacy is possibly the most important reason why interdisciplinary problems have occurred in the course of the oncosimulator's development. However, Stamatakos was able to overcome the reluctance of biologists and clinicians, described as historically the most disturbing factor in the storyline, by convincing them step by step. It started with individual scientists in the ACGT project and is still ongoing in continuing research after ACGT. EU funding seems to be a good environment to meet and collaborate with interdisciplinary-minded scientists and to build up international communities on the edge of emerging research fields such as in silico oncology.

4.3 Coordinating Systems-Oriented Research in a Technological Environment

The quantity of data involved has generated the idea that data-intensive science is a whole new way of doing research (e.g., Mayer-Schonberger and Cuckier 2013; Kitchin 2014). Although the production of such large data stocks has already existed in some domains for some time (e.g., weather prediction, financial markets), biological research has transformed into data-intensive science in the context of Omics and systems-oriented research since the early 2000s (Leonelli 2014, 3). This coincided with the time when technologies for the high-throughput production of genomic data (e.g., DNA sequencing, microarrays) started to become widely used.

The large data stocks have made it necessary to reorganize the storage and management of data. Expectations regarding the use of ICT in systems research to manage large data repositories are currently high and ICT infrastructures are already under way towards realization. The following section focuses on the discrete results or outcomes of the ACGT research project to retrace the current status of ICT in systems-oriented research and to assess the potentials of such an approach. Based on empirical data, we tried to find out what the ACGT members thought to be necessary to maintain an ICT infrastructure and to keep it running and how the scientists evaluate its productivity. Questions related to such issues were included in the questionnaire we used in the interviews described in Sect. 4.1.1. We asked inter alia for the concrete results of the ACGT research project and the reasons why the goal of designing an ICT infrastructure and implementing it into the emerging systems-oriented research community in oncology was not fully reached when the ACGT project was concluded. Many interviewees agreed that the ACGT project was not able to accomplish an ICT infrastructure that can be used in clinical practice because it requires much more effort in terms of financial support and time to be invested than was available in a four-year research project. Therefore, they broadly discussed what lessons they learned regarding the use of ICT for systems-oriented research. The question is: what kind of function will ICT infrastructures have, or, in the eyes of the interviewed ACGT consortium members, are they supposed to have in systems medical research?

4.3.1 The ACGT Project and Its Results

In the proposal of the ACGT research project, its objective was clearly defined: the ACGT consortium aimed at designing and developing an integrated ICT infrastructure that offers tools and techniques for the mining of data from data repositories and the extraction of knowledge from knowledge discovery services (see Sect. 4.1.1). Hence, the project's results can be directly compared to and evaluated by the objective described in the proposal. However, the following section shows that the interviewees consider not only technological innovation, but also indirect outcomes such as gaining experience in the research process as valuable results of the ACGT project.

4.3.1.1 Technological Innovation

As the research guiding objective was the creation of an integrated ICT infrastructure, it is certainly not surprising that the whole endeavor was described as a technological innovation starting from the outset.

"In the beginning, there was nothing," explained an IT-expert in the interview (I3); "so we had an objective and the objective was to create an infrastructure that would allow you to do scientific research over a distributed platform. A platform that would consist of many different institutions that all had their own computational resources that would allow you to do research that was not possible before. But in the beginning we did not have any infrastructure so this Grid infrastructure needs to be created." The computer scientists within the ACGT project started by investigating what type of software was available and what kind of conceptual systems were already built and how existing databases worked. After the basic decisions were made of how to create the ICT platform, the assigned ACGT partners developed different technological components and tools such as the data access services, the clinical trial management system, or the workflow editor (see Sect. 4.1.1). These components were composed as parts of an integrated system. However, many problems occurred when the components designed by different ACGT partners were to be assembled into an integrated architecture.

> The general idea how the components were supposed to interact, that existed already. But whenever we sat down together, when we programmed something together, linked up a few things from various partners, then there was always some kind of problem. And then it sometimes took weeks to find out what the problem was. That was also a reason why ACGT wasn't so successful, because this Grid technology is very, very complex. That means, in the following project we're not taking that kind of approach any more. Instead, we're trying to keep things simple. Because it really may be that in the end, the problem is … if a workflow doesn't run properly because, say, the computer on Crete, that computer's clock is a millisecond ahead of the clock we have here. And then some security alarm went off because it thought that data from the future are coming in—that can't be, so it aborted the process. But you've got to figure that out, and it isn't easy. That can take days and weeks until you've figured out somehow, going through the entire system why one part somewhere seems to think that something isn't working anymore. (I12, IT)[31]

In the course of the ACGT project, the coordination and assembly of the components was continuously presented as an end-to-end demonstration at meetings in front

[31] German original: "Die allgemeine Idee, wie die Komponenten zusammenspielen sollen, gab es halt. Aber es war immer so, wenn wir uns zusammengesetzt haben, irgendwas zusammen programmiert haben, ein paar Sachen von verschiedenen Partnern verknüpft haben, dann gab es immer irgendwo ein Problem. Und dann brauchte man teilweise Wochen, um herauszufinden, woran es lag. Das war auch ein Grund, warum ACGT nicht so erfolgreich war, dass diese Grid-Technologie sehr, sehr komplex ist. Das heißt, im Nachfolgeprojekt haben wir so was auch nicht mehr, sondern versuchen das einfacher zu machen. Weil es wirklich sein kann, dass im Endeffekt das Problem daran … wenn ein Workflow nicht durchläuft, weil die Uhr, die irgendwie der Rechner auf Kreta hat, irgendwie eine Millisekunde vor der Uhr läuft, die wir hier haben. Und dann knallte irgendwas mit der Security. Weil der meint, da kommen irgendwie Daten aus der Zukunft—kann nicht sein, und bricht ab. Und da muss man halt erst mal drauf kommen. Das kann halt Tage und Wochen dauern, bis man dann irgendwie durch das gesamte System herausgefunden hat, warum irgendeine Stelle meint, dass jetzt irgendwas nicht mehr funktioniert." (I12, IT)

of reviewers assigned by the EU commission.[32] These demonstrations were adapted to a scenario-based development process in which a number of scenarios were created. Essentially, they can be described as a sequence of activities conducted by a clinician who is willing to use the ACGT platform in his or her clinical trial. The sequence followed the established procedures of data handling in a clinical trial, that is, access to heterogeneous data, use of various tools for data analysis, and invocation of appropriate tools for visualizing and interpreting results (e.g., ACGT 2009).

The clinician as the final end-user of the ACGT infrastructure was in focus of the scenarios. However, it was often stressed in the interviews that the ACGT project was a research and development project (R&D project). After only four years of research, the developed infrastructure was not ready for regular use in clinical practice and many of the interviewed members of the ACGT consortium did not initially anticipate that by the end a sustainable infrastructure would exist that could be of use to the oncological community. In their view, the ultimate objective of the ACGT project was just to prove the concept. They wanted to show that developing an ICT infrastructure for clinical systems research in cancer is possible*:* "ACGT was a kind of a proof of concept. As is the case I think of the most of EU projects. You are trying to build something to show that it is possible and of course, you are trying to build up on it in future projects. And try to reuse it. But it's not building a production level system." (I5, IT) According to the quotation, clinical application ("the production level," previous citation) was not the scope of the ACGT project but the proof of concept which means in the first place to develop an infrastructural prototype. This is what the ACGT consortium achieved: the first integral version of an ICT infrastructure was presented as an end-to-end demonstration at the final review meeting held in Heraklion (Crete) in September 2010.

Concerning the technological outcomes, it can be said that not the infrastructure, but the individual tools such as the clinical trial management system called OpTiMA or the oncosimulator were the most concrete technological achievements of the ACGT project. Many of those components hold the potential for further use in follow-up projects. For example, the security tool named the Custodix Anonimisation Tool that supports anonymization and pseudonymization of different types of data, designed by the software development company CUSTODIX, is already reused and extended in follow-up projects.[33] The integrated ACGT infrastructure itself broke down several months after the research project had ended. The reason was a very practical one: the technical partners switched off the server capacities for the ACGT infrastructure one after another, and the ACGT computing network that was built all over Europe broke down.

---

[32] End-to-end demonstration basically means that the assembly of components into a system is demonstrated by creating workflows using the system from end to end.

[33] The successor CATS is a versatile service platform for de-identification and pseudonymization which can easily be integrated into high-volume data workflows and is, for example, applied in the ACGT follow-up research projects p-medicine (www.p-medicine.eu) and INTEGRATE (www.fp7-integrate.eu). Both websites accessed September 15, 2014.

4.3.1.2 Experience

The ACGT platform did not persist; however, the project instigated research on ICT infrastructures in the biomedical domain by many former members of the ACGT consortium. In reference to the broader research field, the interviewees highlighted in particular the experience and the knowledge they gained in the research process as a valuable outcome. The ACGT project was "a very good basis for the things that we are doing now" (I2, IT) said an interviewee who is currently working in one of the follow-up EU projects. The ACGT project therefore seemed to be a starting point of promising research that was worth being pursued in future work.

> So in a general point of view, I suppose that we gained a lot of understanding on how difficult it can be to create an infrastructure that is very technological, at a very bleeding edge, advanced, and apply that to a setting that has no clear understanding of the computer science ideas behind it. And so, there were a lot of difficulties that we had to overcome. But, you know, during the course of the project, we also gained a lot of understanding on how we could cope with those difficulties and how we could sort of fix the underlying challenges. That is one. And the other thing from our own personal perspective is that we created something new. We created an architecture that we still use, not in the same types of projects that we did with ACGT, but we are now applying the same type of research to other projects. It allows us to continue the research that we have done and extend on it. So that is good from our perspective. I suppose, but it is guessing, I suppose that from the clinical point of view, there is a better understanding of how computer science can help the research in a clinical setting. Especially also on subjects that have to do with, let's say, the genetic backgrounds and everything that has to do with proteomics, the Omic types of research. We are not a part of p-medicine, once again, but I suppose that the people in p-medicine have a clear understanding of how they could continue with the work, the results that were produced in ACGT. And how you can build upon that and get your own science further and better. (I3, IT)

Assessing ACGT's impact on future research, the cited interviewee underlined that the scientists were gaining a deeper understanding of the theoretical challenges and practical obstacles of developing an ICT infrastructure. This means that the awareness of the problems was created by practical experience in the first place. This approach was expressed by others as well. For instance, biomedical experts pointed out that they learned more about the possibilities and limits of ICT. Some of them considered for the first time the ethical–legal requirements that are indispensable when designing clinico-genomic trials and having a continuous access to data-sharing platforms. Computer scientists, on their side, got an inside view into daily clinical workflows, the amount of information that can be generated from genomic data, and how sensitive these data are in legal and ethical terms.

This experience was described as being important for future research in follow-up projects: "A lot of things were used in p-medicine. Yes. And mostly the experience was used, which is, in terms of time, huge. The biggest thing a lot of times is the experience of what the problems were rather than the actual building of the tool, because the actual building of the tool doesn't take that long if you know exactly what you need" (I9, BioMed).

Looking at ACGT as seen and assessed by the interviewees, it can be concluded that the concrete results of the project were primarily technologically defined.

However, only individual components such as the clinical trial management system, the security tool, or the workflow editor were positively assessed as having the potential for further use, but not the integrated architecture of the ICT infrastructure. Yet, personal outcomes were especially discussed to be as important as the technical ones. The ACGT members appreciated that they were gaining experience in the emerging field of systems-oriented research in oncology. They mentioned in particular that they deeply explored technological and theoretical concepts (e.g., Grid computing, ontologies) and gained practical experience in interdisciplinary work (e.g., clinical workflows, data protection standards) or got opportunities for interdisciplinary networking. These more indirect outcomes were regarded as having a crucial impact on future research in systems-oriented research.

4.3.2 The Sustainability of ICT Infrastructures

Several months after the ACGT project was finished, the integral ACGT infrastructure was shut down. Most of the former ACGT members were not willing to provide server capacities for an indefinite time for an R&D project that was already terminated. In addition, the services were often needed and reused in follow-up projects. This is not unusual in research projects, as one IT expert explained: "The fact that it kept on running before it was finally turned off, that it was in sleep mode, in Halbernet mode, that's what's unusual. The individual institutes, there's no way they can achieve that simply because they're research institutes" (I14).[34] Accordingly, the interviewees collectively agreed that one of the basic problems why the ACGT platform did not succeed to be used in the clinic was the lack of sustainability with regard to server capacities after the completion of ACGT. Claims for sustainability were often expressed in the interviews. We therefore aim at exploring how sustainability affects the potential of ICT in systems-oriented research. The analysis shows that sustainability is defined not only for server capacities, but for different objects and different contexts. In the following sections, we take a closer look at those objects and contexts that, in the eyes of the interviewed scientists, need to be sustainable to keep ICT infrastructures running and implement ICT into systems research.

4.3.2.1 Technological Sustainability

To keep ICT infrastructures running and finally to implement ICT into systems-oriented research, the ACGT interviewees regarded sustainability defined in technological terms as sine qua non. In this context, the technical design of the prototype itself was criticized by some interviewees. They discussed why the ACGT prototype

[34] German original: "Dass es, bevor es endgültig abgeschaltet wurde, so lange noch im Schlafmodus, im Halbernet-Modus, weiter gelaufen ist, ist eher das Ungewöhnliche. Die einzelnen Institute, die können das gar nicht leisten, weil es halt Forschungsinstitute sind." (I14)]

was in their view not the best solution to build up a user platform for the systems-oriented research community in oncology. One reason given was the software used that was still in its infancy.

> A lot of the technology that was used within ACGT was not mature enough. But those pieces were not generated within ACGT. So that is unfortunate, because you are basically building on something that is not mature yet. But you are trying to assess if even this immature technology can be applied to your context. So that is the research project and I think that the results that came out of ACGT were very enlightening. Because a lot of progress has been made on understanding what did not work and what did work. So you can use that in the next project. You now use the things that worked and try different things for the things that did not work. But you are always faced with pieces of the technology that are outside your power. Right so, in ACGT, there was a lot of Grid software that was used, that was developed in other projects like Aggie or Cern, in other European research projects that were primarily focused on creating this Grid infrastructure. So there is nothing you can do about this. (I3, IT)

As the quotation indicates, in particular, the software used came under criticism for its immaturity and complexity. The Grid software is complex as it ties the technical components of the infrastructure closely together.[35] At the same time, the infrastructure was distributed. This means that each technical partner of the ACGT project was requested to offer one or more server(s) that would then be connected with the servers of the other ACGT partners. Hence, the system as a whole and not only the parts of it had to be maintained for its sustainability. As one interviewee put it, "a key question that was set in the beginning of the ACGT was the following: is there value in such a setting of employing a Grid infrastructure? Which I think it is one of those cases where you spend a lot of effort in trying to find the answer and the answer at the end of the day is, it is probably not" (I7, IT). At the same time, Grid technology is more and more being replaced by Cloud technology as computers are becoming faster and cluster computing is no longer necessary. Of course, computer scientists are familiar with such technological developments in which one approach is replaced by another. Looking into their daily work, IT specialists seek cutting-edge technologies such as Grid or Cloud computing in order to use these technologies for their designated application, such as designing an ICT platform for systems-oriented research in cancer.

Seen from today's perspective, the ACGT project was not only working with immature technology, but the Grid computing approach itself was questioned and even soon outdated. Drawing on this argument, one of the interviewees explained that the concepts are still the same although the underlying software might change.

> [W]e are not talking about Grid systems anymore, but we are talking about Cloud systems. And it is a subtle change in approach, but the technology questions are still the same. […] It mostly boils from the awareness that the Grid technology that we created was far too immature. So they ripped out portions of it and they took different portions and integrated that and now it is called the Cloud systems instead of the Grid system. But the concepts are

[35] Grid computing is based on the collection of computer resources from multiple locations to work on a common goal or project.

> still the same. And from my understanding, once again I am not part of p-medicine, but from my understanding they are now applying Cloud technology in p-medicine. That is a logical approach. It makes sense. (I3, IT)

However, the interviewees coincidently said that the ACGT platform was not implemented in systems-oriented research in oncology because it was still a prototype. Many interviewees expressed that the final, but crucial step was not reached during the project: the step from experiment into practice. Hence, the prototype needs to be converted into a production system that can be regularly used by clinicians who are not familiar with high-performance computing. To be ready for customers, engineering of the research software is necessary. This means that the software has to be tested and consolidated, documented, and, finally, certified.

> Research software lacks things that a real application has, like error tolerance, user interfaces, menus or manuals, well, completely normal trivial things that are totally uninteresting for a research project. For example, you show the prototype in a review, that's a proof of concept. You say, this is what we have in mind, this is how it's supposed to work. This is how it would work, fundamentally speaking. That works now. But to be able to sell it, practically as a system, well, quite a lot is still missing, namely software engineering. That means that you have a test department of your own. That means that there are people on staff who really test things from morning to night, checking the whole thing for bugs. There are people for documentation. You don't have them, either, in a research project. The deliverables where you could say, well, a lot of text was produced about the tools, they're for real end users who weren't involved in the project, hardly comprehensible or useless. Well, those are things that are really missing and that take a whole lot of time. And in the world of research, that often isn't so clear. (I14, IT)[36]

From the quotation it follows that scientists coming from university are often not familiar with the requirements of a tool expected to be ready for application on the market. Another interviewee outlined that he realized in the course of the project that it was impossible to establish the ACGT infrastructure for clinical use because of the lack of financial support and time to be invested into software engineering and marketing to achieve marketability.

> It's a vision… and I was ambitious together with a number of other people. Not everybody, but a number of other people. But at the same time, you need to be aware of what the reality is and what life is. And having gone through close interaction of what had caBIG achieved in the United States, I had discussions and I had meetings with the director of caBIG,

[36] German original: "Forschungssoftware fehlt für eine richtige Anwendung solche Sachen wie Fehlertolerabilität, Benutzerschnittstellen, Menüs oder Handbücher, also ganz normale triviale Sachen, die für ein Forschungsprojekt völlig uninteressant sind. Den Prototyp zeigt man zum Beispiel in einem Review, das ist ein Proof of Concept. Man sagt halt, wir stellen uns das so vor, so sollte das laufen. So würde das grundsätzlich gehen. Das funktioniert jetzt. Aber um so etwas quasi als System verkaufen zu können, da fehlt noch wirklich relativ viel, nämlich Software Engineering. Sprich, dass man ein eigenes Test-Department hat. Sprich, dass halt Leute da sind, die wirklich von morgens bis abends nur testen, das Ganze nach Bucks durchschauen. Es gibt Leute, die dokumentieren. So was fehlt auch in einem Forschungsprojekt. Die Deliverables, wo man sagen könnte, es wurde ja viel Text produziert über die Tools, die sind für richtige Endnutzer, die nicht in dem Projekt drin waren, kaum zu verstehen oder unbrauchbar. Also das sind Sachen, die wirklich fehlen und die sehr viel Zeit brauchen. Und das ist oftmals in der Wissenschaft nicht so klar." (I14, IT)

> Professor Buetow, etc. They had a structure. They had offices. They had a marketing director. They had scientific directors. They were functioning as a kind of a company whose task was to develop, to further develop, to open new directions for additional work. But also to make sure that there is support for the community to publicize, to market, etc. And they had the 10 million minimum per year to support their functioning, etc. When you compare that to a European R&D project, although the ambition and the vision was there and I think were supported very heavily, very nicely through our reviewers [...], we realized that it cannot happen. It is very rare that you see a European R&D project, because it is an R&D and not a development project that you end up with a fully functioning infrastructure and the reason for that is that… there are three reasons. Because up to the very end you are exploring scientific and technical issues so you are doing research at various levels. The second is the fact that in European R&D projects, you develop proof of concepts and not production quality systems, the third is that very often you see research groups, once they have reached the proof of concept prototype and published, they lose interest in making it in production quality and production ready system. (I7, IT)

The quotation again referred to ACGT as an R&D project entering new research territories at various scientific and technological levels but not the market. However, several private companies that are usually familiar with the adaptation of products to the market were integrated into the project. Would it have been possible that these companies focus on marketability or how do they define their role in R&D projects such as ACGT? One of the interviewees explained that companies pursue their own interests why or while they seek to be involved in academic research. Essentially, they participate in order to understand trends in future research and to be involved in innovative developments. "First of all, for us it is a kind of an early warning system. We get to listen to academics and what they think is the next big thing although often we find that we tell academia where things are going. It is good, because these things definitely give you a very good platform to project yourself and be seen as an avant-garde company so that you are involved in new things, state of the art things" (I4, IT).

This interviewee assigned the potential of trend-setting innovations to academic research, although companies seem to play a role in the second attempt. Another interviewee of an internationally oriented enterprise pointed directly to the commercial sector and how this would influence his own work.

> The alternative would be that you let industry make a decision. So you go, for example, to Microsoft, and tell them this is my problem and please advise me. Then Microsoft will create a Windows Cloud or Windows Grid or something like that. But it will not give you the opportunity of influencing the decisions that are going to be made there. So you are basically forced to swallow the decisions that Microsoft would have made if they decide to build something like this. Therefore, you have to conform to what they did. Whereas in research projects, there is still a possibility of saying to people who have developed the technology, 'the decision that you made there is maybe appropriate for your line of research, but it is not appropriate for my line of research so please can we talk. Can we figure out a way of trying to solve this?' (I3, IT)

From this it follows that research projects, in particular R&D projects, open up space for new trends and approaches in research. In fact, ACGT was one of the first projects exploring how Grid technology could be applied for doing oncological research. Even if the integral ACGT platform was not mature and sustainable enough to reach clinical use, the interviewees collectively agreed that the results of the ACGT research were valuable and necessary and provided a basis for understanding key issues such as data integration or sustainability of ICT infrastructures. At the same time, the innovative processes taking place seem to be open or democratic enough to allow different stakeholders (e.g., academia, industry) to develop and influence landmark decisions for future research. However, neither academia nor industry seems to be willing or able to be responsible for market introduction within an R&D project. This last aspect is extremely important for the translation of systems biology knowledge and tools to applied research such as systems-oriented research in oncology.

4.3.2.2 Financial Sustainability

Another element of sustainability discussed in the interviews of how to make an ICT infrastructure sustainable to keep it running was the funding. A central server as a sustainable facility where the software can be hosted to allow research to continue after an R&D project had ended was one suggestion mentioned in the interviews. However, to host an ICT infrastructure requires continuous financial support and manpower for maintenance tasks. In this context, some of the interviewees stressed that to date researchers (and funders) usually have the mindset that you don't have to pay for Internet use. To solve this problem, a new path of institutionalization is currently being developed in the ACGT follow-up research project p-medicine.

> A structure is established that is going to be dynamic and that will adapt to new circumstances in the future, too. But it's supposed to be a structure where I can continue to do this research. The important thing is to assemble data, to evaluate them, and to put them in a system that can continue to exist independently of EU funding. That means, we're currently trying to develop a business plan where we, for example … a very simple example. If you take OpTiMA, it's structured like a modular system. If I take this Trial Outline Builder, then you can set it up so that I can collect data without having this Trial Outline Builder. You can get a basic module in OpTiMA for free. And then, if somebody wants to have this Trial Outline Builder, then they can buy it via licensing fees etc. If the modules that I can attach to it are so attractive that someone says, that's what I need, then they'd buy it via licensing fees. If I use a data management system at the hospital, I have to pay for that, too. Well, we're trying to establish long-term funding with this kind of ideas for a business plan. (I18, BioMed)[37]

---

[37] German original: "Es wird eine Struktur aufgebaut, die dynamisch sein wird und die sich auch in der Zukunft wieder an neue Gegebenheiten anpassen wird. Aber es soll eine Struktur sein, wo ich diese Forschung weiter betreiben kann. Das Wesentliche ist eben Daten zusammenzuschweißen, die auszuwerten und in ein System einzubringen, was auch weiter bestehen kann unabhängig von einer EU-Förderung. Das heißt, wir versuchen im Moment einen Businessplan zu entwickeln, wo wir dann zum Beispiel … ein ganz einfaches Beispiel. Wenn man OpTiMA hat, das ist ja aufgebaut wie ein modulares System. Wenn ich diesen Trial Outline Builder hole, dann kann man das so machen, dass ich Daten sammeln kann, ohne dass ich diesen Trial Outline Builder habe. So ein Basismodul in OpTiMA gibt es dann umsonst. Und wenn jemand dann aber diesen Trial Outline Builder haben will, dann kann er sich den dazu kaufen über Lizenzgebühren etc. Wenn diese Module, die ich dann da dran hängen kann, so attraktiv sind, dass jemand sagt, das brauche ich, würde er es dazu kaufen über Lizenzgebühren. Wenn ich heute irgendein Data Managementsystem benutze in der Klinik, muss ich auch dafür zahlen. Ja, und über solche Vorstellungen eines Businessplans versuchen wir eine langfristige Finanzierung zu etablieren." (I18, BioMed)

The license fee provides the possibility to afford staff for data management, including data curation, and for advancing the ontology implemented into the ICT infrastructure. The business plan mentioned by the interviewee refers to the institution named the Study, Trial and Research Center (STaRC) which is currently on its way to becoming an innovative center to host and provide a service-oriented clinical research infrastructure based at the University Clinic of the Saarland in Germany.[38] The researchers will have the opportunity to run clinico-genomic trials and do systems-oriented research on the STaRC platform by paying for its use. However, the researchers have to take up the offer by actually using this ICT infrastructure. As clinicians are accustomed to paying for the use of data management systems in the clinic, the license fee, as outlined in the above quotation, will rather be contextualized in data management systems than in Web-based services.

Most of the interviewees are positive about future research because the results achieved in the ACGT project will be taken to the next level of realization in the follow-up research projects. It thus appears logically that the institutionalization of STaRC breaks new ground in different directions, not only in research but also in academic mindsets to create financial sustainability.

4.3.2.3 Social Sustainability

To break new ground in research and to put innovation into practice always has a social dimension. It requires scientists who change or widen their mode of thought and of doing research. At least the latter aspect is deeply embedded in social interactions as one interviewee of the biomedical domain outlined. Before participating in the ACGT project, the interviewee did mostly clinical research and only worked together with clinicians. Today, he is working with researchers of different disciplines to translate systems-oriented approaches into clinical practice. In the following citation, he is convinced that only cohesive interdisciplinary teams will be able to improve survival rates and progress in health care.

> I have the feeling that you can really make things happen here if you can get everyone with a say in the matter to the table. And it really isn't just the medics who can treat patients in the end. In the future, they'll need IT people. They'll need systems biology. They'll need ethicists and lawyers. They'll need basic research. They'll need the bioinformatics people. And in the future, you'll only be able to help a patient if you have a cohesive team like that. Take pediatric oncology: in the last 30, 40 years, we've achieved a really steep increase in survival rates. We achieved that by working together, doing prospective clinical studies, and gaining new knowledge to improve therapies. And that worked, for purely, … well, clinical considerations. Then, molecular biology was added to characterize patients better. And nonetheless, we're stuck when it comes to certain groups that we can't get healthy. That means that the steep increase in improving survival rates has been turning into a plateau for

---

[38] STaRC was founded by Norbert Graf, the director of the Clinic for Pediatric Oncology and Hematology at the University Clinic of the Saarland and the former quality manager in the ACGT project. Start-up financing is provided by the federal state government of the Saarland and the European Union in the course of the research project p-medicine in the 7th Framework Program. See STaRC, http://eu-starc.eu. Accessed October 3, 2014.

about the last 10 years. Suddenly, we can't improve a certain survival rate and we don't know why one patient is relapsing and another one isn't, because our current knowledge is the same for both. That means we're lacking information. And we have to get that information from these approaches. That's the only way, namely by putting together really all the data about the patients that you have. By developing disease models with the systems biology approach and then combining them and finding out individually for each patient what the best treatment is. And that's the reason why that's a very important development for clinicians. (I18, BioMed)[39]

Of course, for the interviewee cited it was easy to meet and collaborate across disciplines as research projects targeting the development of an ICT infrastructure are interdisciplinary positioned. Many of the interviewed consortium members depicted the ACGT project as a starting point for continuous interdisciplinary cooperation. "So we actually formed some kind of team" (18, IT), said an interviewee who collaborated with nearly all of his former ACGT colleagues afterwards. Someone else stressed that not only collaboration between people but also between the institutions are of vital importance and require continuation. Here, the impact of ICT becomes evident as continuous relations between research centers are not based on personal relations any more.

The cooperation was crucial I think, not only between the persons, but the centers. The information was written down, but it would challenge a similar group of people to go through all the information and learn the same lessons. It is true that you write down the information, but to really read it and use it all would probably take one year on its own, all that amount of information. So in terms of time it was very good that the same people worked on it, because they had both the experience and the access to the same tools. I mean the biggest thing with ACGT is that within four years it created a link between centers and people that never spoke with each other before. Some of them did, but a lot of them didn't. (I9, BioMed)

[39] German original: "Ich habe das Gefühl, dass man hier tatsächlich etwas bewegen kann, wenn man alle die Leute, die was zu sagen haben, an einen Tisch bringt. Und es sind eben nicht nur die Mediziner, die am Ende Patienten behandeln können. Die brauchen die IT in Zukunft. Die brauchen die Systembiologie. Die brauchen Ethiker und Juristen. Die brauchen Basic Research. Sie brauchen die Bioinformatiker. Und nur, wenn man so ein geschlossenes Team hat, wird man tatsächlich in Zukunft einem Patienten helfen können. Wenn ich mir die Kinderonkologie angucke, dann haben wir in den letzten 30, 40 Jahren einen ganz steilen Anstieg von Überlebensraten bekommen. Das haben wir dadurch bekommen, dass wir zusammengearbeitet haben, prospektiv klinische Studien gemacht haben, und neue Erkenntnisse gewonnen haben, um Therapien zu verbessern. Und das ging rein aus … ja, klinischen Überlegungen. Dann kamen molekularbiologische Überlegungen dazu, um einen Patienten besser zu charakterisieren. Und wir bleiben trotzdem bei bestimmten Gruppen hängen, die wir nicht gesund bekommen. Das heißt, diesen steilen Anstieg in der Verbesserung von Überlebensraten geht seit ungefähr 10 Jahren in ein Plateau über. Wir können plötzlich eine bestimmte Überlebensrate nicht verbessern und wissen gar nicht, warum der eine Patient rezidiviert und der andere nicht, weil unser heutiges Wissen für beide gleich ist. Das heißt, uns fehlen Informationen. Und die Informationen müssen wir aus diesen Ansätzen bekommen. Das ist die einzige Möglichkeit, indem man wirklich alle Daten von Patienten, die man hat, zusammenbringt. Indem man aus dem systembiologischer Ansatz Krankheitsmodelle entwickelt und das dann kombiniert und individuell für den einzelnen Patienten rausfindet, was ist wohl die beste Behandlung für ihn ist. Und das ist der Grund, warum das für den Kliniker eine ganz wichtige Entwicklung ist." (I18, BioMed)

These collaborations are lasting, because they are based on joint research objectives or "a common vision" (I8, IT) as one interviewee explained. Finally, the collaborating researchers are becoming more interdisciplinary-oriented and open-minded. "I think that practically all people participating in ACGT had de facto to become more multidisciplinary otherwise such a project, which is by definition a very strong multidisciplinary project, would not ever come to a successful end. I can say that we all enjoyed this opening to new areas, to new knowledge, the sharing of knowledge and interaction. It is a new window to the future somehow" (ebd.). As a result, new interdisciplinary scientific communities have emerged that are assembling around research objectives that can only be approached by interdisciplinary collaboration. As the ACGT consortium shows, because of the interdisciplinary and international approach, ICT has become an integral part of those new scientific communities such as the systems-oriented research community in oncology.

The EU commission has reacted to interdisciplinary community building by defining, for instance, the Virtual Physiological Human (VPH) as a core target of the 7th Framework Program which pursues patient-specific computer models and their applications in personalized health care (Kohl and Noble 2009). Within the frame of this program, about 30 systems-oriented research projects were funded. Their goals mainly addressed technological achievements, including data collection, management, and integration as well as processing and curation of data. Furthermore, reductionist and integrative modeling of pathophysiological processes and, finally, presentation, deployment, and end-user applications were under study.[40] However, references to translation were continuously included as nearly all of these projects dealt with challenges relating to patient-specific, multiscale modeling and the implementation of models and software in clinical environments. Here, simulation, data handling, scientific visualization, and community building were in the focus. Previously identified limitations in ontology annotation and inadequate tools to secure wider sharing of models and data (authentication, authorization, etc.) have also being addressed.[41]

Furthermore, the Virtual Physiological Human Network of Excellence (VPH NoE) was established and funded in the frame of FP 7.[42] The network aimed at connecting the various VPH projects and fostering the development of educational, training, and career structures for those researchers involved in VPH-related science, technology, and medicine. VPH study groups, educational meetings, and training events as well as the VPH conference series to be held every 2 years to showcase the best of VPH research were set up. In addition, the VPH NoE supported the emerging community by building up services freely available to researchers,

---

[40] VPH projects of FP 7 that are identified by the interviewees as follow-up projects of ACGT are, for example, p-medicine, www.p-medicine.de, INTEGRATE, http://www.fp7-integrate.eu/, TUMOR, http://tumor-project.deu/, ContraCantrum, http://www.contra cancrum.eu/. All websites accessed July 3, 2014.

[41] VPH Network of Excellence, Newsletter No 8, Sept 2012, VPH_NoEnews_N8_34p.pdf, 10ff.

[42] Virtual Physiological Human Network of Excellence, http://vph-portal.eu. Accessed June 15, 2014.

for example, developing common standards, open source software, and freely accessible data and model repositories in the context of systems research.

To find mechanisms and strategies that enable the VPH community to continue to profit from the legacy of the VPH NoE beyond the runtime of the EU-funded network, the Virtual Physiological Human Institute for Integrative Biomedical Research (in short VPH Institute) was established and founded as an international nonprofit organization incorporated in Belgium in 2011.[43] Its mission is to ensure that the endeavor of the Virtual Physiological Human will be fully realized, universally adopted, and effectively used both in research and in the clinic. The VPH Institute has continued the work of the VPH NoE in many respects, including the running of the VPH conference series and the management of the VPH Portal after the VPH NoE had finished. To date, the VPH Institute represents over 67 public and private institutions active in VPH research, including many academic, clinical, and industrial key players in the area of in silico medicine.

To sum up, the activities of the VPH NoE show that EU science policy initiated the development of a strong interdisciplinary and Europe-wide scientific community in the field of systems biology for future research.[44] However, the European Union funded the network only for about five years (June 2008 to March 2013). After this funding ended, the community repositioned itself by founding the VPH Institute as an independent, nonuniversity institution. This development shows that interdisciplinary networks and research initiatives require an institutional host and permanent funding to survive. However, the universities have not yet taken up the task to fill this gap and new paths to support and foster systems research in the future are already set up that are more independent of university institutions and public funding.

4.3.3 Concluding Remarks

The broader analysis of results and outcomes of the ACGT project reveals that not all the goals of ACGT stated in the original research proposal could be realized. The reasons rest not only on a gap between too high expectations and reality of a four-year lasting research project. Rather, they refer to an underlying tension of exploring epistemic concepts and developing practice-oriented products. In particular the last aspect is closely connected to sustainability which is brought up in the interviews with regard to different objects and contexts. It thus appears important that the development of innovative products requires a research context in which concepts and practices (e.g., programming, defining parameters) are jointly developed to connect epistemic concepts with practice-oriented problems systematically.

[43] VPH Institute, www.vph-institute.org. Accessed June 23, 2014.

[44] The impact of science policy on scientific developments is further discussed in Chap. 5.

The first insight of our analysis suggests that an infrastructure in a field of application needs in the first place technological sustainability. The frequently mentioned criticism regarding the Grid software shows that the technical design of the infrastructure has to provide an essential basis for sustainability. Furthermore, it is obligatory that an infrastructure in order to continue requires the conversion of a prototype into a production system that can be used by clinicians who are not familiar with high-performance computing. The ACGT platform was a prototype as the crucial steps of software engineering were still missing in terms of testing, consolidating, documenting, and certifying the software developed. The interviewees addressed the issue that university scientists are often not familiar with the systematic development of an innovative tool toward a product ready for the market. In this regard, there was uncertainty of the product maturity that could be expected at the end of the ACGT project. Other interviewees observed that academic researchers, once they have reached the proof of concept and published it, have not much interest in converting it into a production system. Here, the neglected gap between university and the market become apparent. Private companies, on the other hand, do not primarily participate in academic research to take care of marketability of research results but to understand better the trends in future research and to be involved in innovative developments. Hence, neither ACGT participants coming from academia nor industry seemed to be willing or able to be responsible for market introduction of the ICT infrastructure developed. This attitude of both academia and industry has an important impact on the translation of systems biological knowledge and tools to applied research contexts such as systems medicine.

The second insight of our analysis is that the financial sustainability of the ACGT infrastructure was not given because the technical partners were not willing to provide server capacities for an indefinite time after the ACGT project had ended. At universities, computing resources are generally limited and only used in ongoing research projects. The lack of server capacity is partially due to the Grid technology itself as the distributed servers (and technical partners) have to stay in connection to run the integrated platform. Servers where the software after the completion of a research project is hosted require continuous financial support and manpower for maintenance. Despite intense discussions on how to solve this financial problem, the ACGT consortium was not able to solve this issue.

The third insight of our analysis implies that a sustainable infrastructure requires powerful funding bodies that are able to provide a long-term perspective in terms of institutional sustainability. This is a decisive aspect for the development of systems-oriented approaches in general as new interdisciplinary scientific communities have emerged that assemble around research objectives that can only be approached by interdisciplinary collaboration. As the VPH NoE illustrates, the interdisciplinary community building is broadly funded by the European Commission. However, EU funding is time-limited and systems-oriented communities still lack institutionalization at universities. They have already reacted to this situation by founding, for example, the VPH Institute which is independent of university and public funding. Yet, it is still one of the basic challenges in systems-oriented research to find a host to institutionalize ICT infrastructures. The case of ACGT has shown that R&D projects would not have the institutional power to build up sustainable structures. Hence, it is expected that these new forms of institutionalization may serve as a sustainable host for ICT infrastructures. Time will tell if newly founded institutions such as the VPH Institute or STaRC are able to become powerful enough to coordinate ICT in systems-oriented research, at least at the national or even European level.

The last insight is a very obvious one: an infrastructure to be consistent has to be adapted in the field of application. As already broadly discussed, the ACGT platform itself was not integrated into clinical practice until the end of the project. Hence, we show in the last section what the ACGT consortium basically expected from ICT in systems-oriented research in oncology.

4.4 Impact of ICT on Systems-Oriented Research in Oncology

Asked for the relevance of ICT, all interviewees expressed high expectations regarding the use of ICT infrastructures in systems-oriented research in oncology and in research on other diseases. In their view, ICT is indispensable because of the development towards data-intensive science and the requirement in the medical domain to translate knowledge from one research field to the other.

> Infrastructures could be a breakthrough. Because something that really manages the integration of the laboratory knowledge with clinical trials and so on. The knowledge out there might have much more power than what we have now. The thing is, especially in the field of cancer, that the amount of information that we are accumulating every day is huge. But the amount of information that we can use is really … and that we translate to the clinic safely is very little. (I9, BioMed)

Another interviewee from the biomedical domain pointed out that in many trials in the past, only clinical data were taken into consideration that were relevant in order to compare drug A versus drug B. Now, more and more data will be used for diagnosis, prognosis, and prediction of drug response and so on. By comparing them and doing experiments on them (e.g., next-generation sequencing), the amount and the complexity of data increase even more. In the laboratories, biologists are already using the results from clinical trials to set up laboratory experiments or when they investigate the functions of a gene, they routinely include experiments testing effects on drug applications as well. "These things are already happening. What it is not, a lot of times they are not happening in a structured way. Projects like ACGT help structuring the process of things that are already happening, but not in a structured way" (I9, BioMed). According to this view, ICT infrastructures provide a framework to administer large amounts of clinical and laboratory data and to create interoperability between those heterogeneous data sources.

> It is one of the options to improve health care, not only cancer research, but everything. So it is one of the logical next steps that you would take if that technology that we have lying around can very much improve care and research. I think it's a good approach to test

> that. We don't know it for sure. Everyone thinks that it will and everyone has a good feeling about it, but to be honest, we don't know it one hundred percent for sure. I mean there is so much data in care that you want to use in research. That is already available there. There is so much knowledge generated in research that actually has a very hard time to find its way back into care. And what all these projects try to do is to reconcile, to bring those two worlds closer together. So that researchers get more data, get more patients in their trials, and on the other hand, care providers get more direct input, assistance by mining these data to have new guidelines for treatment. They can get this immediately into the systems in form of the decision support. They also get more feedback about their personal patient if someone has done research on data and they found something weird. Then they can get feedback about it. So they have to benefit. The translational research is real. Both partners can benefit from each other and the only way in my view is to do it through ICT. (I11, IT)

The quotation stresses the impact of ICT on translational research by aligning clinical and laboratory research and designing "the structured way" (previous quotation) how clinical practice and laboratory research relate to each other. However, the decisive point is that some procedures and tasks would not be performed without ICT support. For example, the semantic and syntactic integration of different data types based on corresponding standards is regarded as necessary (see Sect. 4.1.3). ICT infrastructures can therefore be described as catalysts of doing translational research. Quite similar, the term "breakthrough" was used in the interviews to describe the impact of ICT on translational processes from the laboratory to the clinic and vice versa (see previous quotation).

However, the description of ICT as a catalyst was only one picture in the interviews. An opposing picture is outlined by the following citation.

> To translate and do it safely, you do need a huge process that can be accelerated if you have good integration of data and you are using the standards and so on. [...] It could be a huge facilitator if you really have a good platform. The discovery might come anyway, but it takes ten years instead of one year if you don't have a good infrastructure. (I9, BioMed)

Here, the term used, "facilitator" (previous citation), indicates that ICT should ease and accelerate research activities. However, according to this interpretation, ICT is not indispensable for translational research as "the discovery might come anyway" (previous citation). Hence, the assigned tasks of ICT infrastructures (e.g., providing access and making data shareable) are not regarded as being part of the original research process. ICT infrastructures rather appear as a data management system and in this sense as a service facility.

As a result, there are two opposing concepts of how ICT functions in systems medical research outlined in the interviews. The first picture (ICT as a catalyst) reflects our second hypothesis outlined in the beginning of this chapter that the application of ICT enables doing systems-oriented research as some research activities would not be possible without ICT support. The second picture (ICT as a facilitator) mirrors the more popular understanding, namely that the ICT infrastructure as a service facility are not taking part in research processes. We discuss in the last section, what kind of function and role ICT infrastructures may in fact play in systems-oriented research in the future.

4.4.1 Conclusion

The ACGT project had the ambitious goal of designing an ICT infrastructure in support of systems-oriented approaches in oncology and implementing it into the emerging scientific community. The integrated platform aimed at offering tools and techniques for the distributed mining of autonomous data repositories and extraction of knowledge by knowledge discovery services.

However, as discussed in Sect. 4.1, the development of the ACGT infrastructure was accompanied by considerable challenges coming from different angles. (1) The overarching task of data integration had to be tackled by considering syntax, semantics, and data acquisition contexts; (2) many technological problems occurred and had to be solved on an ad hoc basis in the course of its development as basic standards for integration processes did not yet prevail; and (3) the necessity to work together within different disciplines required not only individual skills but also elaborated strategies of project management to arrange interdisciplinary collaboration.

As shown in Sect. 4.3, the ACGT project was a pioneering project and the interviewed consortium members often referred to the status of ACGT as an R&D project. It was stressed that it represented one of the first approaches as to how Grid technology could be applied to support and facilitate medical research. Therefore, the consortium had to start conceptually from scratch. The initial goal was to explore whether the Grid technology would be adaptive to the needs and demands of the systems-oriented research community in oncology. At the same time, the interviewees agreed that an infrastructure that can be used in clinical practice requires much more effort in terms of financial support and time to be invested into software engineering. However, the implementation to the clinic was far from realization as the ACGT participants—neither from academia nor from industry—were not able to take over the responsibility to steer this process of marketability in the lifespan of the project. Hence, many of them in the beginning did not believe that it was possible to develop an ICT infrastructure that would be approved by the oncological research community after four years of research.

What was finally developed was a core set of technological components. They were assembled to develop an architectural prototype that was presented as an end-to-end demonstration at the final review meeting. However, the Grid technology was criticized as too complex and too immature by some interviewees and it was consequently replaced in follow-up projects by Cloud technology. In addition to the technological tools and services developed in the course of the project, many interviewees referred to more indirect outcomes. They had a lot to say about gaining experience in the emerging field of ICT in the medical domain. In this context, the ACGT project was often evaluated as the beginning of making a career in interdisciplinary research merging ICT, medicine, and systems biology and as the beginning of lasting interdisciplinary networking. These indirect benefits were even more positively assessed compared to technological achievements as impact on research developments was rather connected to experience and networking than to technological innovation. In particular, gaining experience provides the basis for long-acting achievements in a research field which was often connected to reach impact on research developments. In this context, the work on semantics and in particular on metadata definitions for describing data and the capabilities of tools was highlighted. This work of the ACGT consortium was valued as very important because it influences standardization processes in the research field. They were acknowledged as indispensable prerequisites for the adaptation of ICT infrastructures into clinical practice. By referring to the submission of the ACGT Master Ontology to the Open Biomedical Ontology Foundry, it was claimed in the interviews that the initial foundational work was already done in the domain of metadata definitions that has been capitalized in follow-up projects. However, the ACGT MO was not approved by the quality assurance of the OBO Foundry and, as already broadly discussed, the ACGT platform itself was not integrated into clinical practice.

The OBO Foundry is a good example of standardization efforts in systems medicine. These processes have a crucial impact on the coordination of systems medicine in an ICT environment. As described in Sect. 4.1.3, technological tools and services for the overarching task of data integration are consistently developed on the basis of ICT standards: the ICT formats of existing tools work as standards for the new ones. Because the tools based on ICT standards address syntactic as well as semantic integration, ICT deploys the standards for data storing and processing as well as the standards for data quality, annotation, and exchange. ICT-based standards therefore define what counts as reliable and valid data in the research process. Hence, ICT environments collect and integrate not only data on a technological level, they construct at the same time the data used in system-oriented research by assigning significance and meaning to them.

However, it seems as if the interviewees were not conscious of the ICT's profound impact on systems-oriented research. Unambiguously, they appreciated ICT infrastructures as data management systems providing access to and integration of large heterogeneous data stocks. Responsibility for those activities associated with standardization, integration, and management of data have therefore been ascribed to the scope of the ICT infrastructure and not to research institutions or individual scientists any more. In addition, the advantages of using ICT were regarded as being easy to connect and collaborate within the emerging scientific community of international and interdisciplinary range. In line with this perception is the popular picture of the ICT infrastructure as a facilitator or, in other words, as a service facility. Integration, access, and sharing of data are assigned to ICT infrastructures. These tasks are defined in technological terms assessed as general functions of a management system. The characterization of ICT as a facilitator corresponds to the emphasis on standardization and the categorical division of data into structure and content (see Sect. 4.1.3). Although data are split up into one part that is designated for scientific investigation and another part that contains purely technical information, the corresponding responsibilities of collecting and managing data are now separated from researching data. Accordingly, the ICT infrastructure needs to be stable, static, and enduring, whereas the actual use of stored data by scientists is claimed to be dynamic, creative, and proceeding. Not surprisingly, this picture of ICT infrastructures as service facilities was commonly used in the interviews as it is the most popular picture of ICT in science (e.g., Nyrönen et al. 2012).

However, we assumed that, looking from an in-depth perspective of a case study, understanding and modeling of biomedical systems are deeply shaped by ICT and their underlying design and conceptualization. Therefore, we expected to find evidence for this hypothesis as well. In fact, our hypothesis was corroborated by a second picture found in the interviews. Some of the interviewees characterized ICT infrastructures as catalysts for shaping and transforming systems-oriented research. The decisive point of this argument is that some research activities would not be performed without ICT support. Examples given in the interviews were tasks and processes regarding interoperability between heterogeneous data sources or knowledge discovery workflows (e.g., data mining services, the workflow editor, or the oncosimulator). The detailed analysis of the oncosimulator (see Sect. 4.2) has shown that the development of such a systems-oriented research tool in an ICT environment is based on many indicatory decisions (e.g., using a top-down approach, prioritizing particular mathematics) paving the way intrinsically to integrate ICT in systems-oriented research. However, insights into the oncosimulator's conceptual grounding have revealed that ICT not only enables but also restricts doing research at the same time. The exclusive use of mathematically compatible parameters indicated, for example, those kinds of restrictions.

To explore this transition of scientific and technological processes and mechanisms, it may be useful to look at it from a sociological perspective. The Actor-Network-Theory (ANT)[45] conceptualizes society as a completely interwoven sociotechnical web in which both parts, the social and the technical, influence each other mutually. This principle of symmetry between technology and humans rejects both technological determinism and social determinism and analyzes the mechanism of interactions in human–technological networks. Such sociotechnical networks can only exist when human and nonhuman actors (actants)[46] are permanently connected. They are therefore semiotically defined by how they act and are acted on in the networks of practices. The humans and nonhumans have a certain role and perform certain tasks within the network while delegating other roles and tasks to other actants (Latour 1992). In other words, the actants relate their roles and agency to each other. After the network is finally coordinated, it exists as an independent functional unit having agency on its own.

According to the ANT's perspective, ICT infrastructures in systems medicine operate as a nonhuman actant integrated together inter alia with scientists, scientific organizations, and funding organizations into a sociotechnical network. Technical objects such as an ICT infrastructure have a mediating role in the development of a network as they build, maintain, and stabilize the relations between different actants of all types and sizes, whether human or nonhuman. This means that they embody

[45] ANT was mainly developed by Bruno Latour (1987, 2005), Michel Callon (1986) and John Law (1987; Law and Hassard 1999). For a collection of original ANT papers in German see Belliger and Krieger (2006).

[46] To stress the interaction of nonhumans and humans, one term for both is preferred. According to ANT-terminology, they are defined as actants (e.g., Latour 1996). Nonhuman actors may be, for instance, technical artifacts, laboratories, or companies (Callon 1992, 73).

and measure relations between different actants at the same time. However, this does not mean that they are not actants of the sociotechnical network themselves (Akrich 1992, 205f.).

We argue that the ICT's role in the network is systematically to align the different data acquisition contexts with research activities (see Sect. 4.1.3). Therefore, the ICT infrastructure acts as a bridge between laboratory and clinic, molecular data and clinical data, as well as data acquisition and data interpretation. In the end, the interviewees expected that the ICT infrastructures have the potential to pave the way towards systems and personalized medicine; ICT infrastructures are, as one interviewee put it, "a new window to the future." From this it follows that the departure into the new era of systems medicine basically relies on the use of ICT technology: ICT infrastructures are promised to become new powerful actors (or actants, respectively) in upcoming networks in systems medicine.

However, the analysis in Sect. 4.3.2 has shown that ICT infrastructures require a suitable frame to be successfully integrated into systems-oriented research. In the interviews, different aspects of sustainability were outlined as being necessary to maintain an infrastructure over time. With the termination of the ACGT platform in mind, the interviewees drew on new approaches and concepts of how to tackle technological, financial, and social sustainability of ICT infrastructures for systems-oriented research. In addition to an appropriate technological design, it is obligatory that for an ICT platform to continue it has continuous financial support and manpower for maintenance. Providing financing on an ongoing basis is a sensitive issue not yet solved by the European Commission. Even if the use and development of research infrastructures is an overall objective in the 7th EU Framework Program and in its follow-up program Horizon 2020 (see Sect. 5.1), the EU commission usually funds research projects in the start-up phase of ICT infrastructures only. Hence, the original goal of the ACGT consortium to implement the ACGT platform into clinical practice was condemned to failure right from the beginning because of the lack of sustainability afterwards.

However, participating in R&D projects was highly attractive for the interviewees. They valued R&D projects as gateways for setting trends in research and for inter- and transdisciplinary networking. These highly appreciated advantages of participating in EU-funded projects are used to strike a new path of independence from EU money: to secure long-term funding of ICT infrastructures, new institutionalized frames such as a business plan for the Study, Trial and Research Center, STaRC, or the new foundation of the VPH Institute as a nonprofit organization are about to be realized. As these spin-offs are still very young, the connection to EU-funded research is very close. STaRC, for example, is intertwined with the EU project p-medicine and the VPH Institute can be regarded as follow-up of the VPH Network of Excellence. Within these emerging institutions, the key actors are scientific managers often trained and networked in EU-funded research projects. They need to be inter- and transdisciplinary oriented the more that science, medicine, funding, industry, and politics merge. Of course, these scientific managers are still interested in doing cutting-edge research and realizing systems or personalized medicine. However, by stepping outside the academic world, they have to consider more actors and interests coming from different grounds.

References

Abu-Asab MS, Abu-Asab N, Loffredo CA, Clarke R, Amri H (2013) Identifying early events of gene expression in breast cancer with systems biology phylogenetics. Cytogenet Genome Res 139(3):206–214

ACGT (2005) Annex 1—Description of Work, Proposal, unpublished work

ACGT (2009) Demonstrator specifications, deliverable 13.4, unpublished work

Akrich M (1992) The description of technical objects. In: Bijker W, Law J (eds) Shaping technology/ building society: studies in sociotechnical change. MIT Press, Cambridge, pp 205–224

Ankeny RA, Leonelli S (2011) What is so special about model organisms? Stud Hist Philos Sci A 42(2):313–323

Auffray C, Chen Z, Hood L (2009) Systems medicine: the future of medical genomics and healthcare. Genome Med 1(1):2

Belliger A, Krieger D (2006) ANThology. Ein einführendes Handbuch zur Akteur–Netzwerk-Theorie. transcript, Bielefeld

Bernsen HJ, van der Kogel AJ (1999) Antiangiogenic therapy in brain tumor models. J Neurooncol 45(3):247–255

Brazma A, Hingamp P, Quackenbush J, Sherlock G, Spellman P, Stoeckert C et al (2001) Minimum information about a microarray experiment (MIAME)—toward standards for microarray data. Nat Genet 29(4):365–371

Brochhausen M, Blobel B (2011) Architectural approach for providing relations in biomedical terminologies and ontologies. Stud Health Technol Inform 169:739–743

Brochhausen M, Spear AD, Cocos C, Weiler G, Martín L, Anguita A et al (2011) The ACGT Master Ontology and its applications—towards an ontology-driven cancer research and management system. J Biomed Inform 44(1):8–25

Bucur A, Rüping S, Sengstag T, Sfakianakis S, Tsiknakis M, Wegener D (2011) The ACGT project in retrospect: lessons learned and future outlook. Proc Comput Sc 4:1119–1128

Burgoon LD (2007) Clearing the standards landscape: the semantics of terminology and their impact on toxicogenomics. Toxicol Sci 99(2):403–412

Burkhardt H, Smith B (1991) Handbook of metaphysics and ontology. Philosophia, Munich

Callon M (1986) Some elements of a sociology of translation: domestication of the scallops and the fishermen of St Brieuc Bay. In: Law J (ed) Power, action and belief: a new sociology of knowledge. Routledge & Kegan Paul, London, pp 196–233

Callon M (1992) The dynamics of techno-economic networks. In: Coombs R, Saviotti P, Walsh V (eds) Technological change and company strategies. Economic and sociological perspectives. Academic, London, pp 72–102

Coebergh JW, van Veen EB, Vandenbroucke JP, van Diest P, Oosterhuis W (2006) One-time general consent for research on biological samples: opt out system for patients is optimal and endorsed in many countries. BMJ 332(7542):665

Deisboeck TS, Zhang L, Yoon J, Costa J (2009) In silico cancer modeling: is it ready for prime time? Nat Clin Pract Oncol 6(1):34–42

Desmedt C, Di Leo A, de Azambuja E, Larsimont D, Haibe-Kains B, Selleslags J et al (2011) Multifactorial approach to predicting resistance to anthracyclines. J Clin Oncol 29(12): 1578–1586

Dionysiou DD, Stamatakos GS, Uzunoglu NK, Nikita KS, Marioli A (2004) A four-dimensional simulation model of tumour response to radiotherapy in vivo: parametric validation considering radiosensitivity, genetic profile and fractionation. J Theor Biol 230(1):1–20

Edwards PN, Mayernik MS, Batcheller AL, Bowker GC, Borgman CL (2011) Science friction: data, metadata, and collaboration. Soc Stud Sci 41(5):667–690

Field D, Sansone SA, Collis A, Booth T, Dukes P, Gregurick SK et al (2009) Megascience. 'Omics' data sharing. Science 326(5950):234–236

Forgó N, Kollek R, Arning M, Kruegel T, Petersen I (2010) Ethical and legal requirements for transnational genetic research. C.H. Beck, Munich

García-Sancho M (2012) From the genetic to the computer program the historicity of 'data' and 'computation' in the investigations on the nematode worm C. elegans (1963–1998). Stud Hist Philos Biol Biomed Sci 43(1):16–28

Graf N, Hoppe A (2006) What are the expectations of a clinician from in silico oncology? 2nd international advanced research workshop on in silico oncology (IARWISO). Kolympari, Chania, Greece

Graf N, Hoppe A, Georgiadi E, Belleman R, Desmedt C, Dionysiou D et al (2009) 'In silico' oncology for clinical decision making in the context of nephroblastoma. Klin Padiatr 221(3):141–149

Gramelsberger G (2013) Simulation and systems understanding. In: Andersen H, Dieks D, Gonzalez WJ, Uebel T, Wheeler G (eds) New challenges to philosophy of science. Springer, Dordrecht, pp 151–161

Green S, Wolkenhauer O (2012) Integration in action. EMBO Rep 13(9):769–771

Hanahan D, Weinberg RA (2011) Hallmarks of cancer: the next generation. Cell 144(5):646–674

Hanseth O, Monteiro E, Hatling M (1996) Developing information infrastructure: the tension between standardization and flexibility. Sci Technol Hum Values 21(4):407–426

Kitchin R (2014) Big Data, new epistemologies and paradigm shifts. Big Data & Society 1. doi:10.1177/2053951714528481

Kohl P, Noble D (2009) Systems biology and the virtual physiological human. Mol Syst Biol 5:292

Kollek R (2009) Informed consent. Comment on article 6. In: Ten Have HAMJ, Jean MS (eds) Universal declaration on bioethics and human rights. Background, principles and application. UNESCO Publishing, Paris, pp 123–138

Kuiper RA, Schellens JH, Blijham GH, Beijnen JH, Voest EE (1998) Clinical research on antiangiogenic therapy. Pharmacol Res 37(1):1–16

Kyriazis D, Menychtas A, Dionysiou D, Stamatakos G, Varvarigou T (eds) (2008) Clinical trial simulation in grid environments. Conf. Proc. 8th IEEE International Conference on BioInformatics and BioEngineering (BIBE). Athens, 8–10 October 2008. 6 p

Latour B (1987) Science in action: how to follow scientists and engineers through society. Open University Press, Milton Keynes

Latour B (1992) Where are the missing masses? The sociology of a few mundane artifacts. In: Bijker W, Law J (eds) Shaping technology/building society: studies in sociotechnical change. MIT Press, Cambridge, pp 225–259

Latour B (1996) On actor-network theory. A few clarifications. Soziale Welt 47(4):369–382

Latour B (2005) Reassembling the social: an introduction to actor-network-theory. Oxford University Press, Oxford

Law J (1987) Technology and heterogeneous engineering: the case of portuguese expansion. In: Bijker W, Law J (eds) The social construction of technological systems: new directions in the sociology and history of technology. MIT Press, Cambridge, pp 111–134

Law J, Hassard J (eds) (1999) Actor network theory and after. Blackwell and the Sociological Review, Oxford

Lee JK, Coutant C, Kim YC, Qi Y, Theodorescu D, Symmans WF et al (2010) Prospective comparison of clinical and genomic multivariate predictors of response to neoadjuvant chemotherapy in breast cancer. Clin Cancer Res 16(2):711–718

Leonelli S (2010) Packaging data for re-use: databases in model organism biology. In: Howlett P, Morgan MS (eds) How well do facts travel? The dissemination of reliable knowledge. Cambridge University Press, Cambridge, pp 325–348

Leonelli S (2012) Introduction: making sense of data-driven research in the biological and biomedical sciences. Stud Hist Philos Biol Biomed Sci 43(1):1–3

Leonelli S (2014) What difference does quantity make? On the epistemology of big data in biology. Big Data & Society 1. doi:10.1177/2053951714534395

Leonelli S, Ankeny RA (2012) Re-thinking organisms: the impact of databases on model organism biology. Stud Hist Philos Biol Biomed Sci 43(1):29–36

Leonelli S, Diehl AD, Christie KR, Harris MA, Lomax J (2011) How the gene ontology evolves. BMC Bioinformatics 12:325

Lin B, White JT, Lu W, Xie T, Utleg AG, Yan X et al (2005) Evidence for the presence of disease-perturbed networks in prostate cancer cells by genomic and proteomic analyses: a systems approach to disease. Cancer Res 65(8):3081–3091

Mayer-Schonberger V, Cukier K (2013) Big data: a revolution that will transform how we live, work and think. John Murray, London

Meier S, Gehring C (2008) A guide to the integrated application of online data mining tools for the inference of gene functions at the systems level. Biotechnol J 3(11):1375–1387

Meuser M, Nagel U (1991) ExpertInneninterviews—vielfach erprobt, wenig bedacht. Ein Beitrag zur qualitativen Methodendiskussion. In: Garz D, Kraimer K (eds) Qualitativ-empirische Sozialforschung. Konzepte, Methoden, Analysen. Westdeutscher Verlag, Opladen, pp 441–471

Michelson S, Sehgal A, Friedrich C (2006) In silico prediction of clinical efficacy. Curr Opin Biotechnol 17(6):666–670

Nyrönen TH, Laitinen J, Tourunen O, Sternkopf D, Laurikainen R, Öster P et al (2012) Delivering ICT infrastructure for biomedical research. Proc. WICSA/ECSA Companion Volume. pp 37–44

O'Malley MA, Soyer OS (2012) The roles of integration in molecular systems biology. Stud Hist Philos Biol Biomed Sci 43(1):58–68

Rakyan VK, Down TA, Balding DJ, Beck S (2011) Epigenome-wide association studies for common human diseases. Nat Rev Genet 12(8):529–541

Rubin DL, Shah NH, Noy NF (2008) Biomedical ontologies: a functional perspective. Brief Bioinform 9(1):75–90

Sanga S, Frieboes HB, Zheng X, Gatenby R, Bearer EL, Cristini V (2007) Predictive oncology: a review of multidisciplinary, multiscale in silico modeling linking phenotype, morphology and growth. Neuroimage 37(Suppl 1):S120–S134

Stamatakos G (2011) In silico oncology part I: clinically oriented cancer multilevel modeling based on discrete event simulation. In: Deisboeck T, Stamatakos G (eds) Multiscale cancer modelling. CRC, Boca Raton, pp 407–436

Stamatakos GS, Antipas VP, Uzunoglu NK (2006a) A spatiotemporal, patient individualized simulation model of solid tumor response to chemotherapy in vivo. The paradigm of glioblastoma multiforme treated by temozolomide. IEEE Trans Biomed Eng 53(8):1467–1477

Stamatakos GS, Antipas VP, Uzunoglu NK, Dale RG (2006b) A four dimensional computer simulation model of the in vivo response to radiotherapy of glioblastoma multiforme. Studies on the effect of clonogenic cell density. Br J Radiol 79:389–400

Stamatakos GS, Dionysiou DD, Graf NM, Sofra NA, Desmedt C, Hoppe A et al (2007) The "Oncosimulator". A multilevel, clinically oriented simulation system of tumor growth and organism response to therapeutic schemes. Towards the clinical evaluation of in silico oncology. Conf. Proc. IEEE Eng Med Biol Soc: pp 6629–6632

Stamatakos GS, Dionysiou DD, Zacharaki EI, Mouravliansky NA, Nikita KS, Uzunoglu NK (2002) In silico radiation oncology: combining novel simulation algorithms with current visualization techniques. Conf. Proc. IEEE Special Issue Bioinformatics Adv Challenges 90(11): 1764–1777

Stamatakos GS, Kolokotroni E, Dionysiou D, Veith C, Kim YJ, Franz A et al (2013) In silico oncology: Exploiting clinical studies to clinically adapt and validate multiscale oncosimulators. Conf Proc IEEE Eng Med Biol Soc: 5545–5549

Sujanski W (2001) Heterogeneous database integration in biomedicine. J Biomed Inform 34(4):285–298

Swertz MA, Jansen RC (2007) Beyond standardization: dynamic software infrastructures for systems biology. Nat Rev Genet 8(3):235–243

Symmans WF, Hatzis C, Sotiriou C, Andre F, Peintinger F, Regitnig P et al (2010) Genomic index of sensitivity to endocrine therapy for breast cancer. J Clin Oncol 28(27):4111–4119

Thomas G (2011) A typology for the case study in social science following a review of definition, discourse and structure. Qual Inq 17(6):511–521

Tomita M (2001) Whole-cell simulation. A grand challenge of the 21st century. Trends Biotechnol 19(6):205–210

Tsiknakis M, Kafetzopoulos D, Potamias G, Analyti A, Marias K, Manganas A (2006) Building a European biomedical grid on cancer: the ACGT integrated project. Stud Health Technol Inform 120:247–258

Van't Veer LJ, Dai H, van de Vijver MJ, He YD, Hart AA, Mao M et al (2002) Gene expression profiling predicts clinical outcome of breast cancer. Nature 415(6871):530–536

Werner-Wasik M, Scott CB, Nelson DF, Gaspar LE, Murray KJ, Fischbach JA et al (1996) Final report of a phase I/II trial of hyperfractionated and accelerated hyperfractionated radiation therapy with carmustine for adults with supratentorial malignant gliomas. Radiation Therapy Oncology Group Study 83-02. Cancer 77(8):1535–1543

Wierling C, Herwig R, Lehrach H (2007) Resources, standards and tools for systems biology. Brief Funct Genomic Proteomic 6(3):240–251

Wolkenhauer O, Green S (2013) The search for organizing principles as a cure against reductionism in systems medicine. FEBS J. doi:10.1111/febs.12311

Wolkenhauer O, Auffray C, Jaster R, Steinhoff G, Dammann O (2013) The road from systems biology to systems medicine. Pediatr Res 73(4 Pt 2):502–507

Yin RK (2009) Case study research: design and methods, 4th edn. SAGE, Thousand Oaks

Chapter 5
Science Policy of Systems Biology

Anne Brüninghaus, Imme Petersen, Regine Kollek, and Martin Döring

Abstract In this chapter, we examine the science policy of systems biology and perspectives thereof. Based on interviews with actors from different fields such as science, politics, media, and economy, we contrast the scientists' conceptualization and assessment of systems biology and their perception of science policy with that of societal actors on what systems biology is and how it should be governed.

Discussions in these different fields are interconnected. We therefore highlight interdependences and shared topics where the separate discourses influence and interact with each other. Aspects addressed touch upon the identity of systems biology as a new science, and the effect of further specialization, the similarity of the scientific and public images of what systems biology is, and the sustainability of funding. While participation and inclusion of the general public is seen as an important achievement in politics, media, and public interest groups, it is less important in the scientific perspective. This raises the question of whether it is ascribed an appropriate role.

Keywords Systems biology • Science policy • Science and society • Scientists • Societal actors • Interdependencies

5.1 Systems Biology as a Topic of Science Policy

Systems biology is commonly understood as an emerging interdisciplinary approach in the life sciences. As such, it depends massively on research funding as does any other scientific development. Budgets for systems biology derive to a large part not from universities or research institutions, but—especially regarding personnel cost—from third-party funding. In Germany, those funds are mainly governmental-driven and, as such, are subject to corresponding science policy. In this chapter, we analyze and discuss the social and scientific dynamic interdependencies that result

A. Brüninghaus (✉) • I. Petersen • R. Kollek • M. Döring
Research Centre for Biotechnology, Society and Environment (FSP BIOGUM), University of Hamburg, Hamburg, Germany
e-mail: anne.brueninghaus@uni-hamburg.de

M. Döring et al., *Contextualizing Systems Biology*,
DOI 10.1007/978-3-319-17106-7_5

from this kind of research funding and development. Furthermore, we wanted to know how the discourse on science policy of systems biology is characterized by representatives of science, the media, industry, and public interest groups, and from research funding and science policy.

5.1.1 Systems Biology and the Dynamics of Science Policy

Topics of new approaches in science are not only discussed in science itself, but also in politics and by the public. Questions regarding a science's funding, its expected value for science and society, its applications, implications, and possibly necessary regulations relate to the corresponding science's policy. The latter is usually examined by scientific as well as by societal actors. Relevant societal actors are, for instance, research funding bodies and representatives from science policy and administration, but also by industry, nongovernmental organizations, and public interest groups, as well as the media (comp. Fig. 5.1). The positions of such actors often vary widely, not only regarding the perspective, but also regarding the selection of topics seen as relevant, such as content, research funding, application, chances, and risks of the scientific development.

The discourse on the future direction and funding of science was not always shared between that many actors, as it used to be more one-dimensional: in the past, public opinion mainly placed its trust in the self-regulatory mechanisms of science,

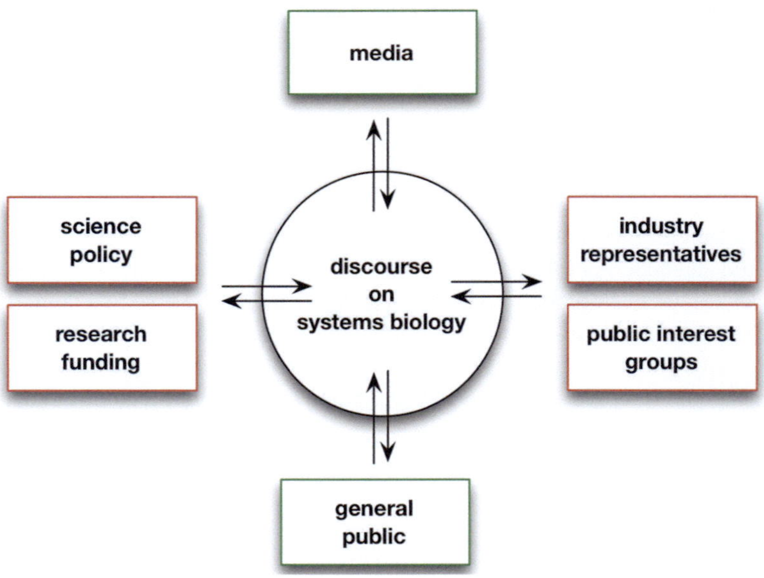

Fig. 5.1 Discourse on systems biology: societal actors

because "[s]cience and technology [were] for the most part removed from attempts to make them topics of public discussion and objects of political regulation"[1] (Bogner 2012, 380). Since then, the relationship between science and the public has been subject to fundamental change: the relationship between scientific practice and public opinion has shifted to a degree that today a "closer coupling"[2] can be found that surfaces in various connections between the two systems (Weingart 2001, 175; see also Dunwoody and Ryan 1985; Sturgis and Allum 2004). Scientific development, for example, make an impact on the daily life of the public, through applications and the public tries to influence the direction of science. In this context, networks of actors in finance, politics, industry, business, and civil society groups interact within the realm of research governance (Peters and Weingart 2009), often by means of the media. Furthermore, Peters et al. (2009) describe the transfer of the direction of science to the political sphere. They discuss that the media play a critical role: the medialization of science is driven by a medialization of politics. Media are increasingly instrumentalized by science to hold priority conflicts and to mobilize public support (Weingart 2001, 244). On the other hand, science is more and more oriented towards political or economic objectives and thus interested in its public image and hence, its perception by the public and the media. This interest has reached a degree that the question arose whether the direction of scientific development is determined by the media (Weingart 2005, 168ff).

As a consequence, the public is not only a passive observer of science, but influences partly (e.g. via the media) its content and directions. Hence, laypersons enter into scientific relationships (compare further, for example, Bogner 2012). In this sense, science, the public, media, politics, the law, industry, and interest groups are not discrete and independent systems but have to be considered as linked and interconnected. The discourse on scientific progress and the orientation of research is thus not determined by science alone but has become subject to the influence of the public and politics, albeit science conceived as a system remains autonomous (cf. Rödder 2009, 33ff).

This is different, however, with regard to its societal legitimization. Here, science does not possess exclusive interpretive authority as the public and politics discuss scientific topics and their worth and merit for society. This is especially true when systems biology is understood as a "technoscience," (Nordmann 2005). Technoscience is characterized by the fact that basic research, technology development, and application are inextricably linked. Inherent in this understanding is that the different subsystems described above are involved in the development of a technoscience (cf. Brüninghaus 2012). As systems biology is seen as at least partly established and increasingly accepted from the perspective of research funding, it is necessary to examine the perceptions and opinions of other relevant actors in order to make statements on its current and possible further development and impacts. In order to put such statements on a sound basis we chose an empirical approach

[1] Original quote: "Wissenschaft und Technik [waren] öffentlicher Problematisierung und politischen Steuerungsversuchen weitgehend entzogen".

[2] Original quote: "engere Kopplung"

that allowed us to include different levels of the discourse on systems biology, as well as important actors and focus on the influence of the different discourses (cf. Bora 2012, 345ff).

In essence, this chapter aims at answering questions regarding science policy of systems biology. We focus on the discourse of science policy in science, media, industry, nongovernmental organizations and public interest groups, research funding, and science policy referring to

- systems biology's funding mechanisms and constraints,
- reactions to its funding,
- its value for scientists, the public, and research funding representatives,
- its application,
- its implications and possibly regulations.

5.1.2 Science Policy and Research Funding of Systems Biology

In Germany, public research funding is mainly driven by the Federal Ministry of Education and Research (BMBF), the German Research Foundation (DFG), and the Helmholtz Association. The BMBF started its first program of systems biology funding in 2001 with a line of funding titled "Systems of Life—Systems Biology" (Systeme des Lebens—Systembiologie). Since then, systems biology was supported financially continuously, for example, in a large collaborative research project on liver cells that started 2004 (HepatoSys); in a project targeting the development of systems biology infrastructure since 2008 (FORSYS); with a focus on medical applications in the 2009 line Medical Systems Biology (Medizinische Systembiologie); in another large collaborative research project that started in 2009 centering around aging processes (GerontoSys), and since 2009 in one that concentrates on new methodologies (SysTec). The BMBF also funds interdisciplinary collaborative projects through a program called ERASysBio (2006–2011). Its primary aim is the development of personalized medicine, and pharmaceutical advances, of treatments for multifactorial diseases, and measurements to increase life expectancy (Rahmenprogramm Gesundheitsforschung 2010).[3]

What follows is a short description of the main funding agencies and instruments in order to provide a background for the better understanding of the statements and claims of the interviewed actors and our interpretations thereof.

The German Research Association (Deutsche Forschungsgemeinschaft; DFG) has no documented guidelines regarding the funding of systems biology due to its nature as a bottom-up organization. It is funding about 80 projects in the area of systems biology.

[3] https://www.erasysbio.net/. Accessed November 15, 2014.

The Helmholtz Association (Helmholtz-Gemeinschaft) has a focus program for systems biology (Helmholtz-Allianz Systembiologie). It aims to "contribute to clarifying the underlying mechanisms in the emergence of complex diseases."[4] Examples are the development of computational mathematical models for cellular processes connected to cancer or heart diseases that are based on data generated by previous experiments and enable a better holistic understanding of processes in human cells. This also includes the "possibility to predict opportunities for targeted intervention when diseases emerge".[5] The funding program includes six research centers within the Helmholtz Association, as well as other universities and research institutes.

The reconstruction of the funding of systems biology by programs of the European Commission is somewhat more complicated. Such programs were started in 1984; first they ran for five, and since 2007 for 7 years. The first seven funding periods were called Framework Programs for Research and Technological Development, abbreviated FP1 through FP7, whereas the current program is named Horizon 2020. The specific objectives and actions vary between funding periods. In FP6 and FP7 the focus was still on technological research, in Horizon 2020 the focus is on innovation-driven research (including support for research infrastructures), developing technologies that support European industries and connecting research results to market, and, finally, benefits to the citizens including research on health, demographic change, food security, energy, and climate as well as secure societies. In FP6 and FP7 (2002–2013) the most important funding instrument was the "Collaborative Research" composed of a minimum of three partners coming from three different EU countries with a typical duration of 3–5 years. The research projects could address basic or applied research. In the context of the "Virtual Physiological Human (VPH)" initiative, more than 30 research projects were funded by the European Union. One of these projects listed is our case study Advanced Clinical-Genomic Trials on Cancer[6] (see Chap. 4) and its various follow-up projects such as p-medicine, INTEGRATE, or VPH Share. Another instrument of FP6 and FP7 were networks of excellence. Such networks were set up to strengthen scientific (and technological) communities in a particular research area through sustainable integration of the research capacities of the participants. As described in Sect. 5.3.2, the "Virtual Physiological Human Network of Excellence" was funded in FP7 to connect the various VPH research projects and to foster the development of educational, training, and career structures in the communities related to VPH and systems biology.[7]

For Horizon 2020 (2014–2020) the funding type terminology has changed according to the general orientation toward innovational research and market-ready

---

[4] Original quote: "einen Beitrag zur Aufklärung der zugrunde liegenden Mechanismen bei der Entstehung von komplexen Erkrankungen zu leisten"

[5] Original quote: "Möglichkeit zur Vorhersage von Möglichkeiten für eine gezielte Intervention bei der Entstehung von Krankheiten."

[6] http://vph-portal.eu/projects. Accessed January 26, 2015.

[7] http://vph-portal.eu/vph-noe-home. Accessed January 26, 2015.

products: (1) "Research and Innovation Action Projects" (RIA) may get 100% funding because they are not close to market. (2) "Innovation Action Projects" (IA) only get 70% funding; they are close to market and especially target small and medium-sized enterprises (SME). (3) "Coordination and Support Action Projects" (CSA) are studies, networking, and distribution of results getting up to 100% funding (but not meant for research). In addition, there are other new funding instruments such as prizes and pre-commercial procurements.[8]

In Horizon 2020, life science research is addressed in the third funding priority of societal changes, in particular in the programs of health and food security, but also in the second funding priority (e.g., biotechnology programs) and in the first pillar (e.g., research infrastructure programs). According to the EU office of the German Federal Ministry of Education and Research (BMBF),[9] the total funding budget for the life sciences has increased from 8 billion euros in FP 7 to 13 billion euros in Horizon 2020.

5.1.3 *Method*

The empirical material and evidence used in this chapter consists of transcribed interviews with actors from different groups that are described below in more detail. In order to capture the different perspectives on science policy, a wide range of actors involved in the discourse were invited for interviews. Included were representatives from science, media, industry, and nongovernmental organizations, public interest groups, research funding, and science policy. The interview method was adapted to the individual interviewee and the actor group to whom he or she belonged.

5.1.3.1 Scientific Actors

The empirical material used in Sect. 5.2 is based on 23 interviews that were conducted with scientists working in systems biology in Germany. The scientists were identified by the following search procedure. First, a literature review was undertaken which used the *PubMed PubReMiner* to locate all scientific reviews available of 10 leading authors. PDF-files of the reviews were downloaded and studied as outlined in Grounded Theory (Glaser and Strauss 1967). The categories brought about by this analysis revolved around the conceptual history of systems biology, explanations of what systems biology is, the assessment of current research undertaken, possible areas of application, basic theoretical concepts applied or defined, and the outline of important future research tasks to be addressed. Results were

---

[8] http://ec.europa.eu/research/participants/data/ref/h2020/grants_manual/amga/h2020-amga_en.pdf. Accessed January 26, 2015.

[9] http://www.nks-lebenswissenschaften.de/de/1075.php. Accessed January 26, 2015.

gathered and correlated to tackle thematic overlaps and divergences. This research was supported and extended by a close reading of edited volumes and introductions to systems biology. Once this conceptual overview was generated, a second round of research started in which the *PubMed PubReMiner* was used to find publications of German scientists: 23 German authors found in the database were located and contacted via e-mail. The sample consisted of heads of institutes, group leaders, and post docs to cover the full range of professional experience and views on systems biology available.

In the meantime, a semi-structured questionnaire was developed based on the literature research undertaken. Five topics were addressed in the interviews: (1) the nature of systems biology (2) the history of systems biology, (3) basic concepts applied in systems biology, (4) national and international differences in doing systems biology, and (5) assessment of possible futures of systems biology as seen through scientists' eyes. Data were taped with a tape recorder, fully transcribed, and analyzed by applying a linguistically (Wetherell et al. 2002, 2003) informed grounded approach (Charmaz 2006; Clarke 2005). This methodological combination was chosen because it productively places emphasis on the deductive development of analytical categories from data and is easy to combine with the analysis of linguistic structures. Once main themes or topics emerged during the process of analyzing data, segments of transcripts were grouped under emergent headings, and significant linguistic structures were analyzed that substantiated ad hoc categories. In doing so, an underlying and saturated semantic network of categories permeating the different topics addressed in the interviews became available. These were then systematized and analyzed from an interpretative point of view.

The second series of interviews is part of our case study of the EU-funded project "Advancing Clinico-Genomic Trials on Cancer: Open Grid Services for Improving Medical Knowledge Discovery (ACGT)" presented in Chap. 4. Here, we refer to 18 interviews conducted with consortium members. Structured questionnaires were used for the interviews. The questionnaire regarding the ACGT research project consisted of four sections addressing the following topics: (1) experiences of scientific and practical cooperation in the project collaboration (in particular interdisciplinary negotiations), (2) experiences regarding the realization of the ACGT infrastructure, (3) judgments regarding the project outcome and science policy, (4) judgments regarding the anticipated profit of ACGT for cancer research and systems biology.

To select the 18 interview partners according to their visibility in the ACGT project, we conducted a bibliometric analysis of the collaborations for internal publications (deliverables) and external publications (peer-reviewed articles, books, conference proceedings). Additionally, we drew on their designated tasks within the project (e.g., work package leadership, project management, quality control). The interviews were digitally recorded, anonymized, and literally transcribed. The empirical results are based on qualitative content analysis by using the software MAXQDA 11. Below, the interview citations are characterized by the professional background of the interviewee.

5.1.3.2 Societal Actors

The governance and policy analyses presented in Sect. 5.3 draw from interviews with German experts. In addition, written documents were analyzed that were deemed relevant to the establishment or evaluation of systems biology. The criteria for selecting of both the interviewees and the documents was their belonging to or close affiliation with the media, industry, public interest groups, research funding, science policy bodies, or administration. For the interviews, actors were identified who possess "technical, processual, and terminological knowledge that relates to [...] [a] specific or professional field of action" (Bogner and Menz 2005, 46). Thus, they can be considered as informants who "possess knowledge that is not available from other sources" (Littig 2008, paragraph 12) and command the power to enforce or champion their own orientations and conceptions within their profession (cf. Bogner and Menz 2005, 46). We thus selected interviewees who do not only have their own informed and specific perspective on the field but who bear, to some degree, an impact on the establishment of systems biology and thus have subtle power (Stehr and Grundmann 2010, 57).

Selection of the interviewees was initially based on a theoretical sampling (Flick 2000, 58). Basis for sampling was a comprehensive media analysis of the German press coverage that enabled us to gather first hints with regard to major players in the field. Other clues came from an analysis of central policy documents and from a formal analysis of persons responsible for science policy in political parties. In selecting interview partners from these fields, we paid attention to maintain political balance in order to avoid bias. Actors from research funding, the science policy, and administrative area were selected on the basis of existing research programs and chosen in order to maintain a balance between the main funding institutions. A similar approach was followed for actors from industry and public interest groups. To represent the media, such authors were contacted who have consistently written about systems biology and/or are known to have a good overview of the life sciences. In selecting the interviewees, we included the results of the document analysis to complete the sample. In total, we interviewed 10 actors from the different fields in Germany.[10]

For the interviews, a semi-structured questionnaire was developed. It consisted of an actor-specific and a general part. The specific section of questions differed depending on the actor area. We asked, for example, for the understanding of systems biology, its state of establishment, research funding, application, its implications, and the role of the general public in science development and governance.

The analysis of the interviews was carried out according to Meuser and Nagel (1991): First, the interviews were fully transcribed and anonymized. As a next step, we paraphrased and sequenced the individual transcripts and created headings

[10] As part of the THCL research project, we also conducted and analyzed interviews with Austrian representatives of media, industry, and public interest groups, research funding, science policy bodies, and administration. These outcomes and a comparison between Germany and Austria will be published elsewhere.

(categories) for individual statements. We then compiled topically similar statements and provided a corresponding main heading for each topic.

In the following three sections we discuss the social and scientific dynamics that result from this kind of research funding and development. Furthermore, we look into how the discourse on science policy of systems biology can be characterized looking at representatives from science, media, industry, and public interest groups, and from research funding and science policy. Thus, Sect. 5.2 discusses science policy from the scientist's view, Sect. 5.3 from the public's perspective, and Sect. 5.4 discusses the interdependencies between scientific actors and the public.

5.2 Scientific Actors' Perceptions of Science Policy

Scientists are an important target group of science policy. Research and development (R&D) programs usually address researchers working at universities whereas some specific programs explicitly include research departments of companies or research organizations as well (e.g., Fraunhofer Organization). The main aim of R&D programs is to ensure the marketability of research and to bring science and industry together. Hence, scientists are requested to contribute to applied research and to establish contacts to industry if they want to participate in such funds. Furthermore, because successful university careers nowadays depend more and more on successfully raising external funds, competition on claiming funds from national and EU R&D funding agencies for doing research is high. To this effect, the landscape of research is changing in terms of research goals. In addition, researchers have to define and to manage their research activities according to funding strategies.

The following section analyzes how—from the scientists' perspective—funding programs are structured and how scientists assess and strategically cope with funding mechanisms. The final section summarizes the findings of both sections, combines them, and reflects on the interdependency of trends in funding and developments in research. The research question we address is threefold. How do funding mechanisms exert an impact on the research landscape in terms of exploitation of research results and development of research fields? What is their impact on establishing research agendas? In what way do these mechanisms contribute to doing research and developing academic careers?

5.2.1 Funding Mechanisms

Let us now turn to the first section in which the reflections of our scientific interview partners are analyzed with regard to funding mechanisms. This is important in order to find out, in which way research activities related to systems biology are or have been established and maintained.

5.2.1.1 Contents of Funding

In the interviews, emphasis is put on R&D programs such as Horizon 2020[11] run by the European Commission (see Sect. 5.1.2). The interviewees stress that the primary focus of R&D funding is research with the future goal of application, but not development of technology, a point that became apparent in the interviews with scientists having worked in the ACGT project on the development of a sustainable ICT infrastructure (see Chap. 4). From a more general perspective, the interviewees refer to research done in the context of such R&D programs in the sense of raising new research questions, defining research objectives, evaluating alternatives, and coming up with innovative solutions to address and answer the research questions and agendas set up by the research agencies.

With regard to new fields of research, one of the key questions is how such fields and targets of research are being selected and defined. This is usually done by national and international funding agencies and programs which in large define topics and substance of such programs via the themes addressed in their calls. Looking at the calls in a research field such as the life sciences from a chronological perspective, they represent common trends in funding domains. As the following quotation shows, these trends exert an impact on current trends and future developments in research. Using the example of the priority of funding research on simulation in systems biology, one member of the ACGT consortium explained who in his view is responsible for this trend in research, which was funded by a series of different funding initiatives of the European Commission (FP6,[12] FP7,[13] Horizon 2020):

> I think, if somebody has to look for reasons, I think, these have to do with the way they are running the procedures in the selection of projects and the design of the research policy. In that sense, I would mainly consider the EU director somewhat responsible for this kind of research. So, if you have a closer look to the genealogy, let's say, to the succession of projects, they tend to have certain lineages with most of the same people being in the same lineages. That means that they do not implement a really open approach in the selection and I suppose this reflects somehow their political agendas. But this of course isn't fair. But at the same time it also reflects some of the procedures in the selection of the projects. In that sense, I can be a bit more specific. There are particular lines of research especially in this particular area we are talking about, systems biology, that tend to do with simulation of particular systems and it appears that they have been pulling money in that kind of research activities, practically paying again and again for the same kind of hypothetical applied research. That is an exercise that is not going to yield significant results. So I would expect this kind of decisive bodies to be more careful with the kind of selection and spending of public or common funds. (I6, BioMed)

Apparently, the interviewee holds the opinion that the funding system is a more or less private and possibly a closed circle. This is indicated by the metaphors used from kinship terminology ("genealogy," "lineage"). A succession of projects related to the respective trend is usually funded. At the same time, scientists who were able

[11] Horizon 2020, ec.europa.eu/programmes/horizon2020/. Accessed January 26, 2015.

[12] FP6, ec.europa.eu/research/fp6/index_en.cfm. Accessed January 26, 2015.

[13] FP7, ec.europa.eu/research/fp7/index_en.cfm. Accessed January 26, 2015.

to integrate themselves in the funding lineage are usually able to receive funding for successive projects in the trend domain. These participating researchers build social networks to apply to and to receive funding together in certain research areas. In particular the ACGT consortium stressed that they were able to build a stable community which succeeded in receiving funding for continuing research together after ACGT had ended. Scientists already participating in one of the funding lineages also support science policy organizations in seeking new trends for funding strategies at the same time. One of the ACGT consortium members provided personal insights of how science policy representatives interact with scientists in the research domain of ICT in the life sciences:

> These so called experts [science policy representatives] are talking to people trying to get their antenna on what is the next big thing, and who the cools are, and who should I speak to in order to draw my next new program? So there is very much dependency on whom you talk to and who will paint the best picture. I will never forget, for example, a statement once by an EU functionary, whose role at the time was to contribute to the work program. I think, this was work program five and they were tasked with writing the ten pages that we receive for the part of ICT in the life sciences. So these poor guys are talking to people like you or me or whoever saying 'Come on guys, help me to write the paper. I've got to write ten pages that will then become the policy booklet of the commission'. And it has to be slightly different from what they wrote four years ago, because otherwise the politicians will say,we have paid for this four years ago!' So they have to come up with new buzzwords. You know, for me this is very silly. I don't know, but it is inevitable. For example, in ACGT we were talking about the Grid. Tell me about the Grid. Now it is about the Cloud. We haven't even solved the problems of the Grid and now it's the Cloud. Why? Because now the Cloud is the next big thing. It is essentially, if you look at it, the same concepts with different clothes. And there is this issue as well. It is like the engine, you know, life goes on, therefore, it has to go on. It has to go, it has to go on. So how do we keep it going on? It is easier to come up with a new dress than to say 'I'm losing my hair or my body is fat. I need to lose weight.' This is more difficult than if I go and buy some fancy clothes and look nice. (I4, IT)

To receive continuing funding for their research, scientists are willing to deliver and follow new buzzwords or, according to the interviewee's metaphor, to present "the same concepts with different clothes." At the same time, the quotation illustrates that the interaction between scientists and science policy representatives leads to scientists contributing or even initiating their own funding programs by delivering new buzzwords for continuous research.

However, scientists sometimes want to give their research a push into a new and different direction. Then they have to find new options of how to receive funding for their upcoming research interests in addition to the established tracks. If the research area is not on the funding agenda yet, one possibility is to try to attract the attention of science policy organizations to set up a new program. However, this is only possible if the respective scientists play a prominent role in the organization. Such individuals may be labeled as science managers exerting political influence to initiate shifts and redeployments in the distribution of funding. To this effect, one interviewee of the systems biology community in Germany described the role of an individual systems biologist in the science policy initiatives of the German Federal Ministry of Education and Research and the European Union to establish systems biology.

> I think it started with these activities that he [the scientist] did for the BMBF [German Federal Ministry of Education and Research] and that then positioned and supported him there. Um, that he acted like a mover and shaker who said, 'we have to press forward with this'. At the time, I was much too naïve, for example [...]. I was the kind of person who, if there was a program, then I applied to take part in it. But I would never have had the idea to go to Brussels and tell them that there should be this kind of program that I would like to have. (I19)[14]

Scientists who proactively approached the science policy organizations in Berlin and Brussels in the view of the interviewee played a crucial role in setting systems biology on the funding agenda. However, science policy representatives of the funding organizations are the necessary teammates to let the funding initiative become real, as the interviewee carried on.

> [...] The BMBF played practically the biggest role. Because I already said I assume that the role that he [the scientist] played in the beginning, that that was also decisively linked to the activities of the BMBF. And this impression that the funding organizations like the BMBF and the EU were actively, that is, not reactively, but actively trying to establish this field. So, the BMBF then said, 'okay, we'll do [it], we'll set up a program. But we'll try to do it not just in Germany, but push for it in parallel at the EU level', and then bringing this conference to Germany and all that. And it isn't ... it wasn't the scientists who were knocking down the doors, it was individual people in the ministries who are really proactive and also incredibly motivating. (I19)[15]

To sum up, the cited quotations show that science policy organizations are positioned in between the political and the scientific spheres. They have to be accountable to the political representatives and have to address the scientific community at the same time. The interviews also illuminate that functional interrelations are deeply interwoven with personal relationships. To receive funding or to make a career in research, scientists align with others to develop stable social networks in order to apply and receive funding. Once they have become part of one of the funding lineages, they may stand a good chance to receiving continuous funding.

---

[14] Original quote: "Ich glaube, das ging los mit diesen Aktivitäten, die er [der Wissenschaftler] fürs BMBF[Bundesministerium für Bildung und Forschung] gemacht hat und die ihn dann dort auch positioniert haben und ihn unterstützt haben. Äh, dass er aufgetreten ist als ein Macher, der gesagt hat, wir müssen das vorantreiben. Da war ich zu dem Zeitpunkt zum Beispiel auch viel zu naiv [...]. Ich war so jemand, der, wenn 's da 'n Programm gibt, dann hab ich mich da beworben. Aber auf die Idee zu kommen nach Brüssel zu fahren und denen zu sagen, dass es so 'n Programm geben sollte, das ich gerne hätte, wär ich damals nicht gekommen." (I19)

[15] Original quote: "[...] Das BMBF hatte fast die größte Rolle. Weil ich ja eben schon gesagt hab, ich vermute, die Rolle, die er [der Wissenschaftler] gespielt hat am Anfang, auch maßgeblich mit den Aktivitäten des BMBF verbunden waren. Und das hat sich, dieser Eindruck, dass die Geldgeber wie BMBF und EU, aktiv, also nicht reaktiv, sondern aktiv darum bemüht [waren], ein Gebiet zu etablieren. Also das BMBF hat dann gesagt 'okay, wir machen [das], wir setzen 'n Programm auf. Aber parallel versuchen wir das nicht nur in Deutschland zu machen, sondern auch auf EU-Ebene voranzutreiben', und dann eben diese Konferenzen nach Deutschland zu holen und so. Und das ist nicht ... das waren nicht die Wissenschaftler, die die Türen eingerannt haben, sondern das sind einzelne Personen in den Ministerien, die wirklich proaktiv sind und auch unheimlich motivierend." (I19)

Beyond that, a few scientists, the so-called science managers and makers, hold a prominent role because they know the network and the structure of the interrelated spheres which to a large part they established themselves. They proactively approach the science policy representatives to have direct influence on funding policy. Social networks between science policy representatives and scientists emerge because they jointly tackle upcoming trends in science and lobby for funding. These successful scientists usually become prominent in scientific networks and in the emerging branch albeit so-called funding lineages clearly display self-referential features.

5.2.1.2 Structure of Funding

National and EU-funding programs mainly support research projects that are expected to be accomplished within a limited budget and timeframe. Generally speaking, a project is defined as a joint enterprise that is carefully planned to achieve a particular aim.[16] Consequently, especially EU-funded projects are structured in multiple tasks that are assigned to different work packages. Such packages are designed to achieve specific goals and they are handled by teams of scientists coming from one or more research sites to deal with the assigned tasks. The work packages are the smallest components of the project and the time schedules, the workflows, and the budgeting of the project are allocated to them.

Many calls for proposals define applied research topics. As such calls target predetermined objectives, the project structure outlined above is particularly suited for applied research. Therefore, the funding structure especially pushes problem-oriented approaches within research domains. This might include specific methodological approaches as well, as one of the German systems biologists explained with regard to funding programs in systems biology:

> [...] nowadays a systems biology project, that is, a proposal with systems biology in the title, requires certain ingredients. You have to use certain high-throughput methods in order to see what's happening. You have to use mathematical models. That's an absolute must today. Without mathematical modeling, it's practically impossible to get funding for a proposal. (I21)[17]

To prioritize research proposals using a specific method or approach is regarded as a necessary requirement to establish new approaches within established research areas. The integrative role of mathematical modeling for systems biology is often brought up in the interviews. The interviewees refer to mathematic modeling as a

[16] See Oxford English Dictionary, www.oed.com/view/Entry/152265?rskey=8tUCpe&result=1#eid. Accessed September 3, 2014.

[17] Original quote: "[...] heutzutage [gehört] zu einer systembiologischen Projektstudie, also zu einem Antrag, der überschrieben ist mit Systembiologie, dass bestimmte Ingredienzien drin sind. Man muss bestimmte Hochdurchsatzmethoden benutzen, damit man ganzheitlich sehen kann, was passiert. Man muss mathematische Modelle verwenden. Das ist heute auch ein absolutes Muss. Ohne mathematische Modellierung können Sie kaum noch 'n Antrag durchkriegen." (I21)

method for integrating knowledge and as a tool for verifying knowledge with regard to consistency.

In systems biology, mathematical modeling is usually used in order to tackle research questions related to medical or biotechnological applications of biological knowledge. This problem-oriented approach is characterized by its interdisciplinary nature. However, crossing disciplinary boundaries in interdisciplinary teams is always challenging (see Sect. 4.1.2); this may be the reason why interdisciplinary research units are seldom found at universities but in larger research institutions that are especially equipped for this type of research, such as the Helmholtz Association. Compared to other countries, it is often stressed in the interviews that the institutionalization of systems biology is relatively poor in Germany. Departments of systems biology where mathematicians, physicists, chemists, and biologists are working under a single roof are strongly requested by the interviewees to establish the systems approach in biology permanently, but they do not exist yet.

One of the essential preconditions to receive funding for establishing systems biology and conducting systems-oriented biomedical research is collaborating in interdisciplinary research projects. Joint projects are usually the first link between the involved disciplines.

> That wouldn't have been possible in the past. All of a sudden, you're directly connected to clinical research, with the people who are basically operating on the ground—that's great! That's, well, I think that's something very important. Scientists are coming together who would certainly never have come together in the past, when systems biology didn't exist as a roof, as a funding roof, well, who wouldn't have been forced to integrate with each other. (I23)[18]

As the quotation shows, the funding structure and its strategies to supporting systems biology—metaphorically described as a roof—coordinate and establish systems biology as an interdisciplinary approach. However, another German systems biologist expressed in the interview that the interests of scientists in interdisciplinary working collaborations are in general limited. In most cases, it is only the higher likelihood to receive funding for their own research that lets scientists get involved with other disciplines. Our interviewee described that he wrote enquiry letters to his colleagues asking for interdisciplinary collaboration before the funding programs for systems biology came up.

> Zero interest. And then, when these programs came up, all of a sudden it worked. That means, you have to give the funding organizations credit for that. (I6)[19]

Another scientist of this interview series goes one step further and wants to turn the scientific culture of biology towards a culture of interdisciplinarity.

[18] Original quote: "Das wär früher nicht möglich gewesen. Auf einmal ist man mit der klinischen Forschung direkt verbunden, mit den Leuten, die sozusagen vor Ort das operieren—toll! Das ist, also das halte ich für was ganz Wichtiges. Wissenschaftler kommen zusammen, die früher mit Sicherheit, wenn es das Dach Systembiologie auch als Förderdach nicht gäbe, äh die nicht gezwungen gewesen wären sich zu integrieren, die wären nie zusammengekommen." (I23)

[19] Original quote: "Null Interesse. Und als dann diese Programme auftauchten, auf einmal funktionierte das. Das heißt, das muss man den Geldgebern anrechnen." (I6)

> A culture of large projects [has yet to] develop in biology. People usually respond to that by saying, 'But they do exist! That Human Genome Project.' But that's of a completely different nature. All you had to do in that project was to sequence. In other words, set up devices here and there. And all of them were independent of one another. So, although that's an example of a large international project in biology, it isn't what I described, where the partners in the various countries with various technologies and various research questions are dependent on each other. (I24)[20]

From this it follows that interdisciplinarity can only develop and prosper in a scientific culture in which disciplinary research is integrated not only in terms of methods but also in terms of common research questions and goals. Interdisciplinary work, using different methods and technology to investigate different research questions, is dependent on a collaborative environment and a common roof. It can be said that the project structure outlined above as a frame for collaboration is a constitutive part of the vision to widen the scientific culture of molecular biology towards interdisciplinary systems biology. However, the last quotation refers to structural aspects of integration only; conceptual integration of research results, for instance, is not addressed. But conceptual integration is needed to pave systems biology's way toward the formation of an independent epistemic culture.[21]

Interdisciplinarity offers not only new prospects, but also points to new challenges. Let us now turn to the constraints and obstacles posed to systems-biological projects and their funding mechanism in general.

5.2.1.3 Constraints by Funding Mechanisms

As described in the previous section, research projects are basically structured in work packages in which consortium partners work together on specific research questions and tasks. The pieces of work have to be aligned and bound together and to be interrelated with other work of the research project and with its objectives. In the interviews it is often stressed that the integration into the whole, the entirety of the overall project, is the most challenging task. It is therefore usually coordinated and supervised by the project management being responsible for the progression of the project and its coherence. However, one of the ACGT consortium members pointed out that most partners do not work enough on what he metaphorically called "the glue".

[20] Original quote: "[I]n der Biologie [muss sich] eine Kultur für Großprojekte entwickeln. Darauf sagen die meistens immer: 'Gibt's doch! Dieses Human Genome Project.' Aber das hat 'ne ganz andere Natur, und zwar hat man dort nur sequenzieren müssen. Das heißt, Geräte sich hierhin gestellt und dahin gestellt. Und alle waren voneinander unabhängig. Also das ist zwar ein Beispiel für ein internationales Großprojekt in der Biologie, aber es ist nicht das, was ich beschrieben hab, wo die Partner in den unterschiedlichen Ländern mit unterschiedlichen Technologien an unterschiedlichen Fragestellungen voneinander abhängig sind." (I24)

[21] Epistemic culture is a prominent concept in Science and Technology Studies referring to the practices and beliefs that constitute a culture's attitude toward knowledge and its way of justifying knowledge claims. Based on this concept, various settings of knowledge production have been identified and distinguished by stressing their contextual aspects (Knorr-Cretina 1999).

[T]he glue is the work that every partner needs to do on top of what they do, on top of their daily business, to make their work stick with the work of the others to have the integrated whole. And in my experience,—I have a lot of experience with EU projects, many years with EU projects—this is a very weak point of EU projects. And the bigger the consortium, the bigger the problem. Because everybody tends to do their own little bit. We tend to scramble together before the reviews to do the general thing and then there is not a lot of thought going into the whole. So generally speaking, I think collaboration is suboptimal I would say. (I4, IT)

Whether an interdisciplinary collaboration in a research project functions, depends on many different factors. In the quote cited above the size of the project is mentioned. Another factor is, again, how experienced the project partners are in interdisciplinary work and how heterogeneous in terms of disciplines the consortium is. The more heterogeneity, the more time the group needs to become acquainted, to get to know each other's strengths, to overcome disciplinary mindsets, and to agree on shared concepts and terminology (see Sect. 4.1.2). In addition, project partners may join the consortium for very different reasons and interests. They also have to be aligned in order to be able to reach the objectives of the project.

I think it is a case of alignment from the project proposal stage on. So usually, if you have had this experience, there are one or two people who have the big idea and then they are scrambling to find partners for the different parts. Now these partners might buy in or they might join the project just because it is good money. And not a lot of thought is put in after the project is accepted for funding. So that is one. I think there are diverging agendas or interests. Not done on purpose, but just because that is the way it is. And the incentive is not enough from the project itself. [...] The universities are pressured to get funding not only by the state. So you need to do competitive bidding. So that means that there is a lot of..., you know, you accept to become a member of a project just for the money rather than the absolute interest. Then there is, I think, also this divergent on the technical level whereby there is not a very close match, or the match is imposed from the top, let's say, from the project structure and you do what you have to do. (I4, IT)

To ensure that the project partners share the project's objectives, or—according to the interviewee's wording—to assure that the glue is working, the project management needs to stick to the project's aim and to communicate the project's goals to the partners. According to the interviewed ACGT consortium members, this is often a problem in EU projects. In particular, uncertainties about the target exploitation of the research project were addressed. Many of the interviewees said that they had no clear expectations in the beginning of the ACGT project regarding its prospective outcomes. Some interviewees would have been satisfied with a research prototype of the developed ICT infrastructure as a final result of the ACGT project and very few expected an implemented infrastructure by the end of the project, which was, according to the project proposal, in fact the explicit target of the ACGT project (see Sect. 4.1.1). One of the ACGT consortium members pointed out that R&D projects need, at least at the EU level, a broader understanding of what exploitation might mean in its specific context:

Because every time you start an R&D project, the first question that they [the European Commission] ask you is 'how are you going to exploit the results?' Of course, you expect

> the results, which could be patented, could be prototype of systems, etc. But there is a need to understand that it is a different… and there are a lot of exploitations that are taken place, of exploiting your individual outcome. Whether it is a technology outcome or whether it is knowledge or expertise, etc. And it is a different issue as I have already said, to try to exploit an infrastructure. An infrastructure should be seen as a service to a community and I think the Commission is now […] realizing that yes, you need on the European level a much more longer termination and dedicated groups who are not only focusing on addressing research questions, but also focusing on making quality production systems and infrastructures that can then be utilized by the wider community. (I7, IT)

The quotation reflects on the definition of exploitation in R&D projects, asking what could be understood as an exploitable result. Another issue, which was raised even more often in the interviews, was the appropriate timeframe necessary to get beyond results and succeed in exploiting them in a different context. As the national and EU funding programs usually limit the duration of a research project to three, four, or sometimes five years, the timeframe was often regarded as too restrictive and too inflexible. In particular, it was criticized that the time allocated by the budget may not match with the project's objectives and that such a mismatch is usually charged to the debit of the project's exploitation. Many of the ACGT members described such a dilemma concerning the expected and realized exploitations of their project. One of them said:

> I see an intrinsic problem in the way European projects are being defined. Because they ask you to focus specifically on research while building a production system. This requires much more effort exactly for implementation, that they wouldn't like to fund anyways, because it would not be research. It would be implementation. So as long as the funding is for research, at the end you end up with a research prototype and then you really need to find ways to turn that into a production system. (I2, IT)

At least in the case of the ACGT project, it seemed very difficult to fulfill the targeted expectations. "We hope that we move closer to a better exploitation through the follow-up projects," the interviewee cited above continued. However, the funding stops after termination and the partners of a current project usually do not know if they will be able to continue their collaborative work. This means that exploiting results is usually something researchers must do outside the time frame of the project. Sometimes follow-up proposals are prepared during the runtime of a project depending on the upcoming calls. But decisions on proposals take time and a smooth transition from project to project is more than unlikely.

> It's difficult for any European project to come out with something as a whole, because it's simply not built for it; the funding mechanism. The only thing that they do is they give you three, four or five years of time to build something and there is no follow-up. The funding stops and it is done. So there is no incentive to actually build something that can last. (I11, IT)

Hence, the ACGT consortium was not successful in building up a lasting infrastructure at the time when the original funding stopped, even though the ACGT consortium was very successful in applying for follow-up funding. The ACGT consortium was able to become part of the funding lineage on ICT in health and many consortium members have therefore been able to continue the collaborations initiated during the lifespan of ACGT still working together on ICT infrastructures, mainly in the domain of systems research in oncology. Nevertheless, in the interviewees' view, not follow-up projects but an extension of the ACGT project would have been the best solution with regard to exploitation.

> The duration of the project was simply too short. It just can't be done in four years. And I think it would actually have been much more practical if the ACGT project had run for eight years. So if people had continued to work in the group, like in the follow-on project p-medicine, that is, entirely new partners, entirely new constellation, entirely new goal. Actually, a lot was thrown away from ACGT. (I15, IT)[22]

> Continuing it in p-medicine, that meant a disruption again. It's true that individual components were used again, but in principle, if people had really wanted a stable environment, then they would have had to say, 'Okay, everyone with any significant involvement will get time and money again to develop precisely the same thing further.' A new system is being created yet again in p-medicine. Based on a slightly different technology. That means that whole procedure is beginning all over again, the definition of the architecture and then agreeing on standards, and so on. That means, another entire loop, and that raises the question again how far you'll get in the end. (I12, IT)[23]

Inspecting the broader picture of EU funding in the field of ICT applied to biomedical questions and problems, several research projects with very similar intentions and goals can be found that have been or are funded simultaneously or consecutively. As the example of ACGT follow-ups illustrate, they may be part of one of the funding lineages we discussed above. However, they may also be part of different funding lineages. In health-oriented programs, for example, medical projects with ICT components are funded, and in programs focused on ICT, technological projects with clinical components are supported. Altogether, conceptual integration of research developments seems necessary to prevent double funding. However, as long as the funding structure is based on timely restricted research projects with no option of extension or integration respectively, double funding will be an intrinsic part of the funding system.

---

[22] Original quote:"Die Projektlaufzeit war einfach zu kurz. In vier Jahren kriegt man das halt nicht hin. Und ich meine, eigentlich wäre es viel praktischer gewesen, wenn das ACGT Projekt acht Jahre gelaufen wäre. Also man dann halt in der Gruppe weiter gearbeitet hätte als wie jetzt das Anschlussprojekt p-medicine, also noch mal komplett neue Partner, komplett neue Zusammenstellung, komplett neue Zielsetzung. Also aus ACGT ist eigentlich sehr viel weggeschmissen worden." (I15, IT)

[23] Original quote: "Die Fortsetzung in p-medicine ist schon ein Schnitt gewesen wieder. Einzelne Komponenten wurden zwar genommen, aber im Prinzip hätte es ja so sein müssen, wenn man wirklich eine stabile Umgebung haben will, dann hätte man sagen müssen, 'So, alle die wesentlich daran beteiligt waren, kriegen jetzt noch mal Zeit und Geld, um genau dieses jetzt weiterzuentwickeln.' In p-medicine entsteht auch wieder ein neues System. Basiert ja auch auf einer leicht anderen Technologie. Das heißt, diese gesamte Prozedur fängt wieder von neuem an, die Definition der Architektur und dann die Einigung auf Standards und so weiter. Das heißt, wieder eine komplette Schleife, wo dann eben auch die Frage ist, wie weit man am Ende kommt." (I12, IT)

5.2.2 Scientists' Reactions to Funding Mechanisms

After having discussed the basics of funding mechanisms in interdisciplinary, systems-oriented research in the life sciences along with constraints exerted by these mechanisms and its consequences we now look at how the scientists react to it.

University careers today depend to a considerable extent on the success of raising external funds. As a rule, applicants for scientific appointments have to prove that they are capable of participating in the competition of receiving funding from national and international funding agencies and programs. The funding either pays one's own job or the one of PhD students and post-docs within one's working group. Hence, successful work and careers of young researchers mostly depend on external funding. Private companies, institutes, and research associations are applying for public funding as well. As discussed in Sect. 4.3.1, they are primarily seeking to be involved in academic research. For them, social networking with academia is important to be at the front line of research, to understand future trends and take them up early, and to be directly involved in innovative developments.

These different groups and interests align together to social networks that apply collectively for money. As illustrated in Chap. 4, the network originated in the ACGT project concentrates, for example, on EU funding to continue its research and established working collaborations. The network has established itself in the funding lineage of ICT for health. "Over time you see the groups that are good keep getting the projects. So I think over time, there is a congregation of the good," said one ACGT consortium member (I4, IT). To be successful in receiving the funding therefore is synonymous with, first, to be good in evaluating trends in science policy and, second, in detecting innovative trends in research that match with the trends in science policy. Using the development of the oncosimulator as an example (see Sect. 4.2), one of the interviewees explained that having a good sense for trends has a lot to do with the right intuition or, in other words, with tacit knowledge (Polanyi 1958).

> But ... and I don't really know why we introduced... I mean in ACGT we introduced knowingly a specific work package through the work of the group of Georgios Stamatakos and other individuals who even then were for years working on a more systemic approach to cancer modeling trying to model and understand cancer evolution as a system, as a phenomenon that evolves rather than simply trying to either define gene signatures or selected elements of information that could be used to predict or validate specific hypothesis. And I think that it was anticipation and of course one does that because, as I said, you have information, you understand or you make a prediction of how things will evolve in the research domain and sometimes you are right, sometimes you are wrong. It seems that we were correct in predicting these dynamics of the field, which gradually through the evolution during the course of ACGT and afterwards... I mean more and more the emphasis both in the ICT domain through the VPH type of projects, but also in the health program, the emphasis is more and more trying to understand living systems as systems. Therefore, systems biology and modeling at various levels of biological complexity, etc. They became very important and as a result, I think, we were we as a group, as a large community of people involved originally in ACGT, we've been very successful after ACGT in various subsequent efforts. (I7, IT)

In silico medicine is one of the prominent targets of the current EU funding program Horizon 2020. It is integrated in different funding domains, such as ICT, health, or emerging technologies. The development of the oncosimulator is therefore placed in a strong funding lineage that was started in FP6 in some of the research projects of the VPH-initiative and has accumulated many grants. As presented in Sect. 4.2.2, the former ACGT consortium member developing the oncosimulator—the In Silico Oncology Group at the Institute of Communication and Computer Systems, National Technical University of Athens—is now participating in numerous projects funded in the EU programs FP7 and Horizon 2020. One of the interviewees explained the success of the In Silico Oncology Group in receiving EU funding by the alignment of research interests with the vision of the future use of systems biology held by the European commission.

> The commission itself has realized that such an approach, the integrity of the systems biology approach, can use in various terms in order to describe the same or almost the same vision. So the vision of the European Commission is absolutely in line with the vision concerning the oncosimulator. And what differentiates the Virtual Physiological Human with in silico medicine is the emphasis to be put on the clinical adaptation and validation of complex biological models dealing with disease primarily. (I8, IT)

From the perspective of the individual scientist, the social networks in the funding lineages are the basis for his or her orientation. What projects are successful in receiving grants and who is successful in submitting proposals? Accordingly, for scientists seeking grants, it is necessary to be acknowledged in the social networks that are established in strong funding lineages.

> A lot of these choices are dictated by who your buddies are and who gets in early on the project consortium as it is forming. And again this ties in… I think this is important. It is very much buddy driven. And social network driven. And why no and why neither and how do I bring somebody in… who do I bring in… so I need to address a specific concern of the call. So when the European Commission is drawing its work programs it is saying, you know, 'I think we should put money in this area.' So then they say, 'Okay, for this area I need to have this and this and that.' So it is kind of like a menu. For me as a proposal writer, I have to fit in. So in order to get my high marks, if I don't have someone who is convincing on exploitation then I've got to bring, let's say, BIOVISTA in. They will write a good one pager on exploitation and I will get good points. You know things like this. (I4, IT)

In particular, those teams are successful in the funding lineages that fulfill the interdisciplinary requirement that is usually part of the calls. One of the strength of the ACGT consortium was, for example, to overcome interdisciplinary problems and to merge into an interdisciplinary team. However, with regard to promoting university careers, the interdisciplinary focus of the social networks is not conducive. Academic careers at universities have to fit into the disciplinary profiles and standards and the national culture of the universities. Hence, instead of international interdisciplinary activities, disciplinary networking in the home country is a crucial factor for successful applications and appointments. One of the interviewees stated that the disciplinary-oriented tenures are an obstacle for interdisciplinary approaches in general.

> At the moment, the situation is that the disciplines and, the impact factor matters and that kind of thing, career concerns, work against such [interdisciplinary] approaches. Well, it isn't motivating for people to work in this field, and that's why it has to be promoted. (I23)[24]

As a result, the individual scientists are somewhat trapped. They are forced to raise as much grant money as possible in order to have a good starting point for a career in science. However, to succeed at least in the EU R&D funding system, they need to be part of and active in interdisciplinary networks. If one tries to make an academic career and to get tenure, this is often counterproductive because in this case one has to be part of national disciplinary social networks.

5.2.3 Concluding Remarks

The final section is dedicated to summarize the findings of the last two sections and to reflect on the interdependency of funding and science. What effects do funding mechanisms have on doing research and making an academic career? According to Norma Morris and Arie Rip (2006), it is undisputed that during the last two decades science policy has increasingly taken over the steering of scientific activities by allocating and distributing funds. The chronological succession of the calls for proposals illustrates scientific trends in funding initiatives. In the interviews, the trends in funding were associated with funding lineages in which certain groups or researchers received their funding. According to the analyses of scholars from the Science and Technology Studies, the biggest impact of science policy is in fact the trendsetting in research areas and the coordination of scientific networking (Reiß et al. 2013, 33). The interviews further illuminated that the scientists are basically willing to accept such trendsetting by funding organizations. Creating new buzzwords was one example of how trendsetting in science policy works and how scientists adapt to it by renaming their concepts, or metaphorically speaking, by dressing the same concepts with new clothes to receive continuous funding.[25]

Concerning the trendsetting, one of the interviewees assigned the responsibility for trends in research solely to the funding organizations, whereas other interviewees underlined the interaction between science policy representatives and scientific actors involved in policy making. Given the example of setting up systems biology programs, it was argued that individual scientists have had the political influence to initiate new trends in science policy and to initiate shifts and reallocations in the distribution of budgets. Ongoing interactions between these politically involved scientists and science policy representatives finally lead to social networks in which

[24] Original quote: "Im Moment ist das so, dass die Disziplinen und die Impactfaktoren-Geschichten und solche Sachen, Karrieresachen, gegen solche [interdisziplinären] Ansätze arbeiten. Also es ist nicht motivierend für Leute in diesem Gebiet zu arbeiten, und deswegen muss man es fördern." (I23)

[25] See also Morris and Rip (2006, 256).

upcoming trends in funding and research are aligned and the distribution of the budget is negotiated.

In the interviews, the decisive framework conditions for receiving funding and, thus, doing research were not negatively evaluated. The interviewees explained that, to support or even establish new approaches or methodologies within a scientific domain such as mathematical modeling, it is necessary to prioritize certain proposals. However, the example also illustrates that the preconditions set by funding organizations definitely restrict the innovative potential of science and corroborate mainstream trends in research.

Currently, applied approaches of systems-oriented research in medicine is one of the most prominent trends on the research funding agenda in the life sciences (see Sect. 5.1.2). Michael Gibbons and his colleagues characterize problem-focused research carried out in a context of application as a new mode of knowledge production. The so-called Mode 2 knowledge is directly generated in its context of application in which the scientific problem arises, methodologies are developed, and outcomes are disseminated and used (Nowotny et al. 2001, 2003; see Sect. 5.3.2). From this it follows that Mode2 knowledge can only be achieved in interdisciplinary teams that work together for certain periods of time on specific problems (Gibbons et al. 1994, 4). Hence, the problem-focused funding agenda is complemented by certain funding mechanisms. First, grants for problem-oriented research are usually not given to support institutions or persons, but to finance time- and budget-limited projects. Second, projects, defined as collaborative undertakings to achieve specific objectives, are a very suitable structure for interdisciplinary research. As interdisciplinary research units are still rare, at least at German universities, grants from funding organizations are the best chance to work on interdisciplinary approaches. To this effect, the funding organizations have a leading role in coordinating and structuring interdisciplinary approaches. At the same time, the funding organizations frame the requirements of interdisciplinary research. Many interviewees stressed that interdisciplinary research needs funding mechanisms adapted to the challenges and conditions of interdisciplinarity. They also made clear that successful interdisciplinary collaboration cannot be achieved in the same time as established disciplinary research.

However, project funding as one of the basic funding instruments for interdisciplinary research was often criticized in the interviews. It seems difficult to ensure that all project partners share the project's vision and objectives, and direct their work toward the common goal. In particular, very heterogeneous interdisciplinary research consortia need more time than groups coming from the same disciplinary background to understand each other's concepts and mindsets. If the time frame of a project is too restrictive, the complexity of these processes often results in dissatisfaction and a lack of exploitation. To extend funding of ongoing research, scientists have only the opportunity to apply for follow-up research projects. Scientists' strategies are to adapt not only to trendsetting in science policy but also to funding instruments, for example, by dressing research they are working on with new clothes in follow-up research proposals.

Even if there are often up-coming calls in the same funding lineage, the discontinuity in funding has a negative impact on the established working collaborations. As a consequence, the scientists have to put a lot of effort into social networking to be continuously present and acknowledged in certain funding lineages. This is a dilemma for the individual scientist who is planning an academic career. He or she is forced to receiving as many grants as possible to get visible in the scientific community and to promote younger fellows. On the other hand, the scientist has to establish herself in social and local networks at the university. At university, the disciplinary orientation, for example, publishing activities in disciplinary high-impact journals, is an important career strategy. Therefore, research in the disciplinary established mainstream is usually given preference over interdisciplinary, risky, or long-lasting research (Reiß et al. 2013, 22). Hence, young scientists find themselves in a double-bind: they are institutionalized in a more or less monodisciplinary, local, and administrative structure, but have to fulfill at the same time the international and interdisciplinary funding requirements. Instead of doing research, becoming a scientist at a university seems to incorporate more and more management skills shaping the actual work context. In conclusion, funding mechanisms, in particular the prioritized funding of projects, have a complex impact on doing research and making an academic career. Generally speaking, time-limited funding programs are suitable to hook up with and give support to new emerging trends in research. However, for individual scientists, following funding lineages such as the ones generated to foster the establishment of systems biology and related approaches is risky, because their influence on important parameters is limited: on their future membership in the network of a funding lineage, on the sustained funding of the lineage, and on the relevance of the research in a funding lineage for their career.

5.3 Societal Actors' Perceptions of Science Policy

In this chapter, we explore the perception and the conceptualizations of systems biology by societal actors. Societal actors are linked to the discourse on systems biology and its science policy, yet without themselves being scientists in the conventional meaning. As representatives of the media, of industry, public interest groups, research funding organizations, administration, and of science policy, they both influence the direction of scientific research, and are affected by science policy.

The discussion on science policy that we analyze in the following sections draws upon themes such as establishment of systems biology, its medical application, and possible implications for science and society. How do different societal actors discuss questions regarding science policy of systems biology and related fields of interest? In their understanding is systems biology already established, or do they perceive it as an approach just emerging? And if so, why? How do they assess the application of systems biology? Which implications and regulations do they deem relevant?

In order to clarify some of the differences that exist with regard to the discourse among scientists, we start with a short section on the conceptualization of systems biology by societal actors in Germany (Sect. 5.3.1). The following sections deal with the establishment of systems biology (Sect. 5.3.2), its application potential (Sect. 5.3.3), and its possible societal implications and eventual regulation (Sect. 5.3.4). These sections are structured according to different actor groups and their perspective on systems biology. The grouping of the individual interview findings is done according to the interviewees' proximity to science policy decisions. Media, industry, and public interest group representatives form the first cluster, research funding and science policy the second (comp. Sect. 5.1 regarding method).

5.3.1 Societal Actors' Understanding of Systems Biology

In the public discussion, there is no predominant understanding or definition of systems biology. This is the result of a first assessment of the discourse on the terminology of systems biology; in this, it resembles the scientific discourse (comp. Sect. 2.1). Thus, the topic of the discourse itself is not clearly defined. However, compared to the scientific actors, industry, media, and public interest group representatives leave even more room for interpretation as they use the term systems biology less specifically. This was made explicit in our interviews as there is, according to one interviewee, "no generally valid definition, as far as I'm aware, but many, many different ones. But in the end, systems biology is mathematical modeling of biological processes on the basis of quantitative biological process—in other words, data. Actually, that's relatively simple"[26] (industry representative).

Other actors try to connect systems biology to existing currents in science by stressing its systemic and integrative nature. "Well, I wouldn't really say a new form of research. I mean, its charm is more in the fact that it brings together the most varied branches of research streams, as it were, and integrates them, and it's precisely that that makes it possible to draw summarizing conclusions and gain new knowledge from them"[27] (industry representative). This interpretation is openly questioned by one representative of the media when he draws a comparison to systems science:

> [W]ell, that's a technical-mathematical description of what's supposedly going on in life processes, but it didn't really have all that much explanatory power, [...] it's simply an attempt to find [...] orientation and meaning in a flood of data [...], well, so living organ-

---

[26] Original quote: "keine allgemeingültige Definition, soweit mir das bekannt ist, sondern viele, viele verschiedene. Aber die Systembiologie ist letztendlich die mathematische Modellierung biologischer Vorgänge auf Basis quantitativer biologischer Prozesse—also Daten. Das ist eigentlich relativ einfach."

[27] Original quote: "[A]lso eine neue Form der Forschung würde ich jetzt eigentlich nicht sagen. Ich meine, sie hat ja eher den Charme, dass sie die unterschiedlichsten Äste sozusagen von Forschungsströmungen zusammenführt und integriert und eben dann ermöglicht, daraus zusammenfassend Schlüsse und neue Erkenntnisse zu ziehen."

isms are somehow biocybernetic systems that respond to their environment in some kind of feedback loops and then reach certain states, and those states change again, and so on. Well, more like a, well, science of the logic of wiring.[28] (media representative)

Still, for most industry representatives, systems biology also carries a certain amount of novelty beyond the connection to existing sciences. Exemplary is the statement that "you don't [have to] understand every little cog [...] any more to arrive at biological understanding, but [you] [...] [can] start to model things precisely because biology and computer science are coming together and then simply compare them with the reality that you observe in an experiment. And that's a pretty interesting way to approach biological systems, after people tried for a long time to simplify model systems to the extent that you could only observe isolated components"[29] (industry representative).

Systems biology is seen here as a discipline providing an example with a process that starts from modeling before going into analysis, an approach that is understood as a novel perspective that might help other areas of research. It may possibly be described as a top-down approach to biological systems.

For industry representatives, the coexistence of the continuity with pre-existing research on the one hand, and the novelty of the approach have pragmatic reasons: systems biology "complements the existing quite well. That is, what we can do quite well already"[30] (industry representative). We were told by one interviewee that one has learned in a variety of different projects that an interdisciplinary cooperation is promising or even indispensable:

Well, experimenters don't like it if theorists make experimental designs for them. [...] There are positive exceptions, too. Well, there are also working groups that have an almost 10-year history together, where theorists from one group and experimenters from the other were systematically paired off. And they've learned that they benefit from it.[31] (industry representative)

---

[28] Original quote: "das ist halt so eine technisch mathematische Beschreibung dessen, was da in Lebensvorgängen abgehen soll, hatte aber nicht so wirklich viel Erklärungskraft, [...] das ist halt der Versuch, in einer Flut von Daten [...] Orientierung und Sinn zu finden [...], also dass praktisch Lebewesen irgendwie biokybernetische Systeme sind, die in irgendwelchen Feedbackschleifen auf die Umwelt reagieren und dann wieder bestimmte Zustände erreichen, die sich dann wieder ändern, und so weiter. Also eher so eine, ja, Verschaltungslogik-Wissenschaft."

[29] Original quote: "man [...] nicht mehr jedes einzelne Rädchen verstehen [muss], um zum biologischen Verständnis zu kommen, sondern [dass man] [...] auch gerade durch das Zusammenwachsen von Biologie und Computerwissenschaften eben Modellierung anfangen [kann] und die dann einfach mit der Realität, die man beobachtet, experimentell abgleichen. Und ist mal ein ganz interessanter Weg, sich biologischen Systemen anzunähern, nachdem man eben lange Zeit eben immer versucht hat, Modellsysteme soweit zu vereinfachen, dass man immer nur isolierte Komponenten eben betrachten konnte."

[30] Original quote: "ergänzt ganz gut das bereits Vorhandene. Also was man schon ganz gut kann."

[31] Original quote: "Also Experimentatoren mögen es nicht, wenn ihnen Theoretiker Versuchspläne machen. [...] Es gibt auch positive Ausnahmen. Also es gibt auch Arbeitsgruppen, die jetzt also schon fast zehn Jahre Geschichte haben, eine gemeinsame, wo es also konsequent Pärchenbildung gibt zwischen Theoretikern von der einen Gruppe und Experimentatoren von der anderen. Und die haben gelernt, dass sie davon profitieren."

Here, the step towards interdisciplinary work is marked as a necessary one if systems biology is to catch up with other natural sciences. Among all, catching up with theory and theoretical reasoning plays an important role in this process. In contrast to (molecular) biology, systems biology contains "Modeling methods coming from mathematics, and that are the standard in physics and in other more technical, or at least non-biological disciplines [...]. But first of all, I think it very clearly has to be organized from the theoretical side"[32] (industry representative).

In retrospect, the emergence of systems biology was interpreted by most actors as consequential, for example, as a "logical further development of what we learned from genome research, that a lot can be seen at the DNA level, but nothing can be understood [...] [and that systems biology] is simply necessary to see how this whole new level of -omics, metabolomics and proteomics, all of them are also benefiting from the systems biology approach"[33] (public interest group representative).

Representatives of science policy in Germany seem to have a more specific definition of systems biology. One interviewee made this explicit as he compared the German understanding with the US-American:

> About looking across the Atlantic, for us, the question is//has always been, what do the Americans mean by systems biology? What do the Europeans mean by systems biology? After all, we in Europe defined the term. Of course, especially against the background, how do we want to assess projects and have reviewers evaluate them if we don't have a uniform definition of what [a] systems-biology research approach means? And if I compare that with America, then we see that there's a different concept especially between the US and Europe—what is systems biology? The Americans define systems biology very broadly. [...] That's why the figures for research funding are so impressive there, because they include a lot of things that we'd consider to be in other areas here.[34] (research funding/administration representative)

---

[32] Original quote: "Modellbildungsmethoden, die also aus der Mathematik kommen, die in der Physik und in anderen eher technischen oder eben nicht biologischen Disziplinen Standard sind [...]. Aber zunächst einmal muss das meiner Meinung nach ganz klar von der theoretischen Seite aufgezogen werden."

[33] Original quote: "logische Weiterentwicklung dessen, was wir aus der Genomforschung gelernt haben, dass einfach auf der Ebene der DNA vieles zu sehen ist, aber nichts verstanden werden kann [...] [und Systembiologie] einfach nötig ist, um dann zu sehen, wie diese ganze Ebene der -omics, die da jetzt kommt, die Metabolomics und Proteomics, die alle profitieren ja auch vom Ansatz der Systembiologie."

[34] Original quote: "Was den Blick über den Teich betrifft, da ist//war für uns immer die Frage, was verstehen die Amerikaner unter Systembiologie? Was verstehen die Europäer unter Systembiologie? Wir haben ja in Europa diesen Begriff definiert. Insbesondere natürlich auch vor dem Hintergrund, wie wollen wir Projekte evaluieren und bewerten lassen von Gutachtern, wenn wir keine einheitliche Definition dessen haben, was [ein] systembiologischer Forschungsansatz bedeutet? Und wenn ich das mit Amerika vergleiche, dann haben wir festgestellt, dass es eine unterschiedliche Auffassung gibt zwischen insbesondere USA und Europa—was ist Systembiologie? Die Amerikaner definieren die Systembiologie sehr breit. Und deshalb sind auch die Forschungsförderungszahlen dort so beeindruckend, weil dort sehr vieles darunter gefasst wird, was wir hier anderen Bereichen zuordnen würden."

Our interviewees' reflection on the definition of systems biology thus seems to be at least partially driven by his awareness of an (apparently) higher funding level connected with this label in the United States. We also found that research funding representatives made a clear differentiation towards synthetic biology, something which was not mentioned in the interviews with members of other groups. Yet, the understanding of the subject matter of systems biology varies widely. We now take a closer look on how different actors discuss science policy regarding systems biology.

5.3.2 Establishment of Systems Biology and Need for Science Policy

The different societal actors did not only have divergent understandings and interpretations with regard to what systems biology is and what it comprises, as documented in the previous section, but also concerning its current state of establishment in science and industry. In fact, the experts do agree in a cautiously optimistic assessment of the scientific progress in systems biology: while they do not expect realization of the grand promises made for systems biology in the near future, such as modeling complete cells or even organs, they look forward to smaller albeit encouraging steps and successes.

5.3.2.1 Media, Industry, and Public Interest Groups

For example, industry representatives expect results from modeling smaller systems: "I'd tend to see [the] next 10 to 20 years more in the simple systems, it isn't all that simple to understand an entire human being"[35] (industry representative). For public interest group representatives, the larger aim plays a role as there is still great hope towards systems biology that leads to some fundamental change in our perspective on life and of the interrelationships between organisms and environment, and expectation for beneficial applications in medicine. One stakeholder put it this way:

> And in this respect, I expect that the possibilities of interfering in organisms and changing the cell metabolism, and producing materials or actually organisms that have new characteristics […]. And of course there will be new knowledge about how life is organized fundamentally, how interactions play out between the environment and the living organism, too.[36] (public interest group representative)

---

[35] Original quote: "[Die] nächsten 10 bis 20 Jahre würde ich eher also in den einfachen Systemen sehen, das ist ja nicht so ganz einfach, einen ganzen Menschen zu verstehen".

[36] Original quote: "Und insofern erwarte ich, dass da eben die Möglichkeiten in Organismen einzugreifen und den Zellstoffwechsel zu verändern, und Stoffe zu produzieren oder eben Organismen zu schaffen, die neue Eigenschaften haben […]. Und natürlich wird es darüber auch neue Erkenntnisse darüber geben, wie Leben grundsätzlich organisiert ist, wie Wechselwirkungen auch zwischen Umwelt und Lebewesen sich abspielen."

Regarding the expectations towards application of systems biology, industry representatives maintain a low profile. For most, systems biology is in the research stage but offers nevertheless interesting perspectives. For them, systems biology is "currently in the research phase [...], but we can already tell that it will be important, an important development in biology, because unlike how it used to be, it enables a kind of holistic observation of cells and cell systems with high-throughput methods"[37] (industry representative). Here, the prevailing view is that some systems approaches in biology indeed begin to reap first recognition; a large-scale deployment, however, is still inconceivable. An industry representative states, "that to this day, systems biology in industry is mostly a hope or a promise. And depending on which company you're talking about, people tend to take these promises more or less seriously. [...] That means especially that hardly any companies are using approaches that really work well to integrate systems biology fully into the company's research workflows that are used to develop pharmaceuticals. And in that way to develop products with which the companies can then recoup their money"[38] (industry representative). Thus, the attitude of industrial professional associations can be described as anticipatory:

> [F]rom the perspective of companies, I think there is relatively little initiative to say now, we'll support systems biology or we'll call for research programs in that area. What we're tending to see, just like in other areas, too, is that opportunities are emerging from basic research, from the classical way of gaining scientific knowledge, new technologies, new analytic platforms, where companies are feeling their way forward cautiously. I mean, it's always about feeling your way forward. It's rarely the case that a new technical perspective pops up on the horizon and then industry jumps on it right away and says: that's what we want.[39] (industry representative)

---

[37] Original quote: "derzeit im Forschungsstadium [...], aber es zeichnet sich schon ab, dass es wichtig wird, eine wichtige Entwicklung in der Biologie wird, weil es eben anders als früher ermöglicht, durch so High-Throughput-Methoden so eine holistische Betrachtung auf Zellen und Zellsysteme ermöglicht."

[38] Original quote: "dass die Systembiologie bis heute in der Industrie im Wesentlichen eine Hoffnung oder ein Versprechen ist. Und in Abhängigkeit davon, über welche Firma man dann spricht, wird diesen Versprechen mehr oder weniger geglaubt. [...] Das bedeutet insbesondere, dass es in kaum einer Firma bisher wirklich gut funktionierende Ansätze gibt, Systembiologie voll zu integrieren in die Forschungsworkflows der Firmen, die eingesetzt werden, um Arzneimittel zu entwickeln. Und damit halt auch Produkte zu entwickeln, mit denen die Firmen dann wieder ihr Geld einspielen können."

[39] Original quote: "auch aus Unternehmenssicht gibt es, glaube ich, relativ wenig Initiative jetzt zu sagen, wir fördern Systembiologie oder wir fordern Forschungsprogramme in dem Bereich. Wir sehen eher, so wie es in anderen Bereichen eben auch ist, dass aus der Grundlagenforschung, aus dem klassischen wissenschaftlichen Erkenntnisgewinn heraus eben Möglichkeiten entstehen, neue Technologien, neue analytische Plattformen, in die sich Unternehmen ja zunächst mal vortasten. Ich meine, das ist auch immer ein Vortasten. Es ist ja selten so, dass da eine neue technische Perspektive am Horizont auftaucht und dann sofort die Industrie drauf springt und sagt, das wollen wir."

Although the immediate application of systems biology currently seems out of reach, industry representatives paint an optimistic picture for the future:

> Well, for years systems biology has been an area that's received government funding, and by now, it has also achieved a certain significance at universities. And that's certainly something that is increasing at the moment and that will continue to become more important. And yes, it has a lot of open connections to other areas of biotechnology. And in that respect, I think it's an area that will still be relevant in the coming years, yes.[40] (industry representative)

We also found a strong agreement between representatives of professional associations and the industry: whereas the former state "a very high potential [...] in questions like that [...] concerning personalized medicine"[41] (public interest group representative), the latter describe: "It's also the case that more and more people are (1) recognizing that that might be the only opportunity the pharmaceutical industry still has to improve its research effectiveness or efficiency. And more and more people are also acknowledging that apparently, it can work, and in individual areas, it really has worked already. [...] So what I expect is that its use will increase massively. Really massively. I mean, by orders of magnitude, possibly by a factor of 10 or 100"[42] (industry representative). An example for the future application of systems biology could be the operation of research service agencies: "There is partly a very marked interest in, well, getting the best overview possible, trying out as much as possible, well, especially companies, big companies are not confining themselves to doing that with their own resources, but trying out research service providers that work in the area [...]. Well, things are happening there, that's clear"[43] (industry representative). Thus, our interviewees from the media agree with the interpretation of systems biology as an emerging approach in science: "Well, I do see that as a major trend, I'd say. So, centralization, coordination, access, networks, big science"[44] (media representative).

---

[40] Original quote: "Systembiologie ist ja seit Jahren ein Zweig eben, der staatlich gefördert wird und auch an den Universitäten einen bestimmten Stellenwert inzwischen hat. Und das ist bestimmt etwas, was im Moment eben im Wachsen ist und auch in seiner Bedeutung eben noch zunehmen wird. Und ja, ganz viele offene Enden hat zu anderen Bereichen in der Biotechnologie. Und insofern denke ich, ist das ein Bereich, der also die nächsten Jahre noch relevant sein wird, ja."

[41] Original quote: "ein sehr hohes Potenzial [...] in so Fragestellungen [...], die die personalisierte Medizin betreffen"

[42] Original quote: "Es ist auch so, dass es mehr und mehr Leute gibt, die erstens erkennen, dass das vielleicht die einzige Chance ist, die die Pharmaindustrie noch hat, um ihre Forschungseffektivität oder—effizienz zu verbessern. Und es gibt mehr und mehr Leute, die also auch sehen, dass es scheinbar funktionieren kann und in einzelnen Feldern auch wirklich schon funktioniert hat. [...] Meine Erwartungshaltung ist schon die, dass also der Einsatz massiv zunehmen wird. Wirklich massiv. Also um Größenordnungen, Faktor 10/100 möglicherweise."

[43] Original quote: "Es gibt teilweise ein sehr ausgeprägtes Interesse daran, also einen möglichst guten Überblick zu bekommen, möglichst viel auszuprobieren, also gerade auch Firmen, große Firmen beschränken sich nicht darauf, das mit eigenen Ressourcen zu machen, sondern testen Forschungsdienstleister, die in dem Bereich aktiv sind [...]. Also da passiert was, ganz klar."

[44] Original quote: "Also das sehe ich schon als einen großen Trend, würde ich jetzt sagen. Also Zentralisierung, Koordinierung, Zugang, Netzwerke, Big Science."

5.3.2.2 Research Funding, Science Policy, and Administration

In the interviews with representatives from research funding agencies, science policy and administration, the experts agreed on one thing: systems biology is seen as an approach that has largely established itself in the research community. Not all promises that were given have already been fulfilled, but important steps have been made towards the initial vision of what systems biology can achieve. The big leap, however, is still to come. Nevertheless, in biological research, medicine, and similar areas, as well as in academic training, systems biology as an approach is perceived as being largely established.

This has created the preconditions for integrating research funding for systems biology in new programs:

> I think that this approach has become established, that it's become the routine. And we're seeing that biologists and physicians are simply integrating this approach […] in many applications for research funding, not only in systems biology, but also in other areas. And that's why I think that in the foreseeable future, it won't be necessary any more for us to promote this approach ourselves, but that we should reorient research funding toward other goals and consider the systems-biology approach to be an integral part of every forward-looking research project.[45] (research funding/administration representative)

Our interviewees underlined that this does not mean that research funding agencies seek to shift funding, but rather that systems biology as a discipline is embedded in different contexts as it is firmly established already as a research approach. It is assumed that the establishment of systems biology will follow the general dynamics of scientific disciplines and their common scheme of disciplinary evolution.

In our interviews, we also found some requirements and expectations with regard to systems biology as they are directed towards applications in medicine and the pharmaceutical industry: "[I] would think that naturally, medicine will benefit from it to an extraordinary degree. We're seeing that research, especially in the field of individualized medicine, has benefited a lot from the funding that we initiated in recent years and are still pushing forward"[46] (research funding/administration representative). Here again, the expectation is not that it will be possible to reach visionary goals such as the modeling of complex biological systems in the near future; it is rather agreed upon that the development of models that will be necessary for medical applications

[45] Original quote: "Ich glaube, dass sich dieser Ansatz etabliert hat, dass er zur Routine geworden ist. Und wir beobachten, dass Biologen und Mediziner […] in vielen Anträgen zur Forschungsförderung nicht nur in der Systembiologie, sondern auch auf anderen Gebieten diesen Ansatz einfach integrieren. Und deshalb denke ich, dass in absehbarer Zukunft es nicht mehr erforderlich sein wird, diesen Ansatz selbst zu fördern, sondern die Forschungsförderung auf andere Ziele auszurichten und den Ansatz der Systembiologie als einen integralen Bestandteil jedes zukunftsweisenden Forschungsprojektes zu betrachten."

[46] Original quote: "[I]ch würde denken, dass die Medizin davon natürlich außerordentlich profitieren wird. Wir beobachten, dass die Forschung insbesondere im Bereich der individualisierten Medizin davon sehr profitiert hat, von der Förderung, die wir in den letzten Jahren angeschoben haben und auch noch anschieben."

still needs a great deal of work, even as "individual compartments are already successful"[47] (research funding/administration representative).

To summarize our findings: the different societal actors perceive systems biology as a discipline that is still on its way towards becoming an established approach or discipline. But there are quite diverging views if one compares the different actors: those related to research funding argue that systems biology is established insofar as it does not need special funding anymore, and that it can be supported by conventional research funding, however, other actors emphasize goals that are not yet reached but also the potential of systems biology to achieve them. Interviewees from all fields agree that the future will see systems biology as an established approach in scientific practice and commercial and medical application and they underline that research for the application of systems biology is still funded and will be funded.

5.3.3 *The Application of Systems Biology*

Research-funding initiatives are often justified with the scientific and technological potential of the emerging scientific field and the theoretical and practical goals to be reached; an important role is also played by the promises and hopes associated with future applications resulting from research. How are aspects of the application of systems biology described in public discourse? How do societal groups assess the importance of systems biology applications in science, medicine, and industry? In seeking answers to theses questions, we also take up considerations on possible paths towards commercialization and aspects of intellectual property.

5.3.3.1 Media, Industry, and Public Interest Group Representatives

For the interviewed members from different societal groups, the value of systems biology surfaces in three areas: basic scientific research, industry, and medical applications. For industry representatives, systems biology is proving its value not "primarily in medicine," but rather "in the laboratory market" or in the "production of resources". By now, the economic impact is determined to be sizable and understood as a "striking business argument"[48]. The use of results from systems research in medicine is likewise a stated goal in the public discourse. Here, the discovery of new drugs and early assessments of their potential play a big role: "And that offers opportunities and risks, it naturally also offers individuals the opportunity to assess early on in the development of a pharmaceutical, what are all the things this active substance does that you wouldn't ordinarily be able to see. And also being able to

[47] Original quote: "Einzelkompatimente bereits erfolgreich sind."

[48] Original quote: "schlagendes wirtschaftliches Argument."

assess earlier on, what is its efficacy/side-effect profile? In other words, to assess opportunities and risks"[49] (industry representative).

This aspect is seen as critical by one representative of a public interest group: He sees a danger in commercialization because systems biology as a scientific approach should primarily be driven by the ambition to gain knowledge instead by an interest in turning it into commercial value. In support of this argument, he notes that possible applications and the development of products often stand in the foreground when systems biology is discussed. Criticism is also directed towards the current state of systems-biology research and its maturity regarding the application in medicine:

> Can I find a better therapy for it? Yes or no? For the patient sitting in front of me, I'd say the answer is: in very few cases yes, in most cases no, so far. Could that change in the future? I'd say we don't know that yet. That research question is still open. There's still hope. And the hope is: more data, better prediction. But whether the prediction comes true, I'd say, well, personally, as a journalist, I'm agnostic. So my opinion is, let's let the researchers figure that out.[50] (media representative)

5.3.3.2 Research Funding, Science Policy, and Administration

In our interviews, representatives from research-funding, science policy and administration described their aims in funding research related to the field of systems biology as a first round of funding was dedicated to the establishment of systems biology as it was perceived as truly new and possibly game-changing:

> It was in 2004 that the BMBF (the German Federal Ministry of Education and Research) started funding systems biology for the first time, that was the systems biology of liver cells. That was a pilot project, and I think that was the first time that a coordinated research and funding measure was initiated in this field in Germany. Of course, the intention behind it had to do with funding policy. For one thing, it was about making this truly new approach, this new methodological approach available to research, too, and to try to integrate this new approach in science, too, to introduce it and to see whether the scientists actually take up this research approach. And later on, the idea was of course to determine whether it was successful, whether research funding in the way we structured it at that time, whether it was actually successful, too. That was certainly the case, and a very large package of research funding emerged.[51] (research funding/administration representative)

---

[49] Original quote: "Und das bietet Chancen und Risiken, das bietet auch für Einzelne natürlich die Chance frühzeitig abzuschätzen in so einer Arzneimittelentwicklung, was tut dieser Wirkstoff so alles, was man normalerweise nicht so ohne Weiteres sehen würde. Damit auch frühzeitiger abschätzen zu können, wie ist er denn in seinem Wirkungs-/Nebenwirkungsprofil? Also um Chancen und Risiken einzuschätzen."

[50] Original quote: "Kann ich dafür eine bessere Therapie finden? Ja oder nein? Für den Patienten, der vor mir sitzt, da würde ich sagen, lautet die Antwort: In ganz wenigen Fällen ja, in den meisten Fällen bisher nein. Könnte das zukünftig anders werden? Ich würde sagen, das wissen wir noch nicht. Die Forschungsfrage ist noch offen. Es gibt noch Hoffnung. Und die Hoffnung heißt mehr Daten, mehr Prognose. Aber ob die Prognose eintritt, würde ich sagen, da bin ich jetzt persönlich als Journalist agnostisch. Also das sage ich mal, lassen wir die Forscher klären"

[51] Original quote: "Es war im Jahr 2004, als die Förderung zur Systembiologie zum ersten Mal gestartet ist durchs BMBF, das war die Systembiologie der Leberzelle. Das war ein Pilotprojekt,

Most interviewees of this group think that funding of applied systems-biology research will yield good results, especially in medicine, within the next few years and look forward to it:

> I would think that naturally, medicine will benefit from it to an extraordinary degree. We're seeing that research, especially in the field of individualized medicine, has benefited a lot from the funding that we initiated in recent years and are still pushing forward. I think that we'll be seeing results in the next few years. I see that in the projects, since we're getting very good results, we will get good results, possibly even breakthroughs in a few small areas. The second area is of course the area of biotechnology, in other words, everything described by the term metabolic engineering, and naturally, systems-biology funding measures play a very decisive role here, too.[52] (research funding/administration representative)

This is similar to how industry representatives judge the situation; they, however, see a need for more funding in the near future. Whereas for other topics (such as the state of establishment) the perspectives of all interviewees were quite similar, with regard to the application potential of current systems biology clear differences between actors' opinions exist: industry representatives emphasize economic interests, and stakeholders of public interest groups point out potential conflicts of interest. The media representatives are skeptical about when systems-biology research can be applied and which projects actually hold commercial value; however, this is in contrast to the more optimistic view of the research-funding agencies. These differences in assessing the state of and potential for application generate a number of issues regarding societal challenges of systems biology, which are discussed in the next section.

5.3.4 *Societal Implications and Regulation*

Important societal actors from funding agencies, administration and industry, as well as many scientists stress the huge application potential of systems-biology research and the results thereof in medicine and biotechnology, as well as the

---

und damit ist zum ersten Mal glaube ich in Deutschland eine koordinierte Forschungs- und Fördermaßnahme auf dem Gebiet gestartet. Die Intention war natürlich förderpolitischer Art. Einmal ging es darum, diesen wirklich neuen Ansatz, diesen neuen methodischen Ansatz auch für die Forschung verfügbar zu machen und zu versuchen, diesen neuen Ansatz auch in die Wissenschaft zu integrieren, hineinzubringen und zu schauen, ob sich die Wissenschaftler tatsächlich auch dieses Forschungsansatzes annehmen. Und im weiteren Verlauf natürlich war festzustellen, ob sich das bewährt, ob sich die Forschungsförderung so, wie wir sie aufgesetzt haben damals, auch tatsächlich dann bewährt. Das war sicher der Fall, und daraus hat sich dann ein sehr umfangreiches Paket der Forschungsförderung ergeben."

[52] Original quote: "Ich würde denken, dass die Medizin davon natürlich außerordentlich profitieren wird. Wir beobachten, dass die Forschung insbesondere im Bereich der individualisierten Medizin davon sehr profitiert hat, von der Förderung, die wir in den letzten Jahren angeschoben haben und auch noch anschieben. Ich denke, da wird es in den nächsten wenigen Jahren zu Ergebnissen kommen. Ich beobachte das in den Projekten, da wir sehr gute Ergebnisse kriegen, werden wir gute Ergebnisse bekommen, möglicherweise sogar Durchbrüche auf einzelnen kleinen Teilbereichen. Der zweite Bereich ist natürlich der Bereich der Biotechnologie, also alles das, was man so mit Metabolic Engineering umschreibt, da spielen natürlich systembiologische Fördermaßnahmen hier auch eine ganz entscheidende Rolle."

putative commercial value that could be generated from these applications. Nobody can really be sure today, whether and to which extent these expectations will come true. However, it is reasonable to assume, that at least some of the putative or anticipated benefits of applied systems biology will be realized and have societal implications. Therefore it may be important to find out what societal actors think about such implications of systems biology and what they expect. This could at least in principle enable policy makers to think about necessary interventions, inasmuch as early interventions "can help to avoid that technologies fail to embed in society and/or help that their positive and negative impacts are better governed and exploited at a much earlier stage" (von Schomberg 2012, 50). Consequently, we identified statements of our interview partners related to possible implications of systems biology and analyzed them with regard to existing societal challenges and controversies.

5.3.4.1 Media, Industry, and Public Interest Group Representatives

In the interviews, few hints are given that point directly towards societal implications: instead, often synthetic biology was brought into the picture when the interviews turned towards the role of technoscientific developments for society. When our interviewees mentioned societal implications, they usually were related to the topic of public access and fair distribution. One public interest group representative states that one does have "very often the impression [...] that technologies are developed, products are developed that aren't actually, let's say, necessarily in the public interest"[53] (public interest group representative). On the one hand, research is understood as meaningful even though resulting inventions are not immediately applicable in practice, however, systems biology could seem to have negative implications for society when industrial and economic interest come into play instead of basic research or medical applications.

Real or perceived negative societal implications often provoke calls for regulation. Well-known examples include stem cell research and genetically engineered crops. In the area of systems biology, our interviewees were reluctant to discuss sensitive issues coming up in fields related to applied systems biology and medicine such as the necessary establishment of large biobanks and databases, eventually comprising personal data. In general, they were cautious to talk about societal implications and regulation. This reluctance, however, does not seem to be due to the sensitivity of the issue but rather stems from the lack of an immediate need to deal with such issues. There seems to be a broad consensus that for systems biology there are no new ethical or societal issues at stake. An exemplary statement from an industry representative argues that ethical concerns would only be relevant, "if it really becomes a topic of discussion in practice. So, if you really have to consider, say, from an entrepreneurial point of view, [...] in case of doubt, it's also a risk in

[53] Original quote: "sehr oft den Eindruck [...], es werden Technologien entwickelt, Produktentwicklung betrieben, die eigentlich nicht im Sinne des, sagen wir mal, des öffentlichen Interesses unbedingt stehen."

terms of an additional regulatory requirement that precisely doesn't result in additional safety, but in more time and effort"[54] (industry representative).

In our interviews, the subject of databases was always linked to the topic of data protection and privacy. However, we found no indication that societal actors were aware of a new quality or challenge introduced by big data storage and processing as it is, for instance, necessary in systems medicine and research related to it. As a business representative remarked, the question of the databases was primarily a technical challenge with the aim of "enabling all research groups to access these resources. Somehow in a way that also conforms to data protection"[55] (industry representative). In contrast, one public interest group representative discusses the lack of transparence that is a reality for patients:

> Well, it's also the case that as a matter of principle, patients are simply asked whether they consent to having the data used in research. But whether they are exploited commercially, [...] whether personal genetic data are even patented, these questions aren't discussed with the patients. And I believe that there should simply be more transparency here. And the [...] level of data storage and anonymization is important, too, of course. And of course, there should be rules about who has access to these data at all and for which purposes.[56] (public interest group representative)

Although it is true that currently, no real need is seen to take regulatory action, such measures are not categorically ruled out for the future: "And I think we will also see, to the extent that these technologies become broadly available, sooner or later they will also be the subject of guidelines and, say, they'll play a role in regulatory frameworks and underlying conditions. [...] I actually don't think there's a need to regulate right now that would go beyond what we have anyway"[57] (industry representative). Again, we found that for systems biology, the consequences are (still) quite unclear and there is a strong feeling that existing regulations for genetic engineering, clinical trials, or data protection are sufficient, inasmuch as no new or

---

[54] Original quote: "wenn es in der Praxis wirklich mal zu einem Thema wird. Also wenn man wirklich auch dann abwägen muss, sagen wir, aus unternehmerischer Sicht betrachtet, ist das [...] im Zweifel auch ein Risiko im Sinne einer zusätzlichen Behördenauflage, die mir eben keine zusätzliche Sicherheit schafft, aber mehr Aufwand."

[55] Original quote: "den Zugriff aller Forschergruppen auf diese Ressourcen zu ermöglichen. In irgendwo einer Art, die dann eben auch datenschutzkonform ist."

[56] Original quote: "Es ist ja auch so, dass Patienten grundsätzlich einfach nur gefragt werden, ob sie damit einverstanden sind, dass die Daten in der Forschung verwendet werden. Ob das dann aber eine wirtschaftliche Verwertung ist, [...] personenbezogene genetische Informationen sogar patentiert werden, diese Fragen werden ja nicht erörtert gegenüber den Patienten. Und ich glaube, da müsste einfach mehr Transparenz vorhanden sein. Und wichtig ist natürlich auch die [...] Ebene der Datenspeicherung, die Anonymisierung. Und es sollte natürlich auch geregelt werden, wer überhaupt Zugriff auf diese Daten zu welchen Zwecken hat."

[57] Original quote: "Und wir werden, denke ich, auch sehen, in dem Maße, wie diese Technologien in der Breite zugänglich werden, werden sie früher oder später auch in Guidelines auftauchen und sagen wir, in regulatorischen Rahmennetzwerken und in Rahmenbedingungen eine Rolle spielen. [...] an sich sehe ich eigentlich momentan keinen Regulierungsbedarf, der über das, was wir ohnehin haben, hinausgehen würde."

enhanced societal effects are expected from systems biology. In contrast to this, synthetic biology evokes much stronger images of possible negative consequences that seem relevant for everyday life:

> Well, I mean, sure there are aspects that extend into classical genetic engineering, but we have a comprehensive legal regulatory framework for that. We have questions concerning biosecurity. But in my opinion, we have a sufficient, at least a sufficient framework for that, too. […] One thing that will certainly play a role in the future, but that doesn't concern systems biology at its core, but more a different area, the topic of synthetic biology. […] Life from the lab, designer organisms, etc. etc., and that will raise the question again, where are the reasonable limits in terms of aspects of security, but also in terms of ethical aspects?[58] (industry representative)

Thus, methods or applications developed in the field of systems biology are deemed to be possible subjects of regulation, yet the discipline itself is free from such restrictions: "Methods that are used in systems biology just as in//well yes, I can image that, but not for systems biology itself, at first. I can certainly imagine applications that aren't in the interest of society"[59] (industry representative). Furthermore, systems biology is not associated with an impact on ethical values as is, for instance, stem cell research: "Likewise, I can naturally imagine systems biology resulting in some kind of abstruse excesses, and especially synthetic biology, too, but first of all, I'd think that that can be managed relatively well—in the area of systems biology as well as in the area of genetics. […] In contrast to, let's say, early stem cell research, systems biology doesn't have the problem that it believes it's dependent on research funding sources that are ethically questionable per se"[60] (industry representative). Again, our interviewees find it difficult to identify problems and concerns. This is most likely due to the difficulty of knowing today what possible implications might surface in the future. None of our interview partners was comfortable with providing concrete examples for negative implications without any further indication that such problems might indeed become reality. Thus, many hopes but few problems or fears are identifiable in the context of systems biology and medicine; the only issue that emerged and may be relevant for regulation was the handling and protection of sensitive data.

---

[58] Original quote: "Also ich meine, klar, da haben wir Aspekte, die in die klassische Gentechnik reinreichen, aber dafür haben wir ja einen umfassenden gesetzlichen Regulierungsrahmen. Wir haben Fragen, die die Biosicherheit betreffen. Aber auch dafür haben wir einen hinreichenden, aus meiner Sicht zumindest einen hinreichenden Rahmen. […] Ein Punkt, der sicherlich für die Zukunft eine Rolle spielt, der aber die Systembiologie im Kern nicht betrifft, sondern eher einen anderen Bereich, das Thema synthetische Biologie. […] Leben aus dem Labor, Designerorganismen, etc. pp, und das wird wieder die Frage aufwerfen, wo sind sozusagen da die aus Sicherheitsaspekten, aber auch aus ethischen Aspekten heraus vertretbaren Grenzen?"

[59] Original quote: "Methoden, die in der Systembiologie eingesetzt werden genauso wie in//also ja, da kann ich mir das vorstellen, bei der Systembiologie selber zunächst einmal nicht. Ich kann mir auch durchaus Anwendungen vorstellen, die nicht mehr im Interesse der Gesellschaft sind."

[60] Original quote: "Genauso kann ich mir natürlich auch bei Systembiologie irgendwelche abstrusen Auswüchse, also gerade bei der synthetischen Biologie sowieso vorstellen, aber ich würde zunächst einmal denken, dass das—also im Bereich Systembiologie genauso wie im Bereich Genetik—relativ gut handhabbar ist. […] Systembiologie hat im Gegensatz zu der—ich sage mal—frühen Stammzellforschung nicht das Problem, dass sie glaubt, angewiesen zu sein auf Quellen oder auf Mittel für ihre Forschung, die per se ethisch bedenklich sind."

In this context, the concept of anticipatory regulation (regulation of future fields of research) surfaces: "That's why I'd see that less in relation to systems biology or synthetic biology, instead, I'd argue strongly for establishing control mechanisms that establish responsible handling of certain research, of sensitive research areas, for example pathogen research and so on. So, similar to medical guidelines"[61] (industry representative). One public interest group representative expresses similar thoughts regarding anticipatory regulation:

> And all these questions and also especially in relation to possible environmental impacts haven't really been discussed so far and should be taken up by the legislature, and they should try first of all to map everything that's actually happening, what's new, and to what extent the current legal provisions are actually sufficient.[62] (public interest group representative)

No immediate measures are called for, but there remains a certain awareness of the fact that should negative implications from systems biology become reality, it would have been better to have taken preventive measures. The guidelines mentioned above are not exactly the strictest option available, and the call for legislative action is brought forward with little urgency. Thus far, the outlooks of the interviewees do not address imminent threats, and not even indirect threats such as possible negative impacts on public opinion. Hence, they do not see a need for proactive or anticipatory regulation.

Furthermore, one stakeholder poses the question of who could develop schemes for dealing with such uncertainty and lack of knowledge:

> Interdisciplinary working groups including civil society should be put in a position to deal with the question, which questions are new, which questions have come up recently, what is the need for regulation? I do think that that is a process that can't really go to the Bundestag (parliament) immediately, where you could say, well, the Bundestag or the government will simply put forward a proposal for a new Genetic Engineering Law, and then it's just about the details. I think that it's actually about a survey, and also about the attempt to take an interdisciplinary look at how to develop reasonable legal provisions here in terms of future developments.[63] (public interest group).

[61] Original quote: "Von daher würde ich das weniger auf die Systembiologie oder synthetische Biologie bezogen sehen, sondern ich würde stark dafür plädieren, dass Kontrollmechanismen etabliert werden, die einen verantwortungsvollen Umgang mit bestimmten Forschungs-, sensiblen Forschungsbereichen, wie zum Beispiel Pathogenforschung und so weiter etablieren. Also ähnlich ärztlichen Leitlinien."

[62] Original quote: "Und all diese Fragestellungen und auch in Bezug eben auf mögliche Umweltauswirkungen sind eigentlich bisher nicht wirklich diskutiert worden und sollten vom Gesetzgeber aufgegriffen werden und sollten also versuchen, hier erst mal abzubilden, was eigentlich alles passiert, was Neues hinzu gekommen ist und inwieweit hier eben auch tatsächlich die derzeitigen gesetzlichen Vorgaben ausreichend sind."

[63] Original quote: "Es müssten interdisziplinäre Arbeitsgruppen auch unter Beteiligung der Zivilgesellschaft dazu in die Lage versetzt werden, sich damit zu befassen, welche Fragestellungen sind neu, welche sind neu dazu gekommen, welchen Regulierungsbedarf gibt es. Ich glaube schon, dass das ein Prozess ist, der nicht jetzt irgendwie sofort in den Bundestag gehen kann, wo man sagen kann, also der Bundestag oder die Bundesregierung macht jetzt einfach einen Vorschlag für ein neues Gentechnikgesetz und dann geht es nur noch um die Details. Ich glaube, hier geht es tatsächlich schon noch mal um eine Bestandsaufnahme und auch den Versuch, interdisziplinär einfach mal zu gucken, wie man auch in Bezug auf die zukünftigen Entwicklungen hier vernünftige gesetzliche Regelungen entwickeln kann."

Today, it is no longer sufficient to assess implications post hoc, but it is understand that new approaches and sciences and their applications require constant assessment.

5.3.4.2 Research Funding, Science Policy, and Administration

Research-funding agencies and science policy, as well as industry emphasize the benefits and possible positive outcomes of systems-biology research and its applications. The possible benefit of systems biology is projected on three areas: basic scientific research, and industrial and medical applications. A research funding representative summarizes this as follows.

> In the end, the benefit lies in advances in knowledge, on the one hand, in other words, the systems-biology research approaches, I'd say, of course have resulted in a very big step in advancing knowledge. And on the other hand, the benefit of systems biology lies in its prospects for innovation, of course, in particular for medicine, and for the chemical industry, too, for the food industry, for the relevant sectors of the economy.[64] (research funding/administration representative)

At the present time, the economical importance is estimated to be substantial and thus seen as a "striking business argument"[65] (research funding representative). This explains the fact that every "business of even only minor significance [...] has systems biology in its portfolio as a research approach[66] (research funding/administration representative).

For systems biology, there is broad consensus that the field does not create outcomes or impacts that would have to come along with ethical concerns. An exemplary quote from an interview with a German industry representative is that ethical concerns would only become relevant "if it really becomes a topic of discussion in practice. So, if you really have to consider, say, from an entrepreneurial point of view, [...] in case of doubt, it's also a risk in terms of an additional regulatory requirement that precisely doesn't result in additional safety, but in more time and effort"[67] (industry representative). Similarly, we found in the interviews with research-funding agencies it is unanimously stressed that "the moment for civil commotion is limited as the research takes place in cell cultures and in a containment

---

[64] Original quote: "Der Nutzen liegt schlussendlich im Erkenntnisfortschritt einerseits, also die systembiologischen Forschungsansätze, ich sage es mal so, haben natürlich zu einem Sprung im Erkenntnisfortschritt geführt. Und andererseits liegt der Nutzen der Systembiologie natürlich in ihren Innovationsperspektiven, für die Medizin insbesondere, und eben auch für die chemische Industrie, für die Ernährungsindustrie, für die einschlägigen Wirtschaftsbranchen."

[65] Original quote: "schlagendes wirtschaftliches Argument."

[66] Original quote: "Unternehmen von auch nur kleinerer Bedeutung [...] Systembiologie als Forschungsansatz in seinem Portfolio hat."

[67] Original quote: "wenn es in der Praxis wirklich mal zu einem Thema wird. Also wenn man wirklich auch dann abwägen muss, sagen wir, aus unternehmerischer Sicht betrachtet, ist das [...] im Zweifel auch ein Risiko im Sinne einer zusätzlichen Behördenauflage, die mir eben keine zusätzliche Sicherheit schafft, aber mehr Aufwand."

environment. Concerns are therefore raised"[68] (research funding/administration representative). Furthermore, it was stated that "I'm not concerned here because of course, I also know that all the actors in the field are well aware of the legal, ethical, and other implications that are relevant. And that is of course routinely part of the reason that research projects in this area are called into question. So I think that's established, and so far I don't know of anything that resulted in major//well, society calling this field of research into question"[69] (research funding/administration representative).

In summary, in the eyes of our interviewees coming from different societal groups, few negative societal implications of system biology are visible at present. Correspondingly there seems to be no acute need for action. In this evaluation of systems biology, no difference between the actors could be found. It is pointed out that a broad discourse involving the public would be helpful to start an anticipatory discussion of advantages and disadvantages, possible consequences, funding strategies, and the handling of the data generated in medical systems research. Desired positive impact and hope draws mainly on medical application. However, although there is wide agreement that an involvement of the public and relevant societal groups in a comprehensive debate would be beneficial, it is stated at the same time that the topic of systems biology is not well known or even accessible to a wider audience. This contradiction is not solved, but the issue is given further attention in the next section.

5.3.5 *Concluding Remarks*

Perceptions and statements of experts from industry, and public interest groups, media, research-funding, administration, and science policy are—regarding systems biology's science policy—relatively homogeneous with the exception of three aspects:

- First, there is no shared interpretation of what systems biology comprises between the different actor groups. The interpretation seems to be rather subjective as we found many diverging variations and no underlying pattern. This is quite similar to our results from examining the scientific discourse, where interpretations range from understanding systems biology as an applied method (comp. Lee et al. 2006) to a focus on mathematic models (comp. Williamson

[68] Original quote: "das Aufruhrpotential begrenzt sei, weil die Arbeit in Zellkulturen und im Containment stattfinde. Bedenken würden deshalb geweckt."

[69] Original quote: "Ich habe da keine Bedenken, weil ich natürlich auch weiß, dass alle Akteure auf diesem Feld die rechtlichen, ethischen und sonstigen Implikationen, die dort relevant sind, sehr im Auge haben. Und das natürlich regelmäßig auch Bestandteil der Hinterfragung von Forschungsprojekten auf diesem Gebiet ist. Also das ist glaube ich eingeführt, und bisher ist mir nichts bekannt, was zu größeren//ja, gesellschaftlichen Hinterfragungen dieses Forschungsfelds geführt hätte."

2005) to a highly integrative, interdisciplinary field of research (Bruggeman and Westerhoff 2007, Kitano 2002). The interviewees seem to accept the obvious inaccuracy of available definitions. This might be a result of the unclear state of establishment of systems biology in Germany, and it can also indicate a lack of a clear and unified understanding of the core of systems biology in science.

- Second, there were clear differences between actors regarding the application of systems biology when it came to questions of a fair distribution of investment and access to knowledge and technology: some raised the point that both funding and access were spread unequally; others didn't seem to share this perspective.
- Third, the establishment of systems biology is partially different: fewer governmentally oriented stakeholders (industry, public interest groups, media) do not understand systems biology as completely established, yet funding stakeholders draw more upon the advanced (but yet not finished) state of establishment. This refers to the very different perspectives on an emerging approach in science.

The interviewed experts are very cautious when it comes to the application of scientific results, the societal implications of systems biology, and its regulation: There is agreement on the importance of the application of scientific results. It is a valid argument for funding basic research. Results from research in systems biology are seen to be relevant for both industrial and medical applications. With regard to industrial application, it was felt that real value is already measurable. Concerning the promise of systems or individualized medicine, applications were not perceived yet and systems biology was understood as not yet (fully) established. It was not expected that systems biology could deliver in the near future grand visions such as modeling complete cells or organs, but many shared a rather positive anticipation of smaller, stepwise successes. Here, the extent of expectations seems to be influenced by the interpretation of systems biology as a science, approach, or applied method.

Societal implications are deemed to be few and immaterial. Thus, the societal actors' perception is scarcely influenced by questions regarding regulation or the necessity of regulation. When it comes to consequences of systems biology for society, topics such as data security and privacy govern the discussion. But we also found that after raising the issue of societal implications, the discussion often turned to synthetic biology, which was, in the context of regulation, often chosen as the example for the application of systems biology. There is an obvious difference in how the discourse is shaped in the two different fields: in the debate on synthetic biology, research results and the handling of the results are a prominent part of discourse, to a degree that stakeholders involved in scientific and technological development such as the J. Craig Venter Institute have started to work on possible strategies for governance (comp. Garfinkel et al. 2007). There is no evidence for similar strategies in systems biology. Following Bogner et al. (2010), we understand that in the framing of the discourse on systems biology, there is no visible role for either risk assessment or ethics. This seems to be somehow in contradiction to the fact that involvement of the public in the discourse is deemed to be necessary by the experts. However, at the same time, the experts express ambivalence towards laypersons taking an active role in discussing research funding, because the topic is highly complex and difficult to grasp. Furthermore, inasmuch as there is no concrete application, there also is no immediate interest for the public. This is quite different when one compares this with the discourse on biobanks (comp. Gottweis and Zatloukal 2007) or on stem cells (Gottweis 2008).

What is not surprising is that actors from the different areas do not disagree with regard to the assessment of application and with regard to the possible societal implications and corresponding need for regulation. Although the actors' perspectives are obviously quite different, they all share the assumption that an application of systems biology is inevitable and desirable, and no implications are currently to be feared. Still, it seems advisable to establish a common ground in an open discussion with all stakeholders in order to make transparent the ongoing development of systems biology.

It is only partly possible to attribute the broad scope of interpretations of the term "systems biology" to the variation of scientific definitions and interpretations. Another source of the observed differences presumably lies in the lack of agreement on the core definition of systems biology in the scientific sphere, resulting from the different research perspectives on systems biology. Based on the premise of understanding systems biology as technoscience that we have introduced earlier, an extended discourse would be necessary to enable the different actors' participation in a fair and meaningful way in a public–scientific discourse. Hence, nonscientists need to be part of the discourse, also and especially for technosciences such as systems biology. To what extent, and whether to include them in the discussion of results, regulation, and/or science policy should be matter of further consideration. Referring to our initial point of the entangled systems of science with public, media, politics, legislation, industry, research funding, and representatives of public interest groups, a close adherence to these premises would mean that all subsystems are interdependent upon one another. It is thus of increased importance to bring the different actors, interpretations, and attitudes together and foster the exchange of perspectives and ideas (see Sect. 5.4) in order to consider the present and the future of systems biology in a concerted and grounded manner.

5.4 Scientific and Public Discourses on Science Policy: Interdependencies

Different actors contribute to the discourse on the science policy of systems biology. As we have argued before, questions regarding funding of systems biology, its value for science and society, its applications, implications, and possible necessary regulations are discussed not only by science policy and funding bodies, but also by administration, industry, nongovernmental organizations, public interest groups, and by scientists. Science policy of systems biology is reflected within in these different actors groups.

Furthermore, the discourse in the different fields of actors are not self-contained. Instead, they refer to and influence each other. The relations between the different

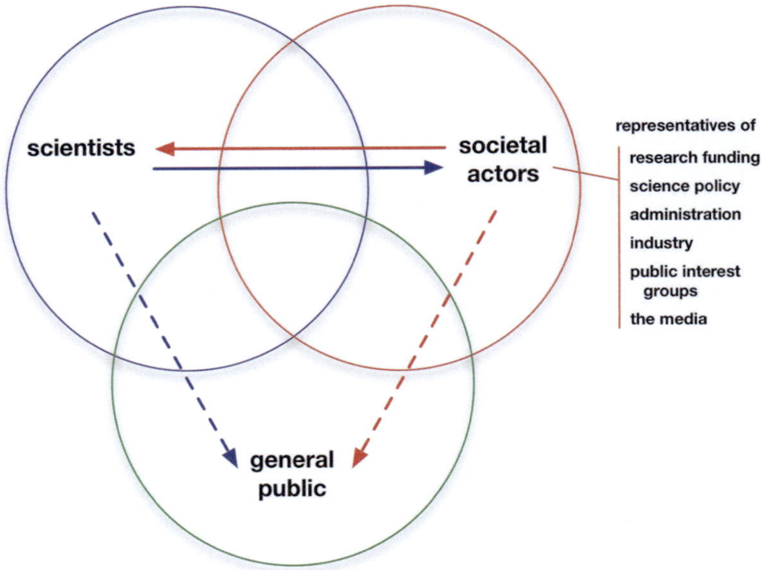

Fig. 5.2 Directions of interdepending discussion on policy of systems biology: availability of data (*solid line*: high availability of data; *dotted line*: low availability of data)

discourses can thus be described as interfaced, interconnected, and even interdependent. They not only consist of direct discussions between actors from different groups (e.g., science and politics) but also take the form of indirect interaction such as acknowledging and referring to discussions in other fields. Two types of communication on science policy exist: the direct communication between individuals and/or groups coming from different fields, and the indirect communication across different fields by referring to position papers, documents, conferences, and the like of other fields.

At this point, we analyze in more detail the mode of indirect communication. How do the different groups of actors influence each other? Where do they diverge and where do they align? We found that interactions between some actors and hence interdependencies between specific groups are much stronger than those between others. We observed, for instance, that the opinion of the general public was not discussed by the scientists in our sample, whereas public actors referred strongly to science (see Fig. 5.2).

The sociologist Peter Weingart describes this relationship between science and the public (including different actor groups) as becoming increasingly interconnected and, as a result, more tightly linked or "coupled" ("engere Kopplung"; Weingart 2001, 175). Often, and especially in resource-intensive, technical disciplines, public funding is a necessary requirement for research. Consequently, the direction of scientific research in systems biology is influenced or even determined significantly by the public and by policy makers, and not by science alone. Therefore, scientific topics develop not only a scientific, but also a social dynamic; the public's opinion influences science policy, which again influences the direction of science, which then is perceived and commentated and annotated by the public, and so on. We define these dynamic, mutual influences as interdependencies that together form an interactive system. This system is highly complex and shows no apparent dominance of a single group. Rather, it allows all individuals to work on and change the texture of the network and thus influence the interdependencies.

In this section, we analyze and discuss some of these interdependencies that emerge from our empirical analysis. By doing this, we go a step further compared to the previous section where we listened to what the different actors had to say; here we want to know how different actor groups frame and discuss the discourse of other groups.

5.4.1 Discursive Interdependencies Between Scientists and Societal Actors

The most significant interdependencies we identified exist between science and science policy (as a subgroup of the public actors; see Fig. 5.1). This is in one sense self-evident, inasmuch as the state is one of the most important sponsors of science. Science policy and funding organizations are important partners for science and research institutions because they negotiate the amount of funding that is going to be allocated to the different sectors of scientific or applied research. Not surprisingly, scientists are motivated and willing to follow the thematic agendas set by funding organizations to receive funding for their research. The role of setting agendas and trends in research is generally assigned to science policy and funding organizations, because they define themes and topics addressed in research programs and calls for proposals. These trends in funding are associated with certain funding lines or clusters of successive projects funded in the trend domain.

Such science policy decisions have far-reaching implications not only for the content of research, but also for the type of research. For instance, *project funding* privileges applied research, because application-oriented topics are usually clearly defined and can—at least in principle—be solved by a structured research agenda and within a limited time frame. Furthermore, problem-oriented systems biology research in medicine often requires, among others, the expertise of biochemists, computer scientists, and mathematicians and the cooperation of experts in interdisciplinary teams. Hence, application-oriented systems biology or medicine projects require interdisciplinary approaches, which then have a greater chance for getting monetary support. As a result, project-oriented funding also privileges problem-oriented or applied research at the expense of theoretical approaches. For example, although many epistemic problems of and in systems biology are not solved or even dealt with yet to a sufficient or even reasonable extent, science policy organizations have prioritized the establishment of mathematic modeling as a methodological approach, because it is expected to be helpful inter alia in elucidating disease mechanisms and defining new targets for the development of new drugs.

Science policy also influences the amount of money that is allocated to a new scientific development. This is closely related to the question of its *establishment* and whether it is progressing into a new phase. Interestingly, for systems biology there are strong differences regarding the state of establishment between different societal actors. They also disagree about the best direction for a science policy for systems biology. Still, most of the public actors agree in assessing systems biology as not established; it is marked as a scientific approach that is still underway and has not yet fulfilled its promises and met its announced aims. Instead, systems biology is described as a science that is still maturing and increasingly growing in importance. Consequently, more funding is needed for the future.

The influence on funding does not only run one way from the societal actors to science; there are also plenty of examples for influences on science policy from the side of science. Even if scientists are basically willing to follow the trendsetting of funding organizations, eminent scientists or science managers may have the power to *influence the initiation, the subject and the direction of research funding*. Systems biology, for example, was first put on the funding agenda because individual scientists proactively approached science policy organizations to set up pertinent funding programs. Ongoing interactions between these politically savvy scientists and science policy agents established social networks in which upcoming trends in funding and research were announced, and where the distribution of funding is negotiated.

Apart from prominent examples of proactively influencing the science policy's agenda, scientists generally care a lot about the distribution of funding as their *careers* crucially depend on the success of raising as much external funding as possible. This is needed to either support one's own position, to promote younger scientists, or to increase one's own reputation. To succeed in the competition of funding, scientists have developed different strategies. First, they refer to the program-related subjects and requirements defined by the funding organizations, which are usually defined in the topics of calls for proposals. Second, scientists communicate with colleagues in their field in order to identify innovative trends in research that may match with the trends in funding. Such networks of scientists develop and submit common proposals, in particular after having had positive experiences in collaborative work. Third, by applying for funding, these collaborative networks adapt to specific and general funding mechanisms of the funding system. Grants are, for example, primarily not given to support institutions or individuals, but to finance research projects that run within a limited budget and time frame. Once a research consortium was successful in a grant application, it may have a good chance for receiving further or even continuous funding. However, when a research project ends, the project partners usually do not know if they will be able to continue their collaboration. They have to invest a lot of effort into networking to maintain the established working relations or to adapt to the altering priorities of funding organizations, either by detecting future trends within their research area or by looking for alternative funding options.

This *impact of science on science policy* and vice versa is also discussed by public actors. They address changes in the research landscape and the establishment of new scientific approaches, as well as dependency of science on funding programs.

According to our findings, for public actors systems biology has become a routine element of biomolecular science. Hence, systems biology appears not in need of special funding, because it becomes part of the life sciences in general. In consequence, the shared perception of public actors is that systems biology will most likely become a basic element of future research programs. Here, emphasis is often put on the interconnection between scientific research and industrial application. For example, the results of fundamental research are seen as basic for the development of new analytical tools that will later be brought into application by industry. The direction of science policy supporting industrial applications thus seems to take up visions and goals of science.

Although the mutual interdependencies between science and science policy are obvious in the interviews, little can be said about the relation between science and *the public*: strikingly, in the scientific discourse, the public plays no role at all. From the scientists' view, there seems to be very little public interest in systems biology. The public too does not seem to be very interested in systems biology. Both observations are in strong contrast to the policymakers' explicit aim to increase public involvement in science and science policy. One factor identified in our interviews that limits potential public involvement is that systems biology in the social actors' discourse is perceived as highly complex, and thus as difficult to understand. Still, it is commonly stressed that the (layperson) public has to be involved in the discussion on science policy and implications of systems biology in the near future. The basic acknowledgment of systems biology seems to be sufficient until today, but it is emphasized by the interviewees that further and ongoing discussion with laypersons is needed. From our perspective, it is here necessary to reflect on when an involvement of the public is necessary, and to what ends and in what context it is induced: does it help to assess the impact of scientific research, or is the public only included to create support for future funding? Also, not all questions can be fed into the public discourse at a given time without risking an erosion of attention. As a consequence, a careful selection of the most relevant questions where public involvement is needed can help to increase the quality and outcome of it.

5.4.2 Conclusions

Science policy of systems biology must be understood as an interface—and result—of three converging discourses of science, of science policy/experts, and of the general public. Interdependencies resulting from this interface influence and change the research landscape as well as science policy and public perception of systems biology.

Our first point concerns the impact of research funding on the practice of research. Science policy *can increase dynamics in the research landscape*: project funding establishes only a temporal sustenance, and we have found evidence that some scientific and societal actors perceive that funding initiatives do not persist until the discipline is universally acknowledged as being fully established. As a consequence, this creates a momentum in the scientific community suggestion that science has to respond to this situation and to seek new research topics, modify existing ones, or adapt them to new scientific trends in order to gain access to further funding. In following this pattern, science has to change constantly, but it has to be asked whether important questions that result from previous research can and will be followed through. Also, scientists are permanently forced to adapt themselves and their professional biographies to these changing research agendas, leading to a stronger and more active competition and intensified selection. However, such an increasingly differentiated science could be counterproductive, not only for scientific biographies and careers but also for science itself as young academics, are less and less able to overlook what is being done and to follow self-defined research agendas. A new alignment between the goals of funding, the goals of science, and the requirements of sustainable scientific careers could be necessary as science not only needs—at least partially—a long-term perspective but also actors who are able to focus on fundamental and theoretical questions and self-reflection without being continuously occupied and absorbed by grant acquisition and pre-defined projects.

The second point concerns the question whether the type of science policy we observed in our study is *sustainable for science*. From the perspective of scientists, current science policy related to systems biology is seen as being not sustainable enough. Is initial funding truly effective in promoting innovation? Will the lack of mid- and long-term funding exert a detrimental effect on systems biology in the longer term? Almost certainly, systems biology will move on from basic research funding to more applied research, such as systems medicine. Based on our evidence, we must raise some doubts about whether systems biology as a basic science will last for long: did funding enable work needed to establish basic methodologies and core concepts? Or did systems biology mainly focus on pragmatic solutions for medicine and not develop (new) concepts that are transferable to other fields?

The third point of our conclusion concerns the lack of a common definition of systems biology. Could the fact that scientists, science policy, and societal actors do have different ideas about *what systems biology is* lead to conflicts? Does a highly complex field such as systems biology provoke problems with defining its main subject area and content? Such conflicts could perhaps provoke further interest in systems biology not only by science or science policy, but also by the general public. However, if opinions on and definitions of systems biology differ too much, this could also divert interest in the field and undermine public and political support.

Fourth, participation and inclusion of the public is in general seen as an important achievement from the perspective of politics, the media, and public interest groups. From the perspective of science, however, the public plays no relevant role and does not demonstrate interest. Here, we point out that we have found no prominent example where the public has been included in the assessment of systems biology. The question remains whether the public has been assigned an adequate role in the discourse described in this chapter, and, more specifically, whether the public should be more involved in the discourse on systems biology and its science policy. In the next chapter we therefore examine how systems biology is discussed in public and which consequences could arise from this discussion.

References

Bogner A (2012) Wissenschaft und Öffentlichkeit: Von Information zu Partizipation. In: Maasen S, Kaiser M, Reinhart M, Sutter B (eds) Handbuch Wissenschaftssoziologie. Springer, Wiesbaden

Bogner A, Kastenhofer K, Torgeresen H (2010) Antizipierte Technikkontroversen als Governance-Problem. In: Gießler E, Rohracher H (eds) Genomforschung – Politik – Gesellschaft: Perspektiven auf ethische, rechtliche und soziale Aspekte der Genomforschung. VS Verlag, Wiesbaden, pp 69–98

Bogner A, Menz W (2005) Das theoriegenerierende Experteninterview. Erkenntnisinteresse, Wissensformen, Interaktion. In: Bogner A, Littig B, Menz W (eds) Das Experteninterview—Theorie, Methode, Anwendung? 2nd edn. VS Verlag, Wiesbaden, pp 33–70

Bora A (2012) Wissenschaft und Politik. Von Steuerung über Governance zu Regulierung. In: Maasen S, Kaiser M, Reinhart M, Sutter B (eds) Handbuch Wissenschaftssoziologie. VS Verlag, Wiesbaden, pp 341–355

Bruggeman FJ, Westerhoff HV (2007) The nature of systems biology. Trends Microbiol 15(1):45–50

Brüninghaus A (2012) Systembiologie als "Technoscience"? Zur Diskussion einer "emerging technology" in Öffentlichkeit, Medien und Wirtschaft in Deutschland und Österreich. Technikfolgenabschätzung Theorie Praxis 21(2):43–49

Bundesministerium für Bildung und Forschung (BMBF) (2010) Rahmenprogramm Gesundheitsforschung der Bundesregierung. http://www.bmbf.de/pub/Rahmenprogramm_Gesund heitsforschung.pdf. Accessed 13 Nov 2014

Charmaz K (2006) Constructing grounded theory. A practical guide through qualitative analysis. Sage, London

Clarke A (2005) Situational analysis: grounded theory after the postmodern turn. Sage, London

Dunwoody S, Ryan M (1985) Scientific barriers to the popularization of science in the mass media. J Commun 35:26–42

Flick U (2000) Qualitative Forschung—Theorien, Methoden, Anwendung in Psychologie und Sozialwissenschaften. Rowohlt, Reinbek

Garfinkel MS, Endy D, Epstein GL, Friedman RM (2007) Synthetic genomicsloptions for governance. Biosecur Bioterror 5(4):359–362

Glaser BG, Strauss AL (1967) The discovery of grounded theory: strategies for qualitative research. Aldine, Chicago

Gibbons M, Limoges C, Nowotny H, Schwartzman S, Scott P, Trow M (1994) The new production of knowledge. The dynamics of science and research in contemporary societies. Sage, London

Gottweis H (2008) Participation and the new governance of life. Biosocieties 3:265–285

Gottweis H, Zatloukal K (2007) Biobank governance: trends and perspectives. Pathobiology 74:206–211

Kitano H (2002) Computational systems biology. Nature 420(6912):206–210

Knorr-Cretina K (1999) Epistemic cultures: how the sciences make knowledge. Harvard University Press, Cambridge

Lee SY, Lee DY, Kim TY, Kim BH, Lee SJ (2006) Systems biotechnology: combined in silico and omics analyses for the improvement of microorganisms for industrial applications. In: Stephanopoulos G, Rigoutsos I (eds) Systemsb, vol 2, Networks, models, and applications. Oxford University Press, London, pp 193–231

Littig B (2008) Interviews mit Eliten—Interviews mit ExpertInnen: Gibt es Unterschiede? Forum Qual Soc Res 9(3):16

Meuser M, Nagel U (1991) ExpertInneninterviews—vielfach erprobt, wenig bedacht. In: Garz D, Kraimer K (eds) Qualitativ-empirische Sozialforschung: Konzepte, Methoden, Analysen. Westdeutscher Verlag, Opladen, pp 441–471

Morris N, Rip A (2006) Scientists' coping strategies in an evolving research system: the case of life scientists in the UK. Sci Publ Policy 33(4):253–263. doi:10.3152/147154306781778957

Nordmann A (2005) In: Rossmann T, Tropea C (eds). Was ist TechnoWissenschaft? Zum Wandel der Wissenschaftskultur am Beispiel von Nanoforschung und Bionik. Bionik. Berlin. pp 209–218
Nowotny H, Gibbons M, Scott P (2001) Re-thinking science. Knowledge and the public in an age of uncertainty. Polity, Oxford
Nowotny H, Scott P, Gibbons M (2003) Mode 2 revisited: the new production of knowledge. Polity, Oxford
Peters D, Weingart P (2009) Governance of and through science and numbers: categories tools and technologies—preface. Minerva 47(3):241–242
Peters HP, Brossard D, de Cheveigné S, Dunwoody S, Heinrichs H, Jung A, Kallfass M, Miller S, Petersen I, Tsuchida S, Cain A, Paquez A-S (2009) Medialisierung der Wissenschaft und ihre Relevanz für das Verhältnis zur Politik Medienorientierung biomedizinischer Forscher im internationalen Vergleich: die Schnittstelle von Wissenschaft & Journalismus und ihre politische Relevanz. Forschungszentrum Jülich, Jülich, pp 9–43
Polanyi M (1958) Personal knowledge. University of Chicago Press, Chicago, IL
Reiß T, Vignola-Gagné E, Kukk P (2013) Thematische Schwerpunktbildung in den Life Sciences durch systemimmanente Prozesse. Fraunhofer, Karlsruhe
Rödder S (2009) Wahrhaft Sichtbar—Humangenomforscher in der Öffentlichkeit. In: Bora A, Maasen S, Reinhardt C, Wehling P (eds) Reihe Wissenschafts- und Technikforschung. Nomos, Baden-Baden
Stehr N, Grundmann R (2010) Expertenwissen. Die Kultur und die Macht von Experten, Beratern und Ratgebern. Velbrück Wissenschaft, Weilerswist
Sturgis P, Allum N (2004) Science in society: re-evaluating the deficit model of public attitudes. Public Underst Sci 13(1):55–74
von Schomberg R (2012) Prospects for technology assessment in a framework of responsible research and innovation. In: Dusseldorp M, Beecroft R (eds) Technikfolgen abschätzen lehren. Bildungspotenziale transdisziplinärer Methoden. Springer, Wiesbaden, pp 39–62
Weingart P (2001) Die Stunde der Wahrheit? Zum Verhältnis der Wissenschaft zu Politik, Wirtschaft und Medien in der Wissensgesellschaft. Velbrück, Weilersvist
Weingart P (2005) Die Wissenschaft der Öffentlichkeit. Velbrück, Weilersvist
Wetherell M, Taylor S, Yates S (eds) (2002) Discourse as data. A guide for analysis. Sage, London
Wetherell M, Taylor S, Yates S (eds) (2003) Discourse theory and practice. A reader. Sage, London
Williamson MP (2005) Systems biology: will it work? Biochem Soc Trans 33(3):503–506

Chapter 6
Systems Biology Goes Public: Representations in German and Austrian Print Media

Anne Brüninghaus, Martin Döring, Regine Kollek, and Imme Petersen

Abstract Media are central for communicating science and its achievements to the public, for the public's discussion of science, and for transferring public opinions and perspectives back into science. In this chapter, we focus on the representations of systems biology in German and Austrian print media. The public perception is analyzed both quantitatively and qualitatively and focuses on the images of systems biology communicated to the public, including its application, research funding, and regulation. These images are derived from an analysis of metaphors that enables us to describe the underlying metaphorical frames and concepts. As we take into account the national differences and compare the public images of systems biology in Germany and Austria, we find some significant differences between both countries in the predominant metaphorical frames. The public image is well reflected in these metaphors, and we suggest that they have an important role in the public understanding of systems biology.

Keywords Systems biology • Media • Public • Science and Society • Metaphor

Media are central for communicating science and its achievements to the public, for discussing research in and by the public, and for transferring public opinions and perspectives back into science. The way this is done has a strong impact on the perception and discussion of science in society. Media representation hence also presents relevant aspects of the public perception of and perspective on systems biology. Metaphorical framings used in these settings influence the perception significantly due to their underlying meaning.

This chapter presents the results of a media analysis of systems biology: it focuses on images and metaphors used to depict systems biology, its approach and goals. We start with a broad analysis of how systems biology is depicted in the media, and discuss three questions: (1) which images of systems biology are communicated to

A. Brüninghaus, Dipl.-Päd (✉) • M. Döring, Ph.D. • R. Kollek, Ph.D. • I. Petersen, Ph.D.
Research Centre for Biotechnology, Society and the Environment (FSP BIOGUM),
University of Hamburg, Hamburg, Germany
e-mail: anne.brueninghaus@uni-hamburg.de; doering@metaphorik.de;
kollek@uni-hamburg.de; imme.petersen@uni-hamburg.de

M. Döring et al., *Contextualizing Systems Biology*,
DOI 10.1007/978-3-319-17106-7_6

the public, (2) what images are used to frame research funding and regulation, and (3) are there national differences between Germany and Austria that stand out?

Following a short introduction into print media as a mediator between science and society (Sect. 6.1), we discuss the relevance of metaphor analysis to identify public images of systems biology in the media, introduce our methodological approach, and describe how we set up the media corpus for our sample (Sect. 6.2). For Germany, we provide an overview of the different metaphorical frames we found in our sample before we continue with a more in-depth analysis of the most important frames and their underlying metaphorical concepts (Sect. 6.3). For Austria, we follow the same approach, starting with an overview followed by the analysis and excerpts for selected metaphorical concepts (Sect. 6.4). Based on this detailed analysis, we compare the public images of systems biology in Germany and Austria, bring attention to similarities and some significant differences between both countries, and develop conclusions and consequences from the use of metaphorical descriptions of systems biology relevant for understanding the current and future role of an emerging approach (Sect. 6.5).

6.1 Print Media as Mediator Between Science and Society

It is without question that media in general have an important function for both science and society today[1] as the relationship among science, media, and the public has changed: media do not only react to scientific developments but take on a more and more proactive role that influences not only society but also science and politics. Knowledge, however, is not simply passed on from scientists to the public. In fact, science, the media, and the public interdepend and influence each other. Thus, there is no one that can be identified as the main actor. Instead, all participating systems act and react to one another. Accordingly, this communication cannot be described as a one-way transfer of knowledge from science to society as there is clear evidence that public perception also influences science, its practice, and direction through the media. As Weingart (2001, 241) argues, feedback resulting from published research and the interaction between public opinion and scientific practice must be taken into consideration by scientists. With the "medialization of science" (Weingart 2005, 28), media do not only communicate scientific results to the public: they define a new relationship between science and society. Science is "construed" in public and becomes "public science". At the same time, science refers to public expectations and demands, thus creating its own, new public, Weingart's "science of public" (ibid.; translation by the authors).

As a result, science becomes more and more visible in the media. Also public opinion is reflected and passed on to science.[2] Our increasingly medialized society is

[1] Rödder (2011) for example, provides an overview of several longitudinal studies of news coverage on the relationship of media and science.

[2] The feedback system also includes politics, industry, and other stakeholders. Its complex internal structure of interrelationships requires a more thorough analysis; compare in more detail Chap. 5.

thus characterized by a greater acceptance and stronger significance of media as an intermediary between different sectors or systems of society.[3] The manner in which scientific information is communicated has a strong influence on its public perception and vice versa. Therefore, media must be considered an important source for the examination of the depicted, communicated, and perceived role and meaning of new developments in science, such as systems biology. This exchange, and the media's mediating function require a critical reflection.

Analysis of print media therefore is a useful tool for examining not only the public perception of science, but also how science is positioned and framed in the media, and how political, scientific, and public discourse interrelates. Our perspective on this circle of interaction is enabled by a print media analysis that acknowledges how scientific concepts or results of systems biology are communicated to the public, but also takes into account how public opinion takes on and reflects systems biology, both regarding its role in shaping the future and its assumed implications. This leads us to the central question of this chapter: how is systems biology addressed and conceptualized in the media? And what role does this representation play for science and the public?

6.2 Metaphors in the Press: Media Images of Systems Biology

Before presenting the different media images of systems biology in the German and Austrian press, we first outline the rationale for using metaphor analysis as a tool for investigating depictions of systems biology in the media. In order to do this, we identified and studied the metaphors used in media language. Metaphor analysis aims at uncovering "patterns of meaning including their ideological attachments" (Fairclough 1989, 119) or "loadings" (Halliday 2001, 190). We use it as a tool to understand interpretations of systems biology that can be found in the media. Drawing from cognitive linguistics, we thus aim to understand how the conceptualization of systems biology underlies and informs public discourse.

6.2.1 Analytical Goals and Methodological Approach

For the present analysis, we used an approach which assumes that a new field such as systems biology is figuratively represented by a set of different metaphors; its perception is diverse and its meaning and purpose have not yet been defined in the public discourse. As such, metaphors not only create and influence conceptual relations but also have an impact on our daily life, be it public or scientific. They help to identify the conceptual frames permeating the news-speak, and thus enable a

[3] Some authors have therefore started using the term media society (comp. in detail Vowe et al. 2008).

deeper analysis and understanding of the possible meanings and interpretations of such a new field, in this case of systems biology. It is important to remember that metaphors take on a double role: they are both a representation of the interpretation of its owner, and a tool to shape the interpretation of others.

Metaphors are an important and irreducible social part of language (see also Chap. 3 and glossary). In a metaphor, meaning is "carried somewhere else"[4]. The classic definition of a linguistic metaphor focuses on words[5] and suggests that in language some words are static and others carry transferred meaning. In its Aristotelian interpretation, the primary function of metaphors is purely ornamental. In modern cognitive linguistics, however, metaphors are perceived as an essential quality of language; they transfer meaning from a source domain to a target domain; thus, they describe or at least indicate conceptual relationships between those domains. Conceptual metaphors were introduced to the wider scientific community by the linguists Lakoff and Johnson (1980), and subsequently developed further into conceptual metaphor theory (e.g. Johnson 1987; Lakoff 1987, 1993; Lakoff and Turner 1989).

Conceptual metaphors help to describe the framework used to think about a topic and to illustrate it. They are an indispensable means of perceiving and understanding the world. According to Lakoff and Johnson (1980) "metaphor is pervasive in everyday life, not just in language, but in thought and action. Our ordinary conceptual system […] is fundamentally metaphorical in nature" (Lakoff and Johnson 1980). They describe conceptual mapping as a transfer from a source to a target domain. This transfer has a central role in explaining and understanding as the unknown is made accessible through the known, by blending into one another: "And this is a result of the massive complex of our culture, language, history, and bodily mechanism that blend to make our world what it is" (Lakoff and Johnson 1987, 104). This mapping is consequently realized in language. Lakoff and Johnson used the example of "LOVE is WAR" to illustrate how we speak (and think) about a very human feeling. For instance, the saying that "he is slowly gaining ground with her" highlights aspects of rivalry, competition, and fighting in a most positive human feeling and makes them explicit. Different submetaphors can also be part of a broader conceptual system as they "jointly provide a coherent understanding of the concept as a whole" (Lakoff and Johnson 1980, 89). Although an analysis of linguistic metaphors might not reveal all conceptual metaphors that structure our mental representation of the world around us, they are seen as an important "source of hypotheses about the structure of abstract concepts" (Casasanto 2009, 143).

Metaphor analysis is thus an instrument to unravel the linguistic and conceptual framework that underlies new areas of interest in society and science. It is a central tool to detect changes in language, and especially to demonstrate the individual and societal meanings assigned to an emerging field such as systems biology. Changes in language used in texts describing scientific research and its results can be induced by a change of perspective that itself might be a result of scientific or technological

[4] Literal translation of the origin greek verb *μετα-φορέω*, (to) carry over.

[5] More correctly: *lexemes*. Lexemes mean the abstract unit of a morphological analysis.

progress. Even scientific texts are not immune to meanings introduced by metaphors; they can "generate new ideas," and be understood as a "productive form of meaning" (Gehring 2009, 81). Therefore we study "metaphors as cognitive *and* social devices, as being anchored in human experience *and* as being anchored in shared cultural experiences" (Nerlich et al. 2002, 93). The use of metaphors also has a societal dimension because they enable us "the control of the world that we make for ourselves to live in" (Richards 1936, 135; cited in Nerlich et al. 2002, 92).

Several studies have already examined the sociocultural role of metaphors. Nelkin (2001) worked on metaphors deriving from science, for example, for genes in public discourse (Nelkin 2001), Kamara (2009) on the language used in green biotechnology. Döring and Nerlich (2004) used metaphor analysis to investigate the public discourse on stem cells in the United Kingdom. With regard to systems biology, Ouzounis and Mazière (2006) reflect on the current use of metaphors in this field, challenging the conception of systems biology.

Investigating the use of metaphors in the media might be able to support a critical reflection of how science sees and positions itself. Three arguments can be brought forward in support of this hypothesis: First, such metaphors were used within the scientific community to describe the new field of systems biology; second, some scientists appropriated the metaphors to explain their practice and thinking to the public; finally, journalists reused them to convey their perspective on science and communicate it to the greater public. Because the media are not only used by scientists in order to spread an opinion about science, but also by politics, the public, or journalists, it can be seen as a means for multidirectional communication between the public, science, and politics.

Our research focuses on the question of how systems biology is conceptualized in the media by using metaphors from different fields of origin and on the meaning that is given to systems biology by the use of these metaphors. To analyze these metaphorically structured patterns of meaning, several steps are necessary (see Box 6.1; compare Döring 2005, 163; Jäkel 1997, 153–154; Jäkel 2003):

Box 6.1: Steps of Metaphor Analysis

1. *Compilation of a text corpus and characterization of text sources* (e.g., for daily papers and journals regarding target group, political periodicity, etc.).
2. *Extraction of linguistic metaphors on a word by word basis.* The relevant lexemes are marked and initially tagged in the corpus and a list of metaphors is compiled.
3. *Identification of a source domain for each metaphor based on the literal use of the lexemes.* The metaphors are tagged in the corpus using the source domain (e.g., for "time is money" the source domain is money/currency).
4. *Lookup of lexemes and identification of semantically similar units.* The meaning of all identified metaphors is checked (simply using a lexicon or for ethical aspects also in co-occurrence databases to check the source domain).

(continued)

Box 6.1: (continued)

5. *Development of conceptual metaphors.* Based on all identified lexemes, a conceptual metaphor is developed and phrased in the form of "target concept *is* source domain". An identification of coherences and connections between the conceptual metaphors prepares the next step of creating a structured overview of all identified metaphors
6. *Development of a construction frame for the conceptual metaphors.* The conceptual metaphors are grouped to create greater units of meaning that help summarize different directions for metaphors as cognitive and social devices.

With these steps, it is possible to identify, structure, and label different conceptual metaphors. However, as in every categorizing approach, some details in difference between metaphors are lost by this procedure. In order to increase transparency and to present our approach, we therefore not only list the construction frames and conceptual metaphors we arrived at, but also depict some exemplary metaphors in detail.

We start with both the analysis of German (Sect. 6.3) and Austrian (Sect. 6.4) media with a brief description of a general conceptual frame and look at conceptual metaphors[6] with linguistic examples that help to illustrate the different perspectives evident in the media discourse.

Following the national analysis, we compare the public discourse in Germany and Austria. Although both are German-speaking countries and thus draw from the same metaphor resources, the two countries are of very different size and have been through different historical and cultural developments. Regarding systems biology, Austria has perhaps the longer history, going back to Bertalanffy presenting his *Allgemeine Systemtheorie* in the early 1930s (comp. Drack et al. 2007 and see Chap. 2 in this book). With regard to today's situation, however, research funding for systems biology in Germany is stronger than in Austria.

6.2.2 Media Sample

For this analysis, we selected printed media for reasons of their high-quality information standard. Our aim was to gather a rich sample with social relevance and a balance between weekly and daily newspapers as well as between political positions that include the most important and influential print media for both Germany and Austria. For Germany, the sample contains three daily newspapers: *die tageszeitung*, *Die Welt*, *Frankfurter Allgemeine Zeitung* and *Süddeutsche Zeitung*. Also, two magazines and weeklies were included in the sample: *Der Spiegel* and *Die Zeit*. Finally, *Spektrum der Wissenschaft* was chosen as a popular science magazine clearly aiming at science communication with the public. For Austria, the sample

[6] A full account of the identified metaphors is in preparation and will be published elsewhere.

includes the daily newspapers, *Der Standard*, *Die Presse*, *Kleine Zeitung*, *Neue Kronen Zeitung*, and *Salzburger Nachrichten*. The weekly newspaper *Profil* completes the sample.

6.3 Systems Biology in the German Press

Two concepts of metaphors that stood out very prominently characterized German press coverage. Metaphors relating to the first concept stem from the domain of orientation and are used both for the subject of research and the scientific discipline of systems biology itself. Examples for the second concept frame the subject of control. As we show, the concepts include different aspects of the promise bound to or associated with systems biology. Starting with the different metaphors we identified—and taking their relative importance and relation into consideration—we built a conceptual frame characterizing the picture of systems biology in German media. We defined the frame as ***Doing SB is Needing Orientation and Getting Control***. It consists of many metaphorical concepts, of which the most important are introduced here.

6.3.1 Systems Biology and the Search for Orientation

The central metaphorical concept ***Doing systems biology is looking for orientation***[7] has two sides: systems biology is still being conceptualized as an emerging science that has not achieved the establishment of a clear direction in public perception. At the same time, it is seen to hold the promise to provide direction, also beyond its immediate goals and boundaries. Several subconcepts represent different aspects of this search for orientation and address topics such as *way*, *mapping*, and *transfer*.

Examples for the *way* metaphor include an article from *Der Spiegel* on March 5, 2008 stating that "The program supports highly innovative projects in Eastern Germany with a total of 45 million Euros. Optical microsystems were selected as well as the research areas [...] medical systems biology [...]. Schavan called the support of these six pilot projects a "***milestone for the promotion of innovation***.""[8] Further examples are that "[i]n the United States, the establishment of the first center for systems biology in Seattle already ***set the course for this***"[9] (*Spektrum* 8/2002), and "[i]n Germany, too, one should continue to pave **the way** ***that has***

[7] Metaphors are set bold and italic.

[8] German original: "Das Programm fördert hochinnovative Projekte in Ostdeutschland mit insgesamt 45 Millionen Euro. Ausgewählt wurden außer optischen Mikrosystemen die Forschungsgebiete [...] medizinische Systembiologie [...]. Schavan bezeichnete die Förderung dieser sechs Pilotprojekte als "**Meilenstein für die Innovationsförderung**.""

[9] German original: "In den Vereinigten Staaten wurden mit der Gründung des ersten Zentrums für Systembiologie in Seattle bereits **die Weichen dafür gestellt**."

already been taken by promoting genome research and effectively bundle knowledge by establishing such centers. ***The first steps in the right direction*** have been taken. But since nobody is able to predict ***how long the path will still be,*** we can only hope that support will not run out of steam halfway through"[10] (*Spektrum* 8/2002).

As the quote cited above and others show, *way* metaphors have moved from their source domain into a development and project context. In short, this indicates that system biology is represented as a scientific area that is still underway and has not yet reached its destination. Systems biology is not perceived as an established field, but as a discipline that has successfully taken some first steps but needs further support. Here, public perception is not so much influenced by scientific progress or individual results, but the focus is more on establishing the discipline itself in the scientific landscape.

On a more reflexive level, the scientific need for orientation surfaces in the metaphorical concept ***Doing systems biology is mapping the biological space***. This refers to the need to get oriented in science by mapping a field to get an overview and put single parts together. This concept is exemplified by metaphors such as "Everything falls within the category of systems biology. It is supposed to examine the new 'inventory lists' of biological systems at a ***higher level***"[11] in *Süddeutsche Zeitung* (January 7, 2004). Another example is "***Complete mapping*** of all disease gene variants of all patient genomes permits the application of dosages and combinations of pharmaceuticals specific to the individual. Each person has all their genome's information on a chip that physicians use for diagnoses and pharmacists for determining dosages"[12] in *Die Welt* (December 29, 2005). And also, "Systems biology, too, is showing progress that makes it possible to bring together the body's metabolic processes ***on a kind of map***: for example, 8,000 chemical signals in the complex network result in the programmed death of a cell"[13] in *Die Welt* (December 23, 2005) or "Mycoplasma pneumonia is systems biology's first model organism: ***Systems biology observes life from a higher vantage point.*** It seeks to understand all molecular processes and to produce computer simulations of them. Its goal is

---

[10] German original: "Auch in Deutschland sollte man **den Weg, der durch die Förderung der Genomforschung bereits eingeschlagen wurde**, weiter ebnen und mit der Schaffung solcher Zentren Wissen effektiv bündeln. **Die ersten Schritte in die richtige Richtung** sind gemacht. Aber da niemand vorherzusagen vermag, wie lange der Weg noch sein wird, bleibt nur zu hoffen, dass der Förderung nicht auf halber Strecke die Luft ausgeht."

[11] German original: "Über allem steht der Begriff der Systembiologie. Sie soll auf einer **übergeordneten Ebene** die neuen ‚Inventarlisten' biologischer Systeme untersuchen."

[12] German original: "Die **Gesamtkartierung** aller Krankheits-Gen-Varianten aller Patientengenome erlaubt die Anwendung von individuumspezifischen Dosierungen und Medikamentenkombinationen. Jeder trägt die gesamte Information seines Genoms auf einem Chip, die beim Arzt zur Diagnose oder Apotheker zur Dosierung abgerufen wird."

[13] German original: "Fortschritte zeigt auch die Systembiologie, die es möglich macht, die Stoffwechselvorgänge des Körpers **in einer Art Karte** zu vereinen: Zum Beispiel führen 8,000 chemische Signale im komplexen Netzwerk zum programmierten Tod einer Zelle."

the virtual cell"[14] in *Frankfurter Allgemeine Zeitung* (December 2, 2009). Mapping can thus be seen as an experimental practice that seeks orientation in an undiscovered field.

In contrast to the metaphorical concept of finding a way, these quotations show that it seems to be a part of systems biology itself to strive for orientation. As systems biology is described as mapping process, concepts such as discovery are invoked: science becomes a tool *to map out* unknown territory, for instance in *genome mapping* which aims at representing the human genome.[15] We saw that the very subject of systems biology itself needs to be defined in public perception, however, systems biology is at the same time seen as holding a promise for discovering and mapping out areas in other life sciences, medicine, and health.

Finally, the metaphorical concept of ***biological processes are transfer*** appears in this context. An example from this domain is *Der Spiegel* reporting "One near-term goal, however, is to ***inject*** genomes into bacteria shells that will transform the single-cell organisms into mini-factories. The era of genetic engineers has already begun"[16] (December 27, 2008). The German connotation of *einschleusen* goes back to water gates which are used in inland waterways. In the quotation they denominate the cell *gates* through which genetic material is inserted into bacteria. Yet, there is also a link to illegal immigration, something that is forbidden and associated with potential negative impact. For systems biology, this metaphor is closely linked with communication and networks: information is often transferred from one domain into another; two areas become linked that were previously set apart (cf. Cellular Networks: Ouzounis and Mazière 2006). When the different connotations are combined, it becomes clear that transfer in systems biology is perceived to have an element of transgression; an element of unpredictability that is bordering risk resonates as existing frontiers are crossed. The ***biological processes are transfer*** metaphor is often used when authors do not clearly distinguish between synthetic biology and systems biology, or the latter is seen as a tool for the former. Thus, synthetic biology as a more exposed scientific discipline is often employed to explain the emerging approach of systems biology.

All three conceptual metaphors on orientation, mapping, and transfer relate to the same domain: that of orientation. However, they highlight that orientation has several meanings in the context of systems biology and that there is a dynamic change in the discipline itself, in its subject, and in the public expectation towards the new development.

---

[14] German original: "Mycoplasma pneumonia ist erster Modellorganismus der Systembiologie: **Die Systembiologie schaut von einer höheren Warte aufs Leben.** Sie will alle molekularen Prozesse verstehen und im Computer simulieren. Ihr Ziel ist die virtuelle Zelle."

[15] See http://www.genome.gov/. Accessed November 15, 2014.

[16] German original: "Ein naheliegendes Ziel aber ist, in Bakterien-Hülsen Genome **einzuschleusen**, die die Einzeller in Mini-Fabriken verwandeln. Die Ära der Geningenieure hat schon begonnen."

6.3.2 Controlling Nature

In the media discourse, systems biology means not only orientation and the search for it, but goes one step beyond: doing systems biology also means to get under control. While mapping points more to the structuring of the field, getting control refers to acquiring a better grasp or more power. This is reflected in further important metaphor that introduces the notion of control: for ***Doing systems biology is getting control*** we identified six subconcepts that concern engineering, machines, industrial production, vessel, tools, and architecture.

As one prominent example, ***biological processes are industrial production*** is used as a metaphor. Systems biology is compared with historic images of industrial production; its subject is associated with the efficient manufacturing of results. It also means that processes can be understood, and regulated in order to optimize or match certain production criteria. As already reported, *Der Spiegel* wrote, "One near-term goal, however, is to ***inject*** genomes into bacteria shells that will transform the single-cell organisms into mini-factories. The era of genetic engineers has already begun" (December 27, 2008), and *Der Spiegel* further reported: "The type of genetic engineering that has so far usually followed the motto, "***inject*** a new gene into an organism and see what happens," is in effect supposed to be replaced by real engineering to design new organisms. Systems biology provides the basis for this: Every genetic network in organisms is to be ***disassembled into individual components, modules that can then be combined in new ways, as is normally possible with technical components.*** The model is the IT sector that started out from individual circuits developed separately and builds processors using standard parts today"[17] (December 27, 2008). We also found a newspaper discussion of a new book authored by philosopher Klaus Mainzer that demonstrated a more reflected perspective on the machine metaphor: "If humans were machines, it would be possible to calculate their lives. This notion has fascinated and frightened people since the Renaissance. In synthetic biology, robotics, and artificial intelligence, it now seems to be becoming reality. But only at first glance. After all, the further researchers decode the interplay of the ***little molecular screws, levers, and cogs*** in our cells, the more clearly they see that the machine metaphor is inappropriate: it has long been replaced by the complex dynamic system. And its ***analysis tool*** is not calculation, but ***computer simulation***. [...] In systems biology, the idea of the dynamic system is the **key**

[17] German original: "An die Stelle einer Gentechnik, die bislang meist nach dem Motto operiert, "**schleuse** ein neues Gen in einen Organismus und schaue, was passiert," soll quasi echte Ingenieurtechnik beim Design neuer Organismen ran. Grundlagen dafür liefert die Systembiologie: Jedes genetische Netzwerk in Organismen soll in **Einzelteile, Module, zerlegt werden, die sich dann wie technische Bauteile standardmäßig neu kombinieren lassen**. Vorbild ist die IT-Branche, die ihren Ausgangspunkt auch von individuell entwickelten einzelnen Schaltkreisen nahm und heute mit Normteilen Prozessoren baut."

to the complexity of life, Mainzer said."[18] (*Frankfurter Allgemeine Zeitung*, November 23, 2010). Describing bacteria as being turned into factories and machine modules by the media suggests that as a start their parts can be analysed and their whole working mechanism be understood. Also, biological parts work together like parts of a machine; if one part of the whole fails, it can be replaced by another part, be it natural or artificial. The well-known image of a machine is no longer sufficient to describe complex processes. Nonetheless, it is still useful to have an image at all to explain a new topic (the complex interplay) drawing from a well-known image (the machine). Even if the discussion highlights the limitations of the machine metaphor with regard to the modeling of biological complexity, it remains within the same framework when replacing the mechanical machine with the computer simulation as an analogy. Perhaps due to the lack of a better analogy, perhaps due to its power in highlighting the strategic aim of systems biology in analyzing smaller parts in order to understand and predict the behavior of larger systems, the mechanical metaphor is reused even when it is limited with regard to what can be said and thought: it is used as it aligns with experiences made previously in other fields.

A further metaphoric concept relates to the aspect of ***Doing Systems Biology is engineering***. Examples relating to this subconcept include the *Frankfurter Allgemeine Zeitung* asking "What do the new biovisionaries have to say? ***Engineering mindset*** flourishing in synthetic and systems biology: Do limits exist for the genome creatures? Germany, the land of the ***bioengineers*** and genome creators?"[19] (November 12, 2008). Another example is *Spektrum der Wissenschaft* reporting "But is the protein really necessary for the life-extending effect? Unequivocally yes, as shown, for example, when its gene was artificially ***switched off***. After all, in the case of the fruit fly, an organism which is quite complex, a lack of food extended their life span only in the presence of the corresponding gene"[20] (Spektrum 10/2006). Further examples include "One of the more than 30,000 molecules of the cell gave a fatal command, all the molecules listened and brought about the division of the entire cell: The ***mini-particle's***

---

[18] German original: "Wäre der Mensch eine Maschine, wäre sein Leben berechenbar. Diese Vorstellung fasziniert und erschreckt die Menschen seit der Renaissance. In der synthetischen Biologie, in der Robotik und der Künstlichen Intelligenz scheint sie nun Wirklichkeit zu werden. Aber nur auf den ersten Blick. Denn je weiter Forscher das Zusammenspiel der **molekularen Schräubchen, Hebelchen und Zahnrädchen** in unseren Zellen entschlüsseln, desto deutlicher sehen sie, dass die Maschinenmetapher hinkt: An ihre Stelle haben sie längst das komplexe dynamische System gestellt. Und dessen **Analyseinstrument** ist nicht die Berechnung, sondern die **Computersimulation**. […] In der Systembiologie ist die Idee des dynamischen Systems der **Schlüssel** zur Komplexität des Lebens, so Mainzer."

[19] German original: "Sprechstunde bei den neuen Biovisionären: **Ingenieursdenken** blüht in Synthetischer und Systembiologie: Gibt es Grenzen für die Genomkreationen? Deutschland, das Land der **Bioingenieure** und Genomschöpfer?"

[20] German original: "Doch ist das Protein für den lebens- verlängernden Effekt auch wirklich notwendig? Eindeutig ja, wie beispielsweise ein künstliches **Ausschalten** seines Gens zeigte. Denn bei immerhin schon so komplexen Organismen wie der Taufliege verlängerte sich die Lebensspanne bei Nahrungsmangel nur, wenn das zugehörige Gen vorhanden war."

mistake can now destroy all 10^{15} cells that make up the human body"[21] (*taz*, February 16, 2006) and "Progress in nanotechnology, stem cell research, systems biology, bionics is already part of the plan to proceed from synthesizing viruses to synthesizing bacteria and higher life forms. [...] Now, viruses are only half organisms; the creation of a synthetic bacterium including a membrane seems to be more complex by orders of magnitude, which is why experienced ***bioengineers*** currently tend to smirk about grandiloquent pronouncements, for example those made by Craig Venter"[22] (*FAZ*, July 6, 2006). These examples indicate the degree to which the thinking not only about synthetic biology but also systems biology is technical and engineering-influenced and how it can be understood in mechanistic terms: you simply have to assemble different components in order to figure out the right solution, and the correct approach for finding that solution is analogous to the physical domain where you can enable and disable circuits in order to test their function and to determine how exactly they need to be put together. In consequence, engineering thus first serves as a source domain for vocabulary that suggests the practicability of a structured approach in the domain of the living as well. Second, it suggests the applicability of existing engineering virtues for getting control, even over the (self-declared) complex subject (of research) of systems biology.

The similarity of the metaphors used to describe systems biology in relation to the concept of ***engineering*** with the manner in which synthetic biology is described in the media suggest that there is significant overlap between the conceptualization in the German media. Even though synthetic biology has a much stronger focus on application, the aspect of engineering, and especially of reverse-engineering, as a method of generating knowledge is shared between the two disciplines (comp., e.g., Boudry and Pigliucci 2013; Gschmeidler and Seiringer 2012).

6.4 Systems Biology in the Austrian Press

Compared to Germany, the representation of systems biology in the Austrian media is substantially different with regard to three points. First, we found a number of confrontation metaphors that range from conflict to war. Second, several metaphors that were often used to characterize the nature of research in systems biology

[21] German original: "Eins der mehr als 30,000 Moleküle der Zelle hat ein fatales Kommando gegeben, alle Moleküle haben gehorcht und die ganze Zelle zum Teilen gebracht: Der Fehler des **Mini-Teilchens** kann jetzt die Gesamtheit der 10^{15} Zellen zerstören, die den menschlichen Körper ausmachen."

[22] German original: "Fortschritte in Nanotechnologie, Stammzellforschung, Systembiologie, Bionik sind bereits eingeplant, um in der Synthese vom Virus zum Bakterium und zu höheren Lebensformen zu schreiten. [...] Nun sind Viren nur halbe Lebewesen, die Erzeugung eines synthetischen Bakteriums samt Membran erscheint um Dimensionen komplexer und läßt erfahrene **Bioingenieure** angesichts großspuriger Ankündigungen, etwa von Craig Venter, derzeit eher schmunzeln."

revolve around the concept of play. And finally, some metaphors that relate strongly to fate and fortune were frequently used when reporting about the effects of systems biology. In short, borrowing from the two most important themes in the media the framing of systems biology in Austria can be summarized as ***Doing systems biology is play and doing science is confrontation***.

6.4.1 Strategic Game or Giving Systems Biology a Try?

Systems biology seems to be associated with role play and gaming, respectively, trying out playfully in the public media. This ***Doing systems biology is play*** becomes visible as ***party game*** or ***theatre***. Reference to playing or acting in the context of systems biology is for instance being made in *Der Standard*: "Classical biologists cannot achieve that alone, agrees Karsten Schürrle of the Society for Chemical Engineering and Biotechnology (DECHEMA) in Frankfurt. The Society is coordinating the ***interplay*** of the working groups involved in the liver cell project across the country. Cell biologists and computer scientists, genetic researchers and control systems engineers, mathematicians and liver specialists must collaborate to ***piece the bio-puzzle together***"[23] (May 12, 2003). Another metaphor refers to interplay: "The new simulation process developed by researchers at the German Cancer Research Center can be used to represent how the genes ***interact*** in this process and thus to determine which molecular targets must be hit in which order for the tumor cells to stop migrating"[24] (*Der Standard*, July 13, 2008). By comparing scientific practice with child's play, innocent and harmless behavior is suggested. Similar to what Kamara (2009) found in his interviews with system biologists, research is also a strategic game for grownups. This is a less benign metaphor and connects closely with the concept of war: games are more than a set of rules; they also require a certain behavior and social interactions. Scientists exploit opportunities. They deal with setbacks by cooperating with money sources, important heads, or sponsors. They use disguise, shepherding, and lobbyism and even team up with rivals as their goal is to win, to strike a decisive hit, a big breakthrough, or a valuable discovery, and be the first to publish it in a high impact journal.

A further aspect of play is exemplified by a quote from an article in *Der Standard* in which play is shifted to a theatre stage: "A spectacular project that will soon

[23] German original: "Klassische Biologen können das alleine nicht leisten, sagt auch Karsten Schürrle von der Gesellschaft für Chemische Technik und Biotechnologie (DECHEMA) in Frankfurt. Dort wird das **Zusammenspiel** der beteiligten Arbeitsgruppen im Leberzell-Projekt bundesweit abgestimmt. Zellbiologen und Informatiker, Genforscher und Regelungstechniker, Mathematiker und Leberspezialisten müssen zusammenarbeiten, um das **Bio-Puzzle zusammenzusetzen**."

[24] German original: "Mit Hilfe des neuen Simulationsverfahrens der Forscher aus dem Deutschen Krebsforschungszentrum lässt sich darstellen, wie die Gene in diesem Prozess **zusammenspielen**, und dadurch ermitteln, welche molekularen Ziele in welcher Abfolge getroffen werden müssen, damit die Tumorzellen aufhören zu wandern."

proceed ***onto the international stage of systems biology*** will be presented at the chemical engineering forum ACHEMA, which will begin this coming Sunday in Frankfurt. The German ministry of research (BMBF) is forking out 50 million euros for the program, which is scheduled to run for 5 years"[25] (May 12, 2003). "Only in recent years have geneticists come to know that those areas of the genetic substance ***play*** a decisive ***role*** in gene control that so far have been called genetic junk because they are not turned into proteins themselves"[26] (*Der Standard*, July 16,2007). This association with the realm of *theatre* suggests that the media conception of systems biology includes the need for science to present itself on stage. Projects are created, developed, and then presented to the public, with the goal of obtaining funding and recognition. This suggests that, following Goffman (2003), it might be appropriate to speak of a backstage and a front stage aspect to systems biology. Reviewing the examples for metaphors on play, it becomes apparent that different aspects are being covered in the media, from the harmless child's play referring to the way scientific progress is seen to be achieved, to the more competition-focused and explicitly political game to achieve first funding, then success, and finally public recognition.

6.4.2 Systems Biology and the Struggle in Science

Another aspect of the media discourse in Austria can be described as ***Doing research is confrontation and war***. The general description of actors in science, and also in the media, is very confrontational: *Der Standard* wrote on August 21, 2006: "The genome research institute ImGuS is being shelved for now despite a financial commitment: it had barely seen the light of day when the elite university in Gugging (previously AIST) claimed ***its first victim***: ImGuS, the planned institute for medical genome research and systems biology will not be realized as conceptualized as a standalone solution on Dr. Bohr-Gasse."[27] Even examples describing collaboration with the aim of adding value through cooperation borrow from war-like metaphors: "***Alliance of disciplines***. The British found this out by going beyond the usual lab experiments. They established an alliance with their highly non-biology colleagues

---

[25] German original: "Auf der Chemietechnikmesse ACHEMA, die am kommenden Sonntag in Frankfurt beginnt, wird ein spektakuläres Projekt vorgestellt, das sich in Kürze **auf die internationale Bühne der Systembiologie** begibt. Das deutsche Forschungsministerium (BMBF) lässt dafür 50 Millionen Euro springen. Das Programm ist auf fünf Jahre angelegt."

[26] German original: "Erst seit wenigen Jahren wissen Genetiker, dass bei der Steuerung der Gene auch jene Bereiche auf der Erbsubstanz eine entscheidende **Rolle spielen**, die bis dato als genetischer Schrott bezeichnet wurde, weil sie selbst nicht in Proteine umgewandelt werden."

[27] German original: "Genomforschungsinstitut ImGuS wird trotz Finanzierungszusage vorerst auf Eis gelegt: Kaum das Licht der Welt erblickt, fordert die Exzellenz-Uni in Gugging (vormals AIST) **ihr erstes Opfer**: ImGuS, das geplante Institut für medizinische Genomforschung und Systembiologie wird in der konzipierten Form als Stand-alone-Lösung in der Dr. Bohrgasse nicht realisiert."

from experimental physics as well as mathematicians and computer scientists."[28] (*Der Standard,* October 7,2007). Not only the interaction within science is confrontational, also the aim of systems biology is being described using a language that is usually reserved for weapons of mass destruction: "The goal of genome research and systems biology is rather to understand disease processes. This knowledge will enable [us] to ***create hard-hitting pharmaceuticals and use them in a targeted fashion***."[29] (*Salzburger Nachrichten,* April 6, 2004). Evidently, there is a large variety of different conflict-type descriptions, sometimes only borrowing from confrontation, sometimes from the realm of war. The language becomes especially tough when the subject of discussion is of monetary nature, but institutions are metaphorically also in greater conflict than issues discussed between individual scientists.

6.4.3 Fateful Science

Finally, a complex of metaphors relates to ***Systems biology is fate***. Here, systems biology takes on the rather challenging task of predicting the future: *Der Standard* wrote that "As small as the object of desire may be, so large are the aspirations: generating computer simulations of all the processes in a cell is truly a Herculean task. The new, aspiring discipline venturing to take on such projects is called systems biology. Sometime in the distant future, virtual cells could be able to ***predict*** what happens when a disease agent enters the cell, a gene is switched off artificially, or a patient takes a medication"[30] (*Der Standard,* May 12, 2003). Not only can the behavior of cells be predicted, but a model built of virtual cells can foretell the behaviour of organisms and their reaction to dramatic interventions. Another article from *Der Standard* states: "In any case, the scientists have gotten very close to the point in time when the **fate of cells** is decided for the first time during embryonic development: embryo or placenta?"[31] (*Der Standard*, January 24, 2011). Here, the fate of cells can

[28] German original: "**Allianz der Fächer**: Herausgefunden haben die Briten das nicht allein über gewöhnliche Laborexperimente. Sie gründeten vielmehr eine Allianz mit ihren sehr nicht biologischen Kollegen aus der experimentellen Physik, Mathematiker und Informatiker."

[29] German original: "Ziel der Genomforschung und der Systembiologie sei es vielmehr, Krankheitsprozesse zu verstehen. Mit diesem Wissen könnten dann auch **schlagkräftige Medikamente geschaffen und zielsicher eingesetzt** werden."

[30] German original: "So klein das Objekt der Begierde auch sein mag, so groß ist der Anspruch: Sämtliche Abläufe in einer Zelle am Computer zu simulieren, ist eine wahre Herkulesaufgabe. Systembiologie heißt die neue, aufstrebende Disziplin, die sich an solche Projekte heranwagt. Irgendwann in ferner Zukunft könnten virtuelle Zellen **vorhersagen**, was passiert, wenn ein Krankheitserreger eindringt, ein Gen künstlich ausgeschaltet wird oder ein Patient ein Medikament schluckt."

[31] German original: "Jedenfalls sind die Wissenschafter dadurch schon ganz nahe an jenen Zeitpunkt heran gekommen, an dem sich das **Schicksal von Zellen** in der Embryonalentwicklung zum ersten Mal entscheidet—Embryo oder Plazenta?"

be interpreted, and broken down into smaller decisions through the insight of systems biology.

On the other end of the spectrum, we found that systems biology itself is seen as an endeavor that requires some faith in (good) fate to believe in: "To Frank Eisenhaber, the head of the Bioinformatics Group at the IMP (Research Institute of Molecular Pathology) in Vienna, systems biology is, on the other hand, more a ***pious hope***. How, he asks, could one speak of such a metascience if we do not even understand the molecular mechanisms in detail?"[32] (*Der Standard*, August 29, 2005).

We found a large number of metaphors that deal with fate, religion, or mysticism, something entirely missing from the German discourse. Indirectly, this echoes the aim of systems biology to provide tools for simulation of organic systems: using these tools to create predictions of how these systems react, the future becomes more conceivable.

6.5 Different Countries: Different Perceptions? Concluding Remarks

The perception of systems biology differs between Germany and Austria. By investigating the use of several metaphors (such as ***Doing systems biology is looking for orientation*** (Germany) and ***Doing systems biology is play*** (Austria)), we have shown that metaphors echo different social experiences through diverse important conceptual framings. We found metaphors related to the source domains of confrontation, play, or fate for Austria. In contrast to this, the main concepts in Germany are orientation and getting control. In the following, we suggest some reasons for how and why the perception of systems biology varies that much in the media of the two countries.

In Germany as in Austria, the **aim** of systems biology is not clearly defined. It is only spoken of in indirect terms. The discourse thus focuses more on the *contents* of science and uses the concept of orientation to explain what approach system biology applies (transfer; mapping the space), and that it is still an emerging science. A further reason for the lack of a defined goal of systems biology is the fact that it is often seen more as an approach (that can be applied in many fields) than a scientific subject era or field in its own right.

In Austria, we found that questions regarding the **aim** of systems biology are often superseded by the quest for funding and infrastructure to start research. The aim itself thus plays a less prominent role as scientists use the media to communicate the deficits in infrastructure. A large number of articles in Austria only speak of the establishment of research programs, or of their dismissal (see also Sect. 5.1).

[32] German original: "Für Frank Eisenhaber, den Leiter der Gruppe Bioinformatik am Wiener IMP (Institut für Molekulare Pathologie), ist die Systembiologie hingegen mehr ein **frommer Wunsch**. Wie könne man von einer derartigen Überwissenschaft sprechen, wenn man noch nicht einmal die molekularen Mechanismen im Einzelnen verstehe?"

In Austria, the focus is not on systems biology's aims, but it is on the (further) development of scientific research and on pragmatic ways of establishing and promoting it.

Thus, in Austria play is used as a metaphor for **describing the method of systems biology**. This comes a bit as a surprise, as the goal is not clearly defined. However, play focuses on two aspects that make it a useful metaphor for scientist: making tactical moves and playing out different positions, and on the other hand making the discipline seem more harmless and innocent. The latter can probably be explained by a conscious use of the metaphor in order to prevent scepticism, because Austria was, for example, very critical regarding green biotechnology.

In Germany, **orientation receives the status of methodology**. This concept demonstrates the need for creating connections between disciplines as well as for finding ways to get an overview about large amounts of data, and to compile them into useful results; play does not have a significant role. Numerous metaphors in the German press are associated with orientation. This indicates that systems biology is in Germany still underway and understands itself as an emerging, moving, and still somewhat elusive discipline.

Fate metaphors are an exception as they relate to a foggy promise of systems biology and a predictability of reactions and cellular processes. However, such descriptions remain vague and cautious. Nevertheless, prediction here claims and replaces the role of fate. In Germany, systems biology is depicted as mastery of nature, looking at the large number and importance of metaphors of **control**. Control is assumed to be an (indirect) aim of systems biology, the method (e.g., play) moves into the background.

The establishment of **systems biology in Austria seems to be a delicate topic**. In this country, financing is an important issue, and open funding for systems biology is very limited compared to Germany. The choice of metaphors highlights how scientists need to fight for funding and resources. In contrast to this, the media touch financial aspects only rarely in Germany.

Based on a linguistic media analysis presented above we were able to show that a number of relevant differences can be identified between Germany and Austria in the public discourse on systems biology. Due to the fact that media are a central element of communication and discussion of scientific outcomes and development, as well as of funding by politics or funding organizations, it is important to ask what the implications of this discourse are, and what it holds for the future of systems biology. We thus now come to our conclusions based on the analysis of the metaphorical concepts.

6.5.1 First, Systems Biology Is Depicted as an Emerging Discipline

In both countries, the media image of systems biology is that of a discipline or an approach that still has to be established. In Austria, where funding is low, the press describes acting in the field of systems biology as conflict and war (e.g., regarding subsidies), and its establishment seems to be indirectly questioned by the media themselves. The choice of metaphors suggests that in Austria, public perception of systems biology lacks an agreed frame and centers on the difficulties and conflicts of finding that framing: systems biology as an approach is still on its way to establish itself. In Germany, the state of establishment is regarded as much more advanced; it is seen as an accepted method of a wider field (the life sciences) that, as a whole, needs to further establish itself firmly and prove its worth. In the German discourse, the media acknowledge an agreed set of goals for systems biology. The need for further establishment refers primarily to the structure and practice of systems biology, and not as much to the general development. Establishment in Germany thus means to demonstrate the value of selected applications by following an established direction rather than setting a new course to follow.

6.5.2 *Second, Systems Biology Is Too Complex to Be Accessible for a Public Media Discussion*

Both, the German and Austrian media, provide different understandings of systems biology. This can be drawn from the fact that articles referring to definitions contain significant differences. No attempt is made to resolve these differences; we found no reference to other, more authoritative definitions and no attempt to clarify what systems biology might be or encompass compared to other scientific approaches in the life sciences. This pattern was observed in both countries. We assume that it is rooted in the complexity of the subject, but that it is, on the other hand, also promoted by a lack of easy to comprehend explanations and explications of what systems biology is and aims to do. A lack of concrete examples for possible application may add to this still nebulous picture. Public comprehension of scientific results seems limited as the spectrum of definitions for systems biology varies ranging from describing it as a new scientific discipline to an auxiliary approach. Referring to the latter understanding, Rheinberger's (2012, 4) definition of systems biology focuses on technology. Therein, systems biology does not primarily refer to biological systems, but rather to the huge amounts of data that are created in laboratories, and to the computation necessary to process this data: "Consequently, we would first and foremost be concerned with the characteristics of a technical system—namely the organization of the biologists' work, and there—with a parallel world of data production and data processing—and less with the characteristics of the organism that this work is devoted to in the end"[33] (ibid). Against the backdrop of such diverse interpretations and understandings, which is complemented by perceptions

[33] German original: "Wir hätten es folglich in erster Linie mit den Eigenschaften eines technischen Systems zu tun—nämlich der Organisation der Arbeit der Biologen und da—mit einer Parallelwelt von Datenproduktion und Datenverarbeitung—und weniger mit den Eigenschaften des Organismus, dem diese Arbeit letztlich gilt."

of systems biology as a completely new science or organization of science, it is impossible to say which conceptualization of systems biology is present or even prevails in the media. This is true for Germany as well as for Austria.

Systems biology—so it seems—is by now such a highly complex science (or an organizational form thereof) that the question arises whether it is still open to any sort of public participation. According to Weingart (2005, 28) this becomes doubtful, once such a stage of complexity is reached. Thus it is questionable whether the public is or will be able to "construe" systems biology as a science or define its expectations towards it. Here, further research could help to understand the influence of media coverage and the way it is done on such a complex scientific development. In that sense, systems biology could be a possible example for an "autonomy of science" that was introduced by Rödder (2011, 838), an autonomy that surfaces as a "mode of communication" (ibid.) and that allows scientists to drive a (relatively) autonomous discourse beyond discursive interventions by the public.

6.5.3 *Third, the Distance Between Science and Public Might Be Increasing*

Following Weingart (2005, 21), the complexity and current absence of concrete applications[34] and corresponding personal stories of scientists, physicians, or even patients increase the distance between science and the public. However, in the case of systems biology it is questionable whether the public actually expresses a "claim to participation, control, and usefulness"[35] (ibid). Systems biology seems to be complex enough and so hard to understand for the public that it is seldom discussed and if it is, then often a flowery and inexact language is used. Hence, for now, the application of systems biology is not linked to social and cultural experiences. This may be another reason why systems biology and even more, its possible applications, are difficult to grasp for the media in both Germany and Austria. Perhaps caused by the embedded nature of systems biology which is always deeply integrated with other, more easily understandable disciplines, there is an apparent lack of personal "stories": it is unclear how systems biology can be applied, for instance, in medicine, biotechnology, or agriculture in order to create practical value. Again, this is not very different between Germany and Austria. Although systems biology has produced results that are accepted by the scientific community, its consequences for everyday life are far less evident; it application possibilities remain vague and without clear examples for the public.

---

[34] One field currently emerging is systems medicine; see Chaps. 1 and 7.

[35] German original: "Anspruch auf Teilhabe, Kontrolle und Nützlichkeit."

References

Boudry M, Pigliucci M (eds) (2013) Philosophy of pseudoscience: reconsidering the demarcation problem. University of Chicago Press, Chicago

Casasanto D (2009) When is a linguistic metaphor a conceptual metaphor? In: Evans V, Pourcel S (eds) New directions in cognitive linguistics. John Benjamins, Amsterdam, pp 127–145

Der Standard (2003) Systembiologie baut Zellen im Computer nach. Die virtuelle Leber im Visier der Forscher. In: Der Standard, 5.12.2003

Der Standard (2005) Zwischen Hype und Hope. Was bringt die Bioinformatik in der Medizin? In: Der Standard, 29.8.2005

Der Standard (2007) Wie aus dem "potenziellen Infogehalt" von Genen eine Wirkung entsteht. Tagung zum Thema Systembiologie in Wien. In: Der Standard, 16.7.2007

Der Standard (2007) Der Schaltplan des Lebens. Um das Innenleben von Zellen besser zu verstehen, sind Biologen eine Allianz mit Mathematikern, Physikern und Informatikern eingegangen. In: Der Standard, 7.10.2007

Der Standard (2008) Warum Zellen zu wandern beginnen. Deutsche Forscher simulierten den Mechanismus, der für Wundheilung, aber auch Metastasenbildung entscheidend ist. In: Der Standard, 13.7.2008

Der Standard (2011) Spätere Aufgaben der Embryonalzellen entscheiden sich früher als gedacht. Transkriptionsfaktor steuert Zelldifferenzierung bereits im Vier- oder Acht-Zell-Stadium. In: Der Standard, 24.1.2011

Döring M (2005) "Wir sind der Deich": Die metaphorisch-diskursive Konstruktion von Natur und Nation. Verlag Dr Kovač, Hamburg

Döring M, Nerlich B (2004) Die metaphorisch-kulturelle Re-Modellierung des Menschenbildes: Die kommunkativen und kognitiven Dimensionen der deutschen und englischen Stammzellendebatte. Zeitschrift für Biopolitik 2:17–29

Drack M, Apfalter W, Pouvreau D (2007) On the making of a systems theory of life: Paul A Weiss and Ludwig von Bertalanffy's conceptual connection. Q Rev Biol 82(4):349–373

Fairclough N (1989) Language and power. Longman, London, New York

Gehring P (2009) Das Bild vom Sprachbild: Die Metapher und das Visuelle. In: Danneberg L, Spoerhase C, Werle D (eds) Bergiffe. Metaphern und Imaginationen in Philosophie und Wissenschaftsgeschichte. Harassowitz, Wiesbaden, pp 81–101

Goffman E (2003) 1959: The presentation of self in everyday life. Doubleday & Company, New York

Gschmeidler B, Seiringer A (2012) "Knight in shining armour" or "Frankenstein's creation"? The coverage of synthetic biology in German-language media. Public Underst Sci 21(2):163–173

Halliday MAK (2001) New ways of meaning. The challenge to applied linguistics. In: Fill A, Mühlhäusler P (eds) The ecolinguistics reader. Language, ecology and environment. Continuum, London, pp 175–202

Jäkel O (1997) Metaphern in abstrakten Diskurs-Domänen: Eine kognitiv-linguistische Untersuchung anhand der Bereiche Geistestätigkeit, Wirtschaft und Wissenschaft. Peter Lang, Frankfurt aM

Jäkel O (2003) Wie Metaphern Wissen schaffen. Verlag Dr Kovač, Hamburg

Johnson M (1987) The body in the mind: the bodily basis of meaning, imagination, and reason. University of Chicago Press, Chicago

Kamara MW (2009) The typology of the game that American, British, and Danish crop and plant scientists play. Minerva 47(4):441–463

Lakoff G (1987) Women, fire, and dangerous things: what categories reveal about the mind. University of Chicago Press, Chicago

Lakoff G (1993) The contemporary theory of metaphor. In: Otony A (ed) Metaphor and thought. Cambridge University Press, Cambridge, pp 202–251

Lakoff G, Johnson M (1980) Metaphors we live by. University of Chicago Press, Chicago

Lakoff G, Johnson M (1987) The metaphorical logic of rape. Metphor Symb Act 2(1):73–79

Lakoff G, Turner M (1989) More than cool reason: a field guide to poetic metaphor. University of Chicago Press, Chicago
Nelkin D (2001) Molecular metaphors: the gene in popular discourse. Nat Rev Genet 2:555–559
Nerlich B, Hamilton C, Rowe V (2002) Conceptualising foot and mouth disease: the socio-cultural role of metaphors, frames and narratives. metaphorik.de 2002(02):90–108
Ouzounis C, Mazière P (2006) Maps, books and other metaphors for systems biology. Biosystems 85(1):6–10
Rheinberger H-J (2012) Biologie – mit System neu definiert. MaxPlanckForschung 2012(1):12–16
Richards IA (1936) The philosophy of rhetoric. Oxford University Press, London
Rödder S (2011) Science and the mass media – "Medialization" as a new perspective on an intricate relationship. Sociol Compass 5(9):834–845
Salzburger Nachrichten (2004) Therapie am PC erproben. Weltkongress der Genomforscher: Spannende neue Perspektiven im Kampf gegen schwere Krankheiten. Kein Optimismus für die Zukunft der Gentherapie. In: Salzburger Nachrichten, 06.04.2004, No 81
Vowe G, Opitz S, Dohle M (2008) Medienpolitische Weichenstellungen in Deutschland – Rückblick und Vorausschau. Medien & Kommunikationswissenschaft 56(2):159–186
Weingart P (2001) Die Stunde der Wahrheit? Zum Verhältnis der Wissenschaft zu Politik, Wirtschaft und Medien in der Wissensgesellschaft. Velbrück Wissenschaft, Weilersvist
Weingart P (2005) Die Wissenschaft der Öffentlichkeit. Velbrück Wissenschaft, Weilersvist

Chapter 7
Back into Future: The Systems Biology to Come

Regine Kollek, Imme Petersen, Martin Döring, and Anne Brüninghaus

Abstract Systems biology is a multidimensional endeavor shaped by cultural and societal factors, as well as by the requirements of scientific practice. By taking up the initial questions of our study, this chapter reveals basic assumptions and constitutive conditions of systems biology, and embeds our findings in a broader scientific and sociocultural context. It first carves out some presuppositions of contemporary science in general, and of systems biology in particular, and reflects them with regard to different paradigms in biology as well as to its past and future developments imagined by systems biologists. Next we discuss the epistemic implications of systems biology's practice, especially its dependence on ICT. Against this background we address the question of whether systems biology should be regarded as an approach or a discipline and offer a new and refreshing answer to this lasting controversy. How science policy pertinent to systems biology is perceived by different actors in the field, and how it shapes systems biology, completes the picture of contextualized scientific development. By referring to public perceptions of systems biology in Germany and Austria and its metaphorical framings in the media, the final section provides a short and speculative outlook on the possible futures of systems biology.

Keywords Presupposition • Reductionism • Holism • Paradigm • Epistemic practice • Systems biology • Scientific discipline

Systems biology is a multidimensional endeavor faced with considerable epistemic, practical, and organizational challenges. On the other hand, it is shaped by many different factors originating in the cultural and social as well as in the scientific or technical spheres. Such spheres or environments create contexts that are constitutive for the establishment and development of systems biology. This general assumption

R. Kollek (✉) • I. Petersen • M. Döring • A. Brüninghaus
Research Centre for Biotechnology, Society and Environment (FSP BIOGUM), University of Hamburg, Hamburg, Germany
e-mail: kollek@uni-hamburg.de

M. Döring et al., *Contextualizing Systems Biology*,
DOI 10.1007/978-3-319-17106-7_7

led to the main questions of this study. How precisely do these environments and factors influence the current and future development of systems biology? How do they shape its concepts, practices, and its possible future implications? And finally: how do they influence the political and public framing of systems biology?

In the preceding chapters of this book we have already carved out a number of factors that affect and shape systems biology, its concepts, its practice, its perceptions, and implications in specific ways. In doing this we aimed for a broad perspective on systems biology and therefore we did not only rely on published material from different fields and sources but also on interviews with scientists and other experts. By exploring this new scientific development from different disciplinary angles and applying mutually complementing but nevertheless distinct theoretical as well as methodical approaches, our results cover a broad spectrum of properties, dynamics, and prospects of systems biology.

In this concluding chapter, we do not merely summarize these results but reflect on them with regard to the overriding question of how this shaping takes place and how the factors and forces involved interact and play together. The aim is to embed our findings in the broader philosophical, social, and cultural context of science. In order to reach this goal we first focus on the (changing) presupposition of systems biology and its implications for science and society (Sect. 7.1) before we—with reference to some results of metaphor analysis in media representations of systems biology—shortly and speculatively dip into the possible futures of this exiting scientific development (Sect. 7.2).

7.1 The Systems Turn in Biology: Presuppositions and Implications

Systems biology has a past, a present, and a future. Its different manifestations over time were and are coined inter alia by implicit presuppositions and epistemic practices[1] as well as by its interactions with society and the general culture. In order to pursue the goal stated above we first carve out some of the presuppositions of systems biology and reflect them with regard to the discourse on concepts and paradigms in biology. The second section focuses on the scientific or epistemic practice of systems biology with special emphasis on the impact ICT has on the structure and content of relevant data sets and on knowledge generation in systems biology in

[1] Scientific or epistemic practice in general denotes knowledge-generating practices, for example, the accumulation, verification, and distribution of knowledge, but also the processes and practices of its justification. Because it is a practice "in which scientists' values for what count as good questions, appropriate methods, and good answers are constructed and negotiated within particular scientific disciplines and communities" it is also a genuine social practice (Sandoval and Morrison 2003, 370). Such practices are inherently epistemic, based on ideas about what kind of knowledge is "true" and justified. This view is supported by current philosophical views of science (e.g., Kuhn 1996) and by sociological studies of professional science (e.g., Latour 1987; Latour and Woolgar 1986).

general. After that we take up the question of what kind of academic entity systems biology is, meaning whether it constitutes an approach, a discipline, or something else, before we turn to the impact of science policy and funding on the establishment and development of systems biology.

7.1.1 Paradigms and Methodologies: Reflecting Presuppositions

In general, this book wants to contribute to a better understanding of the prospects and possible implications associated with systems biology that may support or impede its establishment. In order to do this we also have to explore the presuppositions on which it is based. Such an exploration may help to reveal what is taken into account or what is dismissed, what is made explicit or stays implicit or hidden, or what is been reflected or ignored in the course of knowledge generation following either a more reductionist or holistic approach. In general, presuppositions in science are assumptions made about the nature of reality or parts thereof and which are taken for granted—at least in a certain community—rather than being an explicit part of the content of scientific statements. Usually, basic and secondary presuppositions are distinguished. For instance, one of the very basic presuppositions of the scientific worldview is that the natural world is ontologically, epistemologically, and axiologically accessible to the human mind[2] (Cobern 2000, 238). Another variant of this statement reads: Nature is real, nature is rational, and nature is understandable. The scientific worldview sometimes is also expressed in a number of propositions such as "physical laws influence the real world," "these laws can be detected by observation and experiment," or "all other factors being equal the simplest explanation is best." The latter has also been called the principle of parsimony or "Occam's razor." It was devised by the scholastic philosopher and theologian William of Ockham in the fourteenth century and essentially states that among competing hypotheses, the one with the fewest assumptions should be selected. Furthermore, for each phenomenon, there should only be just one explanation. Such presuppositions reflect our belief in the order and rationality of the world, and our confidence that human beings are able to uncover them. They are inextricable woven into the fabric of science and science education, although they have always also been critically reflected by some scientists as well as philosophers of science. They are nevertheless so much a part of science and everyday practices that they are not consciously noticed by scientists: neither in their daily work nor when they reflect on specific concepts, approaches, or methodologies.

---

[2] Ontology is a branch of philosophy that deals with questions concerning what entities can be said to exist, and how such entities can be grouped, related, or subdivided, whereas epistemology is the branch of philosophy that examines the nature of knowledge, its presuppositions and foundations, and its extent and validity. Axiology refers to the study of the nature of values and value judgments.

However, beyond these basic presuppositions, numerous secondary (or surface) presuppositions are made by individual scientists or groups of scientists about what they expect to find by the help of a method or an experiment, or how a problem should be approached. They can vary between individuals or groups (Cobern 2000, 237). For instance, organisms can be conceptualized as clockwork-like mechanisms by some and as cybernetic machines by others. Some of the presuppositions made by systems biologists were revealed in the analysis of metaphors used to describe basis concepts in systems biology (see Chap. 2). Conceptual metaphors such as, for example, “Life is a machine,” “A system is a structured entity,” or “A model is an integrating entity,” rest on specific perceptions, implicit knowledge, and assumptions. Although they are usually not made explicit they can nevertheless guide experimental strategies or influence what counts as a good explanation in a scientific community or discipline.

In general, a specific set of (primary and secondary) presuppositions is associated with a paradigm.[3] In the context of science, the term paradigm was coined by Thomas Kuhn, a famous historian and philosopher of science. In his book *The Structure of Scientific Revolutions* he defined a scientific paradigm as “universally recognized scientific achievements that, for a time, provide model problems and solutions for a community of practitioners” (Kuhn 1996, 10). A scientific paradigm defines (and controls) what is to be observed and scrutinized, the kind of questions that are supposed to be asked and probed for answers in relation to this subject, how these questions are to be structured, how the results of scientific investigations should be interpreted or an experiment be conducted, and what kind of equipment is available for it. In other words, paradigms “include ways of looking at the world, practices of instrumentation, traditions of research, shared values and beliefs about which questions are considered to be scientific” (De Haro 2013, 7). The geocentric worldview, for instance, was a paradigm in medieval science that was changed to the heliocentric worldview later on. Examples of paradigms in biology include the conviction of a solely unidirectional flow of genetic information that was taken almost as dogma until reverse transcriptase was discovered by David Baltimore in 1969. Another example would be the claim that acquired characteristics could not be inherited which was taken for granted until epigenetic inheritance was described.

What becomes clear is that there is a wide range of ontological, epistemic, and ethical presuppositions woven into any given scientific paradigm (Artigas 2000). Different paradigms can also rest on the same presuppositions. The problem is that neither terminology nor the hierarchy of the terms is agreed upon, and hence they are used inconsistently. Whereas some call reductionism a “metaphysical presupposition” (Marcum 2005, 31), it represents for others “the predominant paradigm of science over the past two centuries” (Ahn et al. 2006, 0709). Independent of such terminological confusion, one of the basic claims of reductionism is that complex

[3] The term paradigm originates in the Greek word *parádeigma*, consisting of *parà* (beside, beyond) and *deiknymi* (to show, to point out). It means “pattern,” “example,” or “sample,” but also “prejudgment,” “worldview,” “belief system,” or “ideology.”

phenomena can be understood by dividing them into smaller, simpler, and thus more tractable units.[4] However, with the advent of systems biology it has been put forward that it is more holistic compared to previous, reductionist approaches[5] or even represents a shift towards a new paradigm in the life sciences (Palsson 2006; Trewavas 2006; Marcum 2008). This has, for instance, already been claimed by Ludwig von Bertalanffy who founded the general systems theory in the 1950s and is one of the pioneers of modern systems thinking in biology.[6] For him, "system" is a new paradigm (in the sense of Thomas Kuhn), which is in contrast to the (then and now) "predominant conceptions focusing on parts instead of interactions and relations" (von Bertalanffy 1972, 415). Although there have been many discussions about the limits of reductionism and the necessity of a more holistic approach in the course of the emergence of systems biology (Gatherer 2010), it has rarely been asked whether or in which way such a systems view in biology is compatible with the reductionist paradigm, or whether they are mutually exclusive. Whereas some scientists and philosophers postulate that the systems paradigm will gradually replace reductionism (Strange 2005; Noble 2008) others think that both paradigms are complementary, have different fields of application (for instance "simple" vs. "complex" diseases) and can coexist in parallel (Ahn et al. 2006, e208). A new paradigm in biology also could—or perhaps should—actively interact with its predecessor, thereby creating new disciplines and subdisciplines (Dronamraju 2006, 41).

Without much doubt, methodological reductionism should be reconcilable with methodological holism (or epistemological antireductionism, Nagel 1998) at least in part of its methodological repertoire (for details see Sects. 2.4 and 2.5). However, it should be kept in mind that the same is not necessarily true with regard to the ontological foundation of the two paradigms and their underlying worldviews. Whereas ontological reductionism assumes that complex phenomena exist of "nothing but" smaller parts and that they can be reduced to smaller elements and their inherent properties, a systems-oriented, holistic perspective (such as ontological antireductionism, Nagel 1998) implies that emergent phenomena exist which cannot be captured on the physical level or be reduced to the workings of lower levels of organization and/or their parts.

Such a change in perspective, or world view, would have far-reaching consequences with respect to different elements and dimensions of science and may—if taken seriously—ultimately question the very foundations of current science. This becomes clear if one reads Ludwig von Bertalanffy in more detail. For him, the systems paradigm comprises three realms: the first "may be circumscribed as systems science, that is, scientific exploration and theory of 'systems' in the various sciences (e.g., physics, biology, psychology, social sciences), and general systems theory as the doctrine of principles applying to all (or defined subclasses of)

[4] In agreement with Derek Gatherer (2010) we do not want to convey the wrong impression that all molecular biologists advocate a "crude" or "naïve" reductionism. Reductionism comes in different variants, which we have outlined in more detail in Chap. 2.

[5] For details see Chap. 1, Sect. 1.1.2, and Chap. 2, Sect. 2.5.

[6] For details see Chap. 2, Sect. 2.3.

systems" (von Bertalanffy 1972, 414). The second is "systems technology [...] including both 'hardware' (control technology, automation, computerization, etc.) and 'software' (application of system concepts and theory in social, ecological, economical, etc., problems)" (ibid., 420). The third realm is "systems philosophy, that is, the reorientation of thought and world view following the introduction of 'system' as a new scientific paradigm (in contrast to the analytic, mechanistic, linear causal paradigm of classical science)" (ibid., 421). Von Bertalanffy's systems philosophy again consists of three elements: First, systems ontology (what is meant by "system" and how are they realized at the various levels of the world of observation). Second, systems epistemology (the investigation of organized wholes comprising many variables requires new categories of interaction, transaction, organization, teleology, and so forth). Furthermore, it has to acknowledge that perception is not a reflection of "real things" or "truth" but an interaction between knower and known, and thus dependent on a multiplicity of factors of a biological, psychological, cultural, and linguistic nature. The third element are values: if reality is (or is perceived as) a hierarchy of organized wholes the image of human beings will be different from what it is in a world of physical particles governed by chance events as the ultimate and only "true" reality (ibid., 421f).

Systems philosophy, especially, the last realm of the systems paradigm, in turn exposes the presupposition of the systems paradigm, at least as it was outlined by von Bertalanffy. For him, it rests on humanistic concerns and therefore "marks a difference to mechanistically oriented system theorists speaking solely in terms of mathematics, feedback, and technology and so giving rise to the fear that systems theory is indeed the ultimate step toward the mechanization and devaluation of man and toward technocratic society" (von Bertalanffy 1972, 423f). Bertalanffy's concern is interesting insofar as it marks and anticipates the two different cornerstones of the current philosophical debates on systems biology: a "bottom-up" approach that tries to reconstruct higher-level properties of complex organisms starting with genes, genomes, and other constituents of the cell, but concentrates on immediate interactions and follows more or less the reductionist causal chain; and on the other hand, a "top-down" approach interested in analyzing downward causal chains and starting with higher-level phenomena that trigger, for instance, cell signaling and gene expression. In contrast to the first causal chain, it is acknowledged that information also flows from proteins to genes and not only the other way around. In current discussions on systems biology, it has often been pointed out that the two approaches are complementary and have to be applied in parallel (Noble 2006). However, to implement these intentions into daily scientific practice does not seem to be easy. Currently, we cannot say which of the two perspectives will gain predominance in systems biology or whether they will coexist in a productive way. It could well be that the turn to a systems view does indeed initiate a paradigm change in biology. It can, however, also not be excluded that a technocratic variant of systems thinking will ultimately prevail. But no matter which path into the future of systems biology will finally be taken, it would be wise to explore further and in more depth its guiding presuppositions.

Some of them have been elucidated by analyzing conceptual metaphors in the preceding chapters of this book; others we have identified await further analysis and reflection. But we have also found that—despite frequent but mostly superficial reference to concepts such as holism or reductionism—there is a relative scarcity of deeper philosophical and epistemic discourse in systems biology. This is not necessarily due to a lack of interest, but mainly caused by a lack of institutionalized opportunities and time in times of "fast science." However, a number of excellent exceptions to this rule exist. For instance, Hans Westerhoff and colleagues (2009) have dealt with presuppositions and principles of systems biology. In their analysis they address two common paradigms (or presuppositions), which are pertinent to much of physics and are sometimes "implemented uncritically in molecular biology and systems biology" (ibid., 3882). At stake is Occam's razor (see above), and the prevalence of minimum energy solutions. They examined in detail whether these paradigms make sense in systems biology and found that they are inapt. Interestingly, they do not think that such reflections and analyses are purely academic exercises because "the implicit disagreement about them is a great burden on the younger developers of this new discipline. Their methodologies are not accepted by their peers or professors from physics or biology, neither their manuscripts finding the complex but real solutions to simple problems, rather than simple apparent solutions to truly complex issues. A clear methodology for Systems Biology is needed" (ibid., 3888). This quotation underlines two important points: first, the importance and practical relevance of a careful and thorough analysis of the presuppositions of systems biology, which not only clarifies its underlying beliefs and assumptions, but also characterizes them as indispensible for further methodological and conceptual development. And second, that time is needed and opportunities have to be created for young and established scientists working in the field to think about, to discuss, and to clarify theoretical questions of systems biology, "slow science" to deepen conceptual reflection.

7.1.2 *Technology and Infrastructure: Shaping Systems Biology in Practice*

Changes in the way of looking at the world or, more specifically, at living entities also involve changes in the way science is done: hence, epistemic practice. Apart from selecting relevant questions, methodologies, and technologies, this also concerns the structure of projects, the social organization of working groups, and the choice of procedures and criteria for justification of facts and findings. As a rule, systems biology aims at analyzing and modeling complex and dynamic phenomena and processes, thereby moving the factors of time, space, and context, which are considered vital for a system-level understanding, into the focus of research. The challenges with regard to technical expertise and workload resulting from such projects by far exceed the ones of traditional experimental research and require methodological and theoretical knowledge from several different disciplines. This is one of the reasons why most systems biology research takes place in large projects that involve biologists and biochemists as well as mathematicians, computer scientists, and members of other disciplines. Because of the complexity of the tasks, the interdisciplinary scope of the working groups, as well as massive dependence on ICT support, establishing systems biology in practice is not an easy task. It has been argued that it needs special structural, institutional, and organizational efforts to cope with it (Hood et al. 2008).

In such contexts the function of technology is often depicted as a mere supportive one. However, such a portrayal may be too simple, and the impact of the reorganization of research due to the demands of systems biology may be much more far-reaching. Evidence supporting this hypothesis was found by exploring the case of systems research in preclinical cancer medicine (see Chap. 4). Such research very fundamentally depends on the large volume and variety of data generated by high-throughput technology from the analysis of molecular processes in cells and organisms. Inevitably, databases and data-processing infrastructures based on ICT have to be established for systems biological and medical research in order to support the processing and integration of data. The impact of such extensive use of and dependence on ICT on the epistemic practice of systems biology is not simply one of speed and scale, although ICT infrastructures certainly increase the amount of data that can be stored and processed at a given time by several orders of magnitude compared to manual handling. Rather, ICT infrastructures have a more fundamental effect on the production and interpretation of data, and hence on the type and content of knowledge produced. For instance, syntactic integration is a necessary precondition for the exchange of data, whereas semantic integration ensures that data are accurately interpreted. Both processes are based on ICT. However, tools and services are constantly under development, whereby recently established standards serve as standards for new developments, which then become standards again for the next generation. As a consequence, ICT-based standards continuously evolve and thus redefine what counts as valid and reliable data in the process of research. Hence, ICT environments and infrastructures not only facilitate storage, processing, and integration of data on a technological level, but at the same time reconstruct the body of data that is or can be used in specific systems-oriented research by assigning significance and meaning to it. Furthermore, for efficient storage, processing, and interpretation of data its division in different categories is needed. Especially the separation into scientific data per se and technical information (metadata) has the effect that the corresponding responsibilities too are separated: whereas the individual scientist or group still is responsible for doing research with (the technically shaped and separated) scientific information, the duty to care for standardization, integration, and management of data and data sets is ascribed to the ICT infrastructure and its providers.

Scientists are usually not aware of the profound impact ICT infrastructures and associated epistemic practices have on the body of data with which they are dealing. In general, such infrastructures are seen as data management systems executing defined tasks, and as service facilities that enable and facilitate collaboration. Such characterization of ICT as mere technical support and facilitator corresponds also to the emphasis that is, for instance, put on standardization. The epistemic impact of computational tools and particularly of community databases on the use of data in experimental biology and on the conceptualization of organisms that underlies those practices has recently been analyzed by Sabina Leonelli and Rachel Ankeny (2012). They especially stress the impact the social context—in this case, the collaborative ethos—had on which data-driven methods were developed and flourished. Furthermore, in systems research, modeling as well as engineering approaches serve as epistemic means that also frame how the objects of research are perceived and conceptualized. Sara Green argues that the use of multiple representational means is an essential part of the dynamic of knowledge generation. She concludes that it is "because of—rather than in spite of—the diversity of constraints of different models that the interlocking use of different epistemic means creates a potential for knowledge production" (Green 2013, 170). Those findings as well as ours point to the fact that the use of extensive ICT support as well as the diversity of approaches and models in interdisciplinary environments creates something completely new that has not been part of biology before. This aspect is corroborated by Miles MacLeod and Nancy Nersessian (2013). They show that doing systems biology is clearly more than a moderate innovation in biology, because the tight interweaving of experimenting and modeling dramatically changes research practices. Systems biology may still be "an approach" but without doubt one that has far-reaching consequences for experimenting and theorizing in the life sciences. Results from empirical social science studies can thus support or question claims of a coming paradigm shift in biology and contribute to clarification of debates lacking contact with the complex reality of interdisciplinarity systems biology research.

7.1.3 Discipline or Approach? The Academic Make-Up of Systems Biology

Behind this background of changing epistemic presupposition and practices the question arises as to how systems biology will develop in the near future. Will it stay an approach or blossom into a fully grown academic discipline? Is it a mere collection of new methods and thus comparable to molecular biology, which is by and large defined by its molecular level of analysis? However, molecular biology introduced the genome as a subject matter and new technological approaches to investigate it (Kirschner 2005, 503). Systems biology also developed new methodologies, but what kind of subject does it have? Is it the system, as Ludwig von Bertalanffy proposed? Or is it a specific perspective on living entities, because a system-level understanding requires, according to Kitano, "a shift in our notion of 'what to look for' in biology" (Kitano 2002, 1662)? Is systems biology on its way to become an academic discipline, which goes beyond former categories? Disciplines are defined by a specific subject area of study and a research tradition incorporating established expertise, practices, methodologies and epistemologies, institutions, projects, communities, conferences, scientific societies, codes of practice, and so on. Hence, they not only involve cognitive, but also social, organizational, practical, and normative factors. Disciplines are therefore places of consolidation and self-assurance, of education and careers and of visibility in the field of academic disciplines and in the public (Knorr-Cetina 1981). The question of how to characterize systems biology therefore is of more than marginal importance for systems biologists, who might search for or constitute a disciplinary identity.

To date, there are already some opinions about how to characterize systems biology and its state of academic development (see also Sect. 1.1.1). For Rainer Breitling, it is a mere "research endeavor" (Breitling 2010, 9). Peter Kohl and his colleague Denis Noble, for instance, put forward that systems biology stands for "an approach to bioresearch, rather than a field or a destination" (Kohl et al. 2010, 25). For Miguel Ángel Medina, "it is commonly accepted that systems biology is an inter-discipline that makes use of principles, knowledge and tools coming from biology, computer sciences, medicine, physics, chemistry and engineering, bridging the gaps between these disciplines" (Medina 2013, 1035). Somewhat more ambitious, Welch and Clegg call it "a grand synthesis" proceeding under the rubric of systems biology (Welch and Clegg 2010, C1288). Other scientists point out that systems biology is more or less physiology in improved clothes. For Denis Nobel, physiology has a major contribution to make to systems biology, because it has developed specific concepts suited to meet the challenge of integrating different data levels a long time ago, whereas the Omics sciences have not achieved this goal yet (Noble 2008). According to Kevin Strange, systems biology as well as physiology aim at understanding the integrated function of complex biological systems (Strange 2005, C968). Hiroaki Kitano (2010), too, advocates systems physiology as an integrated discipline with a functionally centered view. Against this background it has been underlined by several researchers that systems biology should not become a separate discipline but rather be integrated into physiology to create "Integrative Physiology 2.0" (Kuster et al. 2011, 1037). However, others fear that physiology, as an academic discipline, runs the risk of being superseded organizationally and administratively by systems biology (Strange 2005, C968). Taking these statements together it can be said that it is currently not clear how exactly systems biology differs from physiology (Joyner and Pedersen 2011, 1017). In contrast to these positions, others are convinced that systems biology can be distinguished from all other fields in the life sciences. For instance, Yudong Cai and his colleagues advocate the view that systems biology is a "new discipline" that has been created in order to understand how complex interactions in living systems get integrated in nonlinear networks and regulate cell function (Cai et al. 2014). A similar view was put forward some years earlier by Hans Westerhoff and his colleagues (2009). For them, systems biology neither converges with physics, biochemistry, or biology, but rather it is "a science of its own, discovering own fundamental principles" (Westerhoff et al. 2009, 3882).

Obviously, there are uncertainties in two ways. It is unclear whether systems biology is an endeavor, an approach, or a discipline; however, its position in the canon of disciplines, especially its relation to physiology is still under discussion. It remains to be seen, whether systems biology "will become a discipline—and if it does, what kind of discipline it will become" (Calvert and Fujimura 2011, 156). Interestingly, when we explored in interviews (see Chap. 3) how systems biologists themselves conceptualize the past and the future of systems biology, we found that they frame systems biology mainly as an approach, and not as a discipline. However, we also noticed a tension between the conceptual cornerstones of scale of analysis, disciplinary orientation, methodological approach, and theoretical analysis. Definitions of systems biology (and along with them, its disciplinary identity) seem to oscillate between these cornerstones. This fuzzy self-definition and unsolved disciplinary status may partially be due to a lack of a canonical historical account that could function as a reference point for situating and estimating innovations brought about by systems biology. On the other hand, it is counteracted by the current organization of systems biological research.

The question whether systems biology will become a discipline or something else is certainly connected to further questions, for instance, whether a stable academic community can be built and institutionally established. Research in systems biology can hardly be undertaken by single scientists or members of one discipline alone. As we have shown in Chap. 4, there are many hindrances that are stumbling blocks to successful establishment of systems research and corresponding research communities. One of them is interdisciplinarity, inasmuch as misunderstandings between the participating disciplines often occur due to differences in terminology, research practices, and logics, but also to ignorance or mental overload. However, despite all these hindrances, large research consortia that have been established in order to tackle complex scientific tasks succeeded in transcending disciplinary boundaries and finally overcoming the challenges of interdisciplinarity. Once such consortia are settled they involve a core of interdisciplinary-minded people around whom new projects can be designed and built. Targeted program and project funding by national governments and the EU Commission plays a decisive role in this development (see Chap. 5). It considerably contributes to the establishment and growth of a new interdisciplinary and international scientific community of systems biology. Such communities can develop into "epistemic cultures" (Knorr Cetina 1999) characterized by "sets of practices, arrangements and mechanisms bound together by necessity, affinity and historical coincidence which, in a given area of professional expertise, make up how we know what we know" (Knorr Cetina 2007, 363). Keeping this in mind, it could very well be that systems biology never develops into a "classical" discipline, because it probably will always need a considerable amount of methodological, technical, and conceptual input from other disciplines. Furthermore, its activities center around multidisciplinary problems, and not as much on specific questions possibly characteristic for a discipline. This means that in systems biology some centripetal forces, which are necessary for stabilizing disciplines (such as central questions, common research subjects, methodologies, etc.) may still be missing in the future.

Yet systems biology already involves much more than "an approach" which in principle stands for a methodological attitude or theoretical perspective. For instance, some of the large interdisciplinary, systems-oriented project consortia that were established and financed by the European Union or national funding bodies during the last decade did not vanish after the end of the compound project; rather they have sought new institutional homes. One of them is the Virtual Physiological Human (VPH) (see Sect. 4.3.1). In order to stabilize the research community, first the Virtual Physiological Human Network of Excellence (VPH NoE) was founded. Because it was funded by the European Union for only about 5 years, the Virtual Physiological Human Institute for Integrative Biomedical Research (in short VPH Institute) was established next as an international nonprofit organization incorporated in Belgium in 2011, which currently represents over 67 public and private institutions, including many academic, clinical, and industrial key players. The VPH Institute has continued the work of the VPH NoE in many respects, including the running of the VPH conference series and the management of the VPH Portal after the VPH NoE had finished. Obviously, the community repositioned itself as an independent, distributed, and more or less virtual[7] academic institution.

Instead of further contemplating the question of disciplinary identity, the somewhat refreshing conclusions that can be drawn from this example and the arguments discussed above are that with regard to its academic establishment systems biology obviously neither follows the beaten tracks of classical disciplines, nor does it seem to stay simply an approach. Rather it may create something in between, which is different from the concept of a discipline: it neither has disciplinary borders nor does it have to transcend them permanently and work with them constructively in order to fulfill its tasks properly. However, it is also more than and different from a mere approach because in order to stabilize itself it needs at least some technological, financial, and social sustainability and to establish common understandings of what systems biology should be and aim for in order to build up a common identity, history, and academic tradition. Yet, as we have seen, such a sustainability and stability can probably be achieved by means different from the ones classical disciplines employed. What this interpretation of systems biology's identity finally means, and what kind of consequences this has for a systems ontology, epistemology, theory, and methodology, or for academic education in the field, remains to be elucidated.

7.1.4 *Science Policy and Funding: Systems Biology in Society*

New ways of looking at the world in general, and at living entities in particular, affect practices and traditions of research, methodologies, and instrumentation, shared values and beliefs about relevant questions, and so on. However, it has been acknowledged in recent discussions on the concept of paradigm that it involves much more than that. Although the concept first was formulated to describe the features of a discipline or science and what it affects and includes (subjects,

[7] Virtual is meant in the sense of having the attributes of something without sharing its (real or imagined) physical form, but which nevertheless has at least some of the function of its in-reality existing counterpart.

questions, interpretations, etc.), it has now been extended to include not only conceptual and methodological aspects, but also "institutional conditions, governmental constraints and market stimuli that may be supportive of particular paradigms" (De Haro 2013, 7). It therefore is necessary to analyze in more detail how such contextual factors influence and shape systems biology. Against this backdrop we have analyzed in Chap. 5 some features and elements of science policy relevant for the development of systems biology, and hence for a possible paradigm shift in the life sciences. However, inasmuch as scientists, public actors, or the general public may have different interests in and opinions on the development of systems biology and its importance for science and society, science policy cannot be assessed from one perspective only. For this reason we interviewed a number of actors from the groups just mentioned.

From the perspective of scientists, science policy has a profound impact on doing research and developing academic careers. First, because of time-limited funding, science policy can increase dynamics in science because it must permanently seek new research topics or modify existing ones to gain access to further funding. This pressure also extends to scientists as they are forced to adapt themselves and their professional biographies to changing dynamics and agendas. However, this development and the precariousness coming along with it could also be frustrating for scientists and hence counterproductive for scientific development. Counteracting factors may be necessary as science needs actors who are not absorbed by acquisition of new projects and research money but are able to concentrate on central research questions and to engage in productive discourse, which is especially challenging in interdisciplinary projects focusing on complex questions. The second concern of scientists with regard to science policy is whether existing procedures and rules of funding are truly effective in promoting innovation. The lack of mid- and long-term funding could also exert a detrimental effect on systems biology in the longer term. Because systems biology will move on from basic to more applied research, in doing so, will it mainly focus on pragmatic, short-term goals or will it—as a basic science—last long enough in the current funding system to establish basic methodologies and epistemologies? This makes evident how science policy does not relate only to quantitative, but also qualitative, aspects of science.

Beyond these concerns, however, our interview study also showed that scientists do not assign scientific agenda setting solely to funding organizations. Rather, scientific actors themselves may form social networks in which upcoming trends in funding and research are aligned and focused and the distribution of budgets is negotiated. Thus scientific actors can—at least in principle—gain political influence on funding strategies, too, by initiating trends and communicating them to science policy and by influencing shifts and reallocations in the distribution of budgets. This certainly is more difficult in interdisciplinary funding lines where necessary visibility of researchers requires much more effort than in monodisciplinary ones. In addition, the commitment of scientists to such interdisciplinary activities is limited by the fact that they are required to be active in their disciplinary networks as well. Together this makes working and succeeding in an interdisciplinary international community a difficult task.

Opinions of societal actors coming from industry, stakeholder organizations, and public interest groups, the media, as well as research-funding agencies and administration were relatively homogeneous except with regard to the following aspects: first, there was no shared conception of systems biology. Understandings ranged from systems biology as being an applied method to being a highly integrative, interdisciplinary field of research. Compared to scientists, societal actors seemed to be more tolerant with regard to definitions of systems biology and less concerned about their possible inaccuracies. Second, there were clear differences between actors regarding the application potential of systems biology, the fair distribution of investment, and access to knowledge and technology. Third, the stage of establishment of systems biology was partially judged differently: although less governmental-oriented actors (industry, stakeholder representatives, media) do not see systems biology as established, stakeholders from funding organizations draw more upon the advanced (but yet not finished) state of establishment.

The interviewed societal actors are very cautious when it comes to the application of scientific results of systems biology, its societal implications, and possible regulation. Results coming from systems biology were in fact seen to be relevant for industrial and medical applications. With regard to industrial application, it was felt that real value was already measurable. Medical applications, however, were perceived as still less mature. Societal implications were estimated to be few and immaterial; hence regulation-related questions barely influenced the perception of societal actors. This was somewhat different when future applications were addressed, for instance, in medicine. In this case topics such as data security and privacy governed the discussion. With regard to regulation, the development of synthetic biology was often chosen by our interviewees as a comparison for how to deal with systems biology. Despite this, there is empirically an obvious difference in the discourses on systems and synthetic biology: whereas in synthetic biology, research results and their handling are a prominent part of the public discourse, there is no evidence for similar strategies in systems biology. In general, however, all interviewed actors think that application of systems biology is inevitable and desirable, and that they currently fear no negative implications. This coincides with the fact that no immediate interest of the public towards a direct involvement in science policy decisions was observed by our interviewees; they themselves also expressed some ambivalence towards laypersons taking an active role in discussing research funding.

What can be concluded from these findings generated by policy analysis is that systems biology can be understood as an interface of at least three converging fields: science, science policy, and the general public (cf., e.g., Weingart 2001; Peters et al. 2009; Rödder 2009). All fields interact, and hence influence and shape each other in specific ways. In these processes, systems biology is not only fashioned or reorganized superficially, but existing and emerging landscapes of research as well as epistemologies and epistemic practices are deeply molded, framed, and configured by pertinent contextual influences. Thus, the societal context in general and science policy in particular, take an active part in bringing to life a new scientific field. Because of the shaping by policy-related and societal factors systems biology does not only consist of specific presuppositions, ontologies, methodologies, and epistemologies, but also of institutional framework conditions, funding decisions, career plans, legal constraints (or the absence thereof), and market stimuli that support or restrict its development. In this sense, current systems biology is not only a result of internal scientific logic, available technologies, or guiding scientific questions, but also one of social, institutional, and economic decisions and hence, of the societal context and the forces, interests, and powers working in it.

7.2 Public and Cultural Perception of Systems Biology: Outlook

Interestingly, when we conducted our interviews with scientists and societal actors, ethical, social, ecological, or economic implications were rarely mentioned; if that was the case, they were mostly related to current or future fields of application such as synthetic biology or personalized medicine (see Chap. 5). In spite of that, the public certainly has expectations in the potential of systems biology. Such expectations are inter alia conveyed by the media which play a decisive role in the public's discussion of science. They communicate science and its achievements to the public, but also transfer public opinions and perspectives back into science. Representation of systems biology in the media hence expresses relevant aspects of the public perception of and perspective on scientific developments. Because such perspectives and perceptions are crucial for the future societal handling of systems biology we were interested in gaining a deeper understanding of this representation, especially with regard to its cultural and normative colors and undertones. This is why we analyzed the metaphorical framings of systems biology in German and Austrian print media (see Chap. 6). Because we aimed at the elucidation of cultural factors such a comparative analysis is sensible: apart from allowing deeper insights into the process of communication and perception of science, such an analysis also provides some clues to cultural differences in the perception and handling of new scientific developments. Based on our metaphor analysis of written texts from public media, we identified not only common perceptions of systems biology but also a number of relevant differences between Germany and Austria.

First, in both countries, the media image of systems biology is that of an emerging discipline or approach that still has to establish itself. At the same time, the degree of establishment is seen differently: in Austria, public perception of systems biology lacks an agreed framing and focuses on the difficulties thereof. In Germany, systems biology is seen as an accepted method in a wider field (the life sciences) that needs to further establish itself firmly and to prove its relevance and productivity. Second, systems biology is quite complex and thus not very accessible for a public media discussion. Both the German and Austrian media seem to have difficulties agreeing on a common or even consistent understanding of systems biology. We assume this situation is due to the complexity of the field, but also promoted by a lack of easy to comprehend explanations and explications of what systems biology is and aims to do. Third, the distance between science and public might be increasing.

Systems biology obviously is perceived as so complex that it is seldom discussed. If that is indeed the case, often a flowery and inexact language is used. Although systems biology has produced results that are appreciated by the scientific community, its consequences for everyday life are far less evident. Hence, systems biology and its possible applications are difficult to grasp for the media in Germany as well as in Austria.

These findings and conclusions are quite telling as they point to the current—still fragile—state of systems biology's implementation in public perception. However, does metaphor analysis also reveal something about the future development of systems biology? A closer analysis of the metaphors employed by the media revealed striking differences between Germany and Austria: whereas metaphors related to the source domains of confrontation, play, or fate were found in Austria, the main themes in Germany were orientation and getting control (see also Chap. 6). In Germany, the concept of orientation is used in order to explain the methodological approach of system biology (transfer; mapping the space); orientation almost receives the status of methodology. This demonstrates the need for creating connections between disciplines as well as for finding ways to get an overview of large amounts of data, and to compile them into useful results. Furthermore, a large number of metaphors of control and their importance imply that control is assumed to be an (indirect) aim of systems biology. In contrast to this, Austrian media often used play as a metaphor. Play focuses on two aspects that make it a useful metaphor for scientists: it refers to tactical games and the exercise of scientific rivalry, but it also makes the discipline seem to be more harmless and innocent which suggests a possible conscious use of the metaphor in order to prevent skepticism, especially as Austria in the past often was very critical regarding biotechnology.

Based on this analysis one can say that perception of systems biology in the media oscillates between orientation and getting control, and play and confrontation. However, one can also imagine that these two diverging images stand for the two currents existing in parallel in contemporary systems biology: one which is rooted in molecular biology and genetics and aims at mapping genomes and other parts of biological entities, and the other one looking at networks and interactions and probing them in virtual models. At this point we have to acknowledge that the development of new scientific developments is not independent of other societal developments. For instance, during the past few decades, awareness has risen considerably with regard to the interconnectedness of objects and processes in the world in general, and within and between organisms in particular, human beings included. Therefore, it seems that worldviews that put emphasis on the properties and behavior of isolated elements, on determinism, prediction, and control, are questioned and lose persuasive power and ground. This overall development may—in direct and causal terms—have nothing to do with the development of systems biology, which has its own history and emerged at a certain stage of historical, technological and conceptual development in the life sciences. Nevertheless, the emergence of a new systems perspective in biology could also be an expression of an increased awareness of such interconnectedness of all living beings, and of a new modesty with regard to our ability to understand, describe, and control complex and dynamic phenomena and processes.

Such perceptions of the world, of life, and of our ability to understand and describe them by scientific means often unconsciously find their way into our language metaphors used to describe science, its concepts, procedures, or aims. Often they are also represented by cultural icons or symbols. In a certain sense one could also say, that "a culture's icons are a window onto its soul" (Lander 2010, 1). However, such cultural icons are also mediators between the scientific and the public sphere and vice versa. They connect both spheres having their roots for metaphorical representation either in the scientific or in the public context. In the public sphere, cultural icons originally coming from science function as benchmarks in the public understanding of science. Hence, they shape in the long run the public assessment of developments in science and technology. In the scientific sphere, cultural icons originating in the public sphere often influence scientific perception and practice right from the beginning of a newly established research paradigm even though scientists are not necessarily aware of it. As is well known, in and outside science the double helix was considered to be the icon of the genomic age which conveyed the impression that the DNA is the ruler of life (Nelkin and Lindee 1995). The question is which icon will represent the perhaps coming age of a network-oriented, more holistic systems biology. It has been suggested that a hairball of interrelated data on biochemical networks of interactions could serve as some sort of icon for systems biology, or systems biology is symbolized by it (Lander 2010). If that will indeed be the case we would then have to explore what impact such a symbol and what it represents has on our cultural understanding of life, health, organisms, environmental change, and other important subjects. However, this would be the task of new projects.

References

Ahn AC, Tewari M, Poon CS, Phillips RS (2006) The limits of reductionism in medicine: could systems biology offer an alternative? PLoS Med 3(6):e208

Artigas M (2000) The Mind of the Universe. In: Ramos A, George MI (eds) Faith, scholarship, and culture in the 21st century. The Catholic University of America Press, Washington, pp 113–125

Breitling R (2010) What is systems biology? Front Physiol 1(Article 9):1–5

Cai Y, González JV, Liu Z, Huang T (2014) Computational systems biology methods in molecular biology, chemistry biology, molecular biomedicine, and biopharmacy. Biomed Res Int. doi:10.1155/2014/746814

Calvert J, Fujimura JH (2011) Calculating life? Duelling discourses in interdisciplinary systems biology. Stud Hist Philos Biol Biomed Sci 42:155–163

Cobern WW (2000) The nature of science and the role of knowledge and belief. Sci Educ 9:219–246

De Haro S (2013) Science and philosophy: a love-hate relationship. Talk delivered at the conference rethinking liberal education. Amsterdam University College, 15 June 2013. http://philsci-archive.pitt.edu. Accessed 22 Jan 2015

Dronamraju KR (2006) Hybrid paradigms. Endeavour 30(2):41

Gatherer D (2010) So what do we really mean when we say that systems biology is holistic? BMC Syst Biol 4:22

Green S (2013) When one model is not enough: combining epistemic tools in systems biology. Stud Hist Philos Biol Biomed Sci 44(2):170–180

Hood L, Rowen L, Galas DJ, Aitchison JD (2008) Systems biology at the Institute for Systems Biology. Brief Funct Genomic Proteomic 7(4):239–248

Joyner MJ, Pedersen BK (2011) Ten questions about systems biology. J Physiol 589(Pt 5):1017–1030

Kirschner MW (2005) The meaning of systems biology. Cell 121(4):503–504

Kitano H (2002) Systems biology: a brief overview. Science 295(5560):1662–1664

Kitano H (2010) Grand challenges in systems physiology. Front Physiol 1(3):1–2

Knorr-Cetina K (1981) The manufacture of knowledge. an essay on the constructivist and contextual nature of science. Pergamon Press, Oxford

Knorr Cetina K (1999) Epistemic cultures: how the sciences make knowledge. Harvard University Press, Cambridge

Knorr Cetina K (2007) Culture in global knowledge societies: knowledge cultures and epistemic cultures. Interdiscipl Sci Rev 32(4):361–375

Kohl P, Crampin EJ, Quinn TA, Noble D (2010) Systems biology: an approach. Clin Pharmacol Ther 88(1):25–33

Kuhn TS (1996) The structure of scientific revolutions, 3rd edn. University of Chicago Press (first published 1962)

Kuster DW, Merkus D, van der Velden J, Verhoeven AJ, Duncker DJ (2011) 'Integrative Physiology 2.0': integration of systems biology into physiology and its application to cardiovascular homeostasis. J Physiol 589(Pt 5):1037–1045

Lander AD (2010) The edges of understanding. BMC Biol 8:40

Latour B (1987) Science in action. Harvard University Press, Cambridge

Latour B, Woolgar S (1986) Laboratory life: the construction of scientific facts, 2nd edn. Princeton University Press, Princeton

Leonelli S, Ankeny RA (2012) Re-thinking organisms: the impact of databases on model organism biology. Stud Hist Philos Biol Biomed Sci 43:29–36

MacLeod M, Nersessian NJ (2013) The creative industry of integrative systems biology. Mind Soc 12:35–48

Marcum JA (2005) Metaphysical presuppositions and scientific practices: reductionism and organicism in cancer research. Int Stud Phil Sci 19(1):31–45

Marcum JA (2008) Does systems biology represent a Kuhnian paradigm shift? New Phytol 179(3):587–589

Medina MA (2013) Systems biology for molecular life sciences and its impact in biomedicine. Cell Mol Life Sci 70:1035–1053

Nagel T (1998) Reductionism and antireductionism. In: Bock GR, Goode JA (eds) The limits of reductionism in biology. Novartis Foundation Symposium 213. Wiley, Chichester, pp 3–14

Nelkin D, Lindee MS (1995) The DNA mystique. The gene as a cultural icon. Freeman Comp, New York

Noble D (2006) The music of life: biology beyond the genome. Oxford University Press, Oxford

Noble D (2008) Claude Bernard, the first systems biologist, and the future of physiology. Exp Physiol 93(1):16–26

Palsson BO (2006) Systems biology: properties of reconstructed networks. Cambridge University Press, Cambridge

Peters HP, Brossard D, de Cheveigné S, Dunwoody S, Heinrichs H, Jung A, Kallfass M, Miller S, Petersen I, Tsuchida S, Cain A, Paquez A-S (2009) Medialisierung der Wissenschaft und ihre Relevanz für das Verhältnis zur Politik. In: Peters HD (ed) Medienorientierung biomedizinischer Forscher im internationalen Vergleich: die Schnittstelle von Wissenschaft & Journalismus und ihre politische Relevanz. Forschungszentrum Jülich, pp 9–43

Rödder S (2009) Wahrhaft Sichtbar – Humangenomforscher in der Öffentlichkeit. In: Bora A, Maasen S, Reinhardt C, Wehling P (eds) Reihe Wissenschafts- und Technikforschung. Nomos, Baden-Baden

Sandoval WA, Morrison K (2003) High school students' ideas about theories and theory change after a biological inquiry unit. J Res Sci Teach 40(4):369–392
Strange K (2005) The end of 'naive reductionism': rise of systems biology or renaissance of physiology? Am J Physiol Cell Physiol 288:C968–C974
Trewavas A (2006) A brief history of systems biology. Plant Cell 18:2420–2430
Von Bertalanffy L (1972) The history and status of general systems theory. AMJ 15(4):407–426
Weingart P (2001) Die Stunde der Wahrheit? Zum Verhältnis der Wissenschaft zu Politik, Wirtschaft und Medien in der Wissensgesellschaft. Velbrück, Weilersvist
Welch GR, Clegg JS (2010) From protoplasmic theory to cellular systems biology: a 150-year reflection. Am J Physiol Cell Physiol 298(6):C1280–C1290
Westerhoff HV, Winder C, Messiha H, Simeonidis E, Adamczyk M, Verma M, Bruggeman FJ, Dunn W (2009) Systems biology: the elements and principles of life. FEBS Lett 583(24):3882–3890

Glossary

Actant In Actor-Network-Theory, human and non-human actors are both labelled as actants (Latour 1996). They interact to constitute social networks linking together technical and non-technical elements. An actant is enrolled in a network when it gives strength to a position within the network. When a biologist, for instance, argues for the existence of a molecule, the data that prove this existence are enrolled actants.

Actor Actors are endowed with human agency; they have the capacity to act and interact with other actors. Agency implies the power to intervene in the world or to refrain such interventions, with the effect of influencing a specific process or state of affairs (Giddens 1984). Actors can participate in different fields, such as science, society, politics etc.

Actor-Network-Theory (ANT) Actor-Network-Theory is widely acknowledged and used in the field of Science and Technology Studies (↗ Science and Technology Studies) originally created by Bruno Latour (1987, 1992), Michel Callon (1992) and John Law (1987). ANT aims at understanding processes of technological innovation and scientific knowledge creation by considering all surrounding factors. Artefacts, devices and entities are part of social networks in which both parts, the social and the technical, influence each other mutually. This principle of symmetry between technology and humans rejects both technological determinism and social determinism and analyzes the mechanism of interactions in networks. Human and nonhuman actors, both defined as actants (↗ Actant), are semiotically defined by how they act and are acted on in the networks of practices. The important fact here is not that humans and nonhumans are treated symmetrically but that they are defined relationally according to their role and function in the network, and not otherwise. This leads to a relational epistemology which rejects the naive positivist view of objects or actors as existing in themselves prior to any participation in semiotic networks of interactions (including the interactions by which they are observed, named, etc.).

M. Döring et al., *Contextualizing Systems Biology*,
DOI 10.1007/978-3-319-17106-7

Boundary object Boundary object is a widely used concept in Science and Technology Studies (↗ Science and Technology Studies) which was introduced by Susan Star and James Grisemer in 1989. Boundary objects are socially constructed objects that enable communication and coordinated action towards a commonly conceived goal by uniting scientists or other social actors around them. Boundary objects are characterized as both plastic enough to adapt to local needs and constraints of the several parties employing them and robust enough to maintain a common identity across sites. They have different meanings in different social contexts but their structure is common enough to more than one context to make them recognizable, a means of translation. According to Star and Grisemer (1989, 393), the creation and management of boundary objects is key in developing and maintaining coherence across intersecting social contexts.

Clinical trial/clinical study Clinical trials are prospective biomedical or behavioral research studies on patients or healthy volunteers that are designed to answer specific questions about biomedical or behavioral interventions, e.g. drugs, vaccines, biological products, surgery procedures, radiology procedures, devices, behavioral treatments, process-of-care changes or preventive care. *Multicentric trials* are conducted in several locations (e.g. clinical centers). *Clinico-genomic trials* explicitly approach the integration of genomic data with clinical data in medical research.

Cognitive model/idealized cognitive model A cognitive model or an idealized cognitive model is a cognitive representation of knowledge in a semantic frame. This knowledge holds no objective relationship with reality but rather represents an experiential and sometimes individual conceptualization of experience which is mirrored in linguistic structures such as metaphors and metonymy. See ↗ Metaphor, ↗ Metonymy.

Concept/conception A concept is an abstract idea, mental representation or mental symbol that exists in the mind. The terms concept and conception are sometimes used interchangeably. However, a conception may also be more encompassing and detailed than a concept with regard to factors considered and theoretical reflections. In metaphysics, and especially ontology, a concept is a fundamental category of existence.

Conceptual metaphor Conceptual metaphors are used in all kinds of human reasoning. They can be understood as a cognitive configuration or structure which helps to make an abstract domain accessible. Conceptual metaphors develop coherent models or so-called cognitive models which represent experiential simplifications of an even more complex reality and at the same time provide a semantic structure that pervades scientific thought and practice. Metaphor underlies and structures what Michael Polanyi (1958) calls tacit knowledge (↗ Tacit knowledge) in the present context. This means that the analysis of metaphor holds the possibility to reveal and analyze the cognitive patterns and processes used to reason about a scientific problem or concept. In this view, for example science and scientific reasoning should be understood as embodied processes of metaphorical reasoning which transforms knowledge into so-called experientially gained structures. Thus, scientific facts are consequently a product of embodied and experiential metaphorical reasoning which provide a fragmentary but nevertheless important perspective towards reality. See ↗ Metaphor.

Constructivism Constructivism is a philosophical approach that avoids essentialist explanations of events or innovations (e.g. explaining a successful theory by understanding the combinations and interactions of elements that make it successful, rather than saying it is "true" and the others are "false"). According to constructivists the world is independent of human minds, but knowledge of the world is always a human and social construction. Constructivism opposes the philosophy of objectivism, which embraces the belief that a human can come to know the truth about the natural world not mediated by scientific approximations, with different degrees of validity and accuracy. Social constructivism applies the general philosophical constructivism into social settings, wherein groups collaboratively construct knowledge and meanings for one another. According to it, scientific knowledge is constructed by the scientific community, seeking to measure and construct models of the world. See ↗ Objectivism.

Conventional metaphor Figurative language semantically structures the content of the discourse on a certain subject and does not hold a high degree of metaphoricity (↗ Metaphoricity). This aspect becomes visible in the conceptual metaphors used to frame mental activity. Examples are the pervasive text and script metaphors used during the sequencing of the human genome. They became a conventionalized image to convey that the chemical structure of the DNA is something like a text that could be read, understood and—to use another metaphor—be re-written in research. The initial metaphorical mapping gradually disappears and undergoes a process of standardization which finally changes the metaphor into a stand-alone word. See ↗ Metaphor.

Embodiment The notion of embodiment describes in the context of metaphor and metonymy (↗ Metonymy, ↗ Metaphor) the fact that language ascribes meaning to the world by using embodied metaphors. This means that certain ways of knowing the world is based on bodily experiences such as forces, movement or that of being an entity help us to conceptualize abstract and other domains of thought. Thus, metaphors such as "How shall we handle this concept?" metaphorically conceptualize ideas as entities which one can touch, manipulate and shape. Such metaphors are impossible without experiences generated by and through a human body.

Emergentism Emergentism is the philosophical position which states that a property is emergent if it is a novel property of a system or an entity that arises when that system or entity has reached a certain level of complexity and that, even though it exists only insofar as the system or entity exists, it is distinct from the properties of the parts of the system from which it emerges.

Epistemic culture Epistemic culture is a prominent concept in sociology of science introduced by Karin Knorr Cetina (1999). It refers to the practices and beliefs that constitute knowledge and its way of justifying knowledge claims. The emphasis is not on the creation of knowledge, but on the construction of the machineries of knowledge construction. Based on this concept, various cultural attitudes of knowledge production have been identified and distinguished by stressing the differences between scientific fields. See ↗ Epistemology.

Epistemology Epistemology is a branch of philosophy that is concerned with knowledge. The word epistemology is derived from the Greek *epistēmē* meaning "knowledge" and *logos* meaning "study of." The main concerns of epistemology are the definition of knowledge, the sources of knowledge (innate ideas, experience, etc.), the process of acquiring knowledge and the limits of knowledge. Much of the debate in this field has focused on the philosophical analysis of the nature of knowledge and how it relates to connected notions such as truth, belief, and justification.

Experientialist position The experientialist position is based on the assumption that our observations of biological (or other) phenomena are deeply shaped by social, cultural and many other factors. They contribute to form the knowledge and practices with which human beings wend their way through the world.

Frame/framing Framing is an analytical term used in the social sciences and humanities. It comprises a set of theoretical concepts and approaches devoted to analyze the social construction of political, medical, scientific or other issues such as research on embryonic stem cells or the possible uses of synthetic biology. Frames are conceived as mental representations that guide the interpretation of certain issues, help actors to communicate, semantically compress and simplify complex or controversial issues.

Genomics Genomics is part of genetics that applies recombinant DNA, DNA sequencing methods and bioinformatics to sequence, assemble and analyze the structure and the function of genes and genomes and studies their expression and regulation. See ↗ Postgenomics, ↗ Molecular technologies.

Images Images represent empirical outcomes on what systems biology is and compose systems biology in its diverse shapes. They unite concepts, notions, meanings and beliefs. Images stem from different areas, and can therefore be cultural, societal, political or scientific. Accordingly, they include a plurality of different connotations associated with the term "systems biology".

Implications Developments in science can have direct or indirect positive or negative consequences. They can pertain to science itself, but also to policy and politics, economy or society. In order to assess a scientific (or technological) development, its scientific and broader implications need to be analyzed. This is part of Technology Assessment (TA). TA wants to understand the developments of and shifts in science and technology as early as possible. The goal is not to predict, but to identify possible future scenarios with the aim to provide scientists, governments and stakeholders with relevant information which helps them—if necessary—to pro-actively intervene and to design adequate responses to technical, ethical, economic or societal issues resulting from scientific development.

In silico In silico describes the modelling, simulation and visualization of biological and medical processes in computers referring to any application of computer-based technologies.

Interdisciplinarity Interdisciplinarity is commonly defined as integrated research activities of scientists combining or involving two or more academic disciplines or fields of study that are usually considered distinct. However, interdisciplinarity transcend multidisciplinarity that is restricted to scientific work on a shared topic side-by-side without answering crossdisciplinarity research questions and interconnecting disciplinary research methods.

Interoperability Interoperability is the ability of making systems and organizations work together (to inter-operate). Initially, the term was defined for information technology. A broader definition takes into account non-technical (social, political, organizational) factors that impact system to system performance.

Lexeme A lexeme is a specific term used in linguistics that refers to the meaning or semantics of a single word. It depicts a linguistic unit of meaning and neglects the morphological or synthetic function of a word.

Linguistic metaphor Linguistic metaphors refer to a realization of a metaphor on the language or world level. Metaphors such as "we can now read the DNA" first appear on the level of words or sentences. In the course of the analysis and throughout the development of a scientific or public discourse they can become conventional metaphors if used again and again. Furthermore, most of the linguistic metaphors are motivated by conceptual metaphors such as "the DNA is a readable entity". See ↗ Metaphor.

Medialization Medialization describes the relationship between the media and subdomains of modern societies: science, the public, economy and politics. The media do not any longer only report on scientific outcomes, but can be used by all stakeholders to influence discussion on science, its outcomes, direction and policy. Thus, the media can e.g. be instrumentalized by science to hold priority conflicts and to mobilize public support (Weingart 2001, 244). Furthermore, science is more and more oriented towards political or economic objectives and thus focused on the public image in, and the public reception through the media—to a degree that the question arose whether the direction for science would be determined by the media (Weingart 2005, 168ff). Medialisation can thus be central for the justification of science.

Metadata Metadata is data that describes other data. Metadata summarizes basic information about data, which can make finding and working with particular instances of data easier. Examples of very basic metadata of a document are author, date created, date modified and file size. In the context of genomics or systems research metadata are sets of formalized information which can contain details about the data such as their origin, the scope of their utilization, the processes and goals of their production including experimental methods, materials, instruments, protocols. The main purpose of metadata is to assist in the discovery of relevant information by allowing it to be found by specified criteria. Metadata also help to organize electronic resources, provide digital identification, and support archiving and preservation of the data resource.

Metaphor Metaphors must be understood as a ubiquitous phenomenon and constitutive element of language and cognition both in everyday life and science that pervade and structure all kinds of discourses. Based on a mapping process (↗ Metaphorical mapping), metaphors make abstract domains of knowledge via concrete domains of knowledge accessible. Example: "The results taken from her research undermine, at least to some extent, the argument that cancer could be understood as genetically determined." Literally, the sentence does not make sense but, as soon as one reads it, one understands it's meaning immediately.

However, the word undermine at first does not appear to be metaphorical. Its metaphorical content only becomes obvious at second glance and provides an underlying image of arguments as buildings which can be undermined. One example for this mapping process is the metaphorically used language that creatively connects everyday discourse with scientific discourse. Metaphors thus "[…] play […] an essential role in establishing links between scientific language and [experiences taken from; the authors] world" (Kuhn 1993, 539). See also ↗ Conceptual metaphor, ↗ Conventional metaphor, ↗ Linguistic metaphor, ↗ Ontological metaphor, ↗ Orientational metaphor, ↗ Structural metaphor.

Metaphorical mapping A metaphorical mapping connects two cognitive domains. This cognitive coupling of knowledge domains in many cases makes abstracts domains accessible by using a concrete domain of discourse or experience. Thus, one can "dig deep to get data" even though one does not use a shovel or spate in the course of one's scientific analysis. The experience of digging is mapped onto the abstract domain of doing a scientific analysis and makes it understandable. Mapping processes can highlight certain aspects while hiding others. Thus, the previous mapping for example highlights aspects of hard work and the possible use of different tools while it covers aspects of working in an office and using a computer. The critical analysis of highlighting and hiding can be used to uncover blind spots and reframe domains of discourse. This would mean that in medicine, for example, that the war on certain diseases could be reframed in more cooperative terms such as interaction which might lead to more efficient drugs and better therapies.

Metaphoricity Metaphoricity is a term that designates the degree to which a term holds a metaphorical content or not. Degrees of metaphoricity differ between a conventional example such as "This is the way we should follow in our investigation" and a more elaborate one such as "His/her smile is like a blooming flower which longs for the sun". See ↗ Metaphor.

Metonymy/metonymies Metonymy is a figure of speech in which the parts stand for the whole. Typical metonymies are for example "The US say that…" meaning "The government of the United States says that…." Here, the nation stands for the government. Other frequent structures are capital city for government such as "Berlin says that…" meaning "the German government says that…." Metonymies are cognitive structure of meaning making (such as metaphors) but they use contiguity to as a principle of mapping and linguistic metonymies are apervade our everyday language such as metaphors. See ↗ Metaphor.

Mode-2-Knowledge Mode-2-Knowledge is a term from the sociology of science which refers to the production of scientific knowledge. Michael Gibbons and colleagues (1994) argued that a new form of knowledge production began emerging in the mid-twentieth century that was context-driven, problem-focused and interdisciplinary. It involves multidisciplinary teams that are brought together for short periods of time to work on specific problems in the real world. Mode 2 is distinguished from traditional research, labelled Mode 1, which is academic, investigator-initiated and discipline-based knowledge produced in theoretical and/or experimental environments and adapted or transferred into context afterwards.

The implication of the Mode 1/Mode 2 distinction is, that science in Mode 2 can no longer be regarded as an autonomous space, clearly demarcated from others of society. Instead all these domains had become heterogeneous and interdependent, even transgressive to the extent that they sometimes ceased to be distinguishable (Nowotny et al. 2001, 2003). See ↗ Science policy.

Molecular technologies Molecular technologies are the basic tools to study genetic information. They are used to characterize, isolate and manipulate the molecular components of cells and organisms. The most basic molecular technology is the polymerase chain reaction (PCR) deployed to produce multiple identical copies of DNA fragments. Other key technologies include DNA sequencing methods used to determine the order of the four bases (Adenine, Guanine, Cytosine, Thymine) in a strand of DNA, and DNA microarrays which visualize the gene expression of an organism at a particular stage (expression profiling).

Narrative A narrative is a structure of how a story is told, which normally has a start, a middle part and an end. It depicts a sequence of events that recounts an event or something that happened and provides a specific perspective on things that happened. By using a specific order of different structures, it develops an account of developments, events or things and interprets them. Thus, history is made of narratives (stories) as well as everyday talk. The analysis of narratives tries to uncover how certain stories, which have become history, are structured, how they semantically frame certain events and what past perspectives they convey.

Notion In philosophy, a notion is a reflection in the mind of real objects and phenomena in their essential features and relations. Notions are usually described in terms of scope and content. Notions are often created in response to empirical observations (or experiments) of co-varying trends among variables.

Objectivism Objectivism is based on the conviction that scientific descriptions are "true" representations of reality and that reality can be captured by scientific observation and experimentation. Objectivism is in contrast to constructivism (↗ Constructivism).

Omics Omics is a suffix that refers to a specific field of study. The objects of investigation are some sort of pools of biological molecules such as genome, proteome or metabolome. *Omics studies* (e.g. genomics, proteomics or metabolomics) aim at collectively characterizing and quantifying the molecules that translate into the structure, function and dynamics of an organism. See ↗ Genomics, ↗ Postgenomics.

Ontological metaphor An ontological metaphor reifies abstract entities as substances or things and makes them manageable. Consequently, time can become a precious entity such as money (time is money). See ↗ Metaphor.

Ontology Ontologies are formal representations of areas of biological knowledge in which essential terms and concepts are described both by their meaning and their relationship to each other. Ontologies are different from annotations (descriptions of data objects) in that they formalize the meaning of terms through a set of assertions and rules that are collectively known as description logics. Each term or concept within the ontology usually has a term name, a unique identifier that defines its relationships to one or more other terms in the same knowledge domain and sometimes to other domains, a textual definition and a list of synonyms which are classed as being equivalent to the term name. An advantage of ontologies is that the description logic can be used both for querying an information set and for facilitating analyses across information sets that are not traditionally accessible to searching and comparing. See ↗ Semantic interoperability.

Oral history Oral history is an approach that gathers enlived historical accounts by interviewing people. The aim consists in gathering and analyzing people's memories of past events to better understand the different perspectives on historical developments and events.

Orientational metaphor This is a kind of metaphor (↗ Metaphor) that ascribes directionality to an abstract domain. Metaphors such as "He is down again" or "Sales went up" use directionality to conceptualize good or bad.

Personification Personification is a kind of metaphor (↗ Metaphor) that conceptualizes an entity or a thing as a person, which is able to act. Examples are: "Nature can strike back" or "The sea conquers the land".

Philosophy of biology The philosophy of biology emerged as an independent subfield of philosophy of science in the 1960s and 1970s. It deals with epistemological, metaphysical and ethical issues in the context of biology.

Postgenomics Postgenomics refers to any fields of study that is only possible after the genome of an organism was published. Postgenomic research investigates which genes are active at particular times and under different environmental conditions (gene expression), e.g. how genes are transcribed into messenger RNA, the chemical that carries the instructions for forming proteins (transcriptomics), how genes are expressed as proteins (proteomics), and in how they influence the chemicals that control our cellular biochemistry and metabolism (metabolomics). See ↗ Genomics, ↗ Molecular technologies.

Prototype Prototypes are culturally and cognitive knowledge structures which help the individual to categorize the world and its phenomena. They are based on an ideal model of an entity. Thus, a representative example for a bird in the western hemisphere is a sparrow or a robin. Using the concrete example of a sparrow, all kinds of birds compared to the typical sparrowness or robinness. Thus, eagles represent more typical birds due to their "sparrowness" while ostriches or penguins as non-flying birds are conceived as less typical ones. One has, however, to bear in mind that these structures are dynamic and open to change which means that processes of leaning and other kinds of experience change and re-order the structure of prototypes.

Science and Technology Studies (STS) Science and Technology Studies is a relatively new interdisciplinary field of study that examines the creation, development and consequences of science and technology in their cultural, historical and social contexts. STS emerged from the confluence of a variety of disciplines and disciplinary subfields, all of which had developed an interest during the 1960s and 1970s in viewing science and technology as socially embedded enterprises. STS, as practiced in academia today, can be divided into two broad streams or developments. The first consists of research on the nature and practices of science and technology. Studies in this stream approach science and technology as social institutions possessing distinctive structures, commitments, practices and discourses that vary across cultures and change over time. The second stream is more concerned with the impact and control of science and technology, with particular focus on the risks that science and technology may pose to peace, security, community, democracy, environmental sustainability and human values.

Science policy Science policy governs the development and funding of science and research, both basic and applied. It is developed by politics in discussion with other stakeholders, and often evolves from other policies, such as the political goal of the application of technologies or medicine in the field of biotechnology. Science policy of systems biology in Germany e.g. is inter alia driven by the aims for its practical application. For scientists, it is thus central how current research is assessed by funding organizations. Another frequent question for science policy concerns the issue whether the public should be involved more directly in decisions related to the development of science and technology.

Semantic interoperability Semantic interoperability is the ability to automatically interpret the data exchanged meaningfully and accurately in order to produce useful results as defined by the end users of both systems. To achieve the simultaneous transmission of the meaning of the data (semantics), both systems must refer to a common information exchange reference model transmitting shared vocabulary and its associated links to an ontology. See ↗ Interoperability, ↗ Ontology, ↗ Syntactic interoperability.

Standard/standardization Standards are used as a measure, norm or rule to construct uniformities in a given context, discipline or field. Considering the broad application of standards in nearly all spheres of life, standards have different formats and are assigned to realize different goals. Technical standards, for instance, are supposed to maximize compatibility, interoperability (↗ Interoperability), safety, repeatability or quality of technical objects or processes. The process of developing and implementing standards is labelled as standardization. Top-down-standardization is initiated by standard development organizations (e.g. ISO, SEN, HL7). They are usually entitled to develop formal standards for a specific setting. The opposite process is the bottom-up approach, where user communities or industry trigger de facto standards by informal convention or dominant usage. De jure standards are part of legally binding contracts, laws or regulations. Scholars of Science and Technology Studies (↗ Science and Technologies Studies) stress that standards are always socially constructed as standards are usually built collectively and need approval by others to matter.

Structural metaphors Structural metaphors transfer a complete metaphorical pattern on an abstract domain of thought or discourse. Thus, a Ph.D., can become "a journey where you follow different tracks, encounter obstacles, have to go back to the start but finally reach the goal". See ↗ Metaphor.

Syntactic interoperability Syntactic interoperability refers to the ability of two or more computer systems to communicate and exchange data, e.g. by querying one database by another. The tools for packaging and transmission mechanism of data (syntax) are specified data formats and communication protocols. See ↗ Interoperability, ↗ Semantic interoperability.

Tacit knowledge Tacit knowledge is a kind of knowledge that is not formalized in writing, formulas or protocols etc. It is a non-explicit kind of knowledge gained through practical experience and transferred between individuals by practically sharing ways of situated problem solving.

Term Terms are words and compound words that in specific contexts are given specific meanings. These may deviate from the meanings the same words have in other contexts and in everyday language. *Terminology studies* are dealing with the development of terms, their interrelationships and their use.

References

Callon M (1992) The dynamics of techno-economic networks. In: Coombs R, Saviotti P, Walsh V (eds) Technological change and company strategies. Economic and sociological perspectives. Academic Press, London, pp 72–102

Gibbons M, Limoges C, Nowotny H, Schwartzman S, Scott P, Trow M (1994) The new production of knowledge. The dynamics of science and research in contemporary societies. Sage, London

Giddens A (1984) The constitution of society. Outline of the theory of structuration. University of California Press, Berkeley, CA

Kuhn T (1993) Metaphors in science. In: Ortony A (ed) Metaphor and thought, 2nd edn. Cambridge University Press, Cambridge, pp 533–542

Knorr-Cetina K (1999) Epistemic cultures: how the sciences make knowledge. Harvard University Press, Cambridge

Latour B (1987) Science in action: how to follow scientists and engineers through society. Harvard University Press, Cambridge

Latour B (1992) Where are the missing masses? The sociology of a few mundane artifacts. In: Bijker W, Law J (eds) Shaping technology/building society: studies in sociotechnical change. MIT Press, Cambridge, pp 225–259

Latour B (1996) On actor-network theory. A few clarifications. Soziale Welt 47(4):369–382

Law J (1987) Technology and heterogeneous engineering: the case of portuguese expansion. In: Bijker W, Law J (eds) The social construction of technological systems: new directions in the sociology and history of technology. MIT Press, Cambridge, pp 111–134

Nowotny H, Gibbons M, Scott P (2001) Re-thinking science. Knowledge and the public in an age of uncertainty. Polity Press, Oxford

Nowotny H, Scott P, Gibbons M (2003) Mode 2 revisited: the new production of knowledge. Minerva 41:179–194

Polanyi M (1958) Personal knowledge. University of Chicago Press, Chicago, IL

Star S, Griesemer J (1989) Institutional ecology, 'translations' and boundary objects: amateurs and professionals in Berkeley's Museum of Vertebrate Zoology, 1907–39. Soc Stud Sci 19(3):387–420

Weingart P (2001) Die Stunde der Wahrheit? Zum Verhältnis zu Politik, Wirtschaft und Medien in der Wissensgesellschaft. Velbrück, Weilerswist

Weingart P (2005) Die Wissenschaft der Öffentlichkeit. Essays zum Verhältnis von Wissenschaft, Medien und Öffentlichkeit. Velbrück, Weilerswist

Index

M. Döring et al., *Contextualizing Systems Biology*,
DOI 10.1007/978-3-319-17106-7

MIX
Papier aus verantwortungsvollen Quellen
Paper from responsible sources
FSC® C105338

If you have any concerns about our products,
you can contact us on
ProductSafety@springernature.com

In case Publisher is established outside the EU,
the EU authorized representative is:
Springer Nature Customer Service Center GmbH
Europaplatz 3, 69115 Heidelberg, Germany

Printed by Libri Plureos GmbH
in Hamburg, Germany